Nutrients for Sugar Beet Production

Soil–Plant Relationships

Nutrients for Sugar Beet Production

Soil–Plant Relationships

A. Philip Draycott

formerly of Broom's Barn Research Station,
Bury St Edmunds, Suffolk, UK

and

Donald R. Christenson

Professor Emeritus in Crop and Soil Science,
Michigan State University, East Lansing, USA

CABI Publishing

CABI Publishing is a division of CAB International

CABI Publishing
CAB International
Wallingford
Oxon OX10 8DE
UK

Tel: +44 (0)1491 832111
Fax: +44 (0)1491 833508
E-mail: cabi@cabi.org
Web site: www.cabi-publishing.org

CABI Publishing
44 Brattle Street
4th Floor
Cambridge, MA 02138
USA

Tel: +1 617 395 4056
Fax: +1 617 354 6875
E-mail: cabi-nao@cabi.org

A catalogue record for this book is available from the British Library, London, UK.
A catalogue record of this book is available from the Library of Congress, Washington, DC, USA

ISBN 0 85199 623 X

Typeset in 9 pt Palatino by Columns Design Ltd, Reading
Printed and bound in the UK by Cromwell Press, Trowbridge

Contents

About the Authors

Philip Draycott specializes in sugar beet research and development having worked on soil fertility for nearly 20 years at Broom's Barn, the UK centre for research on the crop. *Donald Christenson* is Professor Emeritus in the Crop and Soil Sciences Department at Michigan State University in the USA where sugar beet has been central to his research for over 30 years.

They are respected authorities on sugar beet nutrition in their own countries and internationally through the Institut International de Recherches Betteravières in Europe and the American Society of Sugar Beet Technologists. In this book they combine their knowledge to produce a unique review of their own research and by assembling nearly 1000 references from the world literature.

Preface

About 40 million t of sugar are produced annually from beet, a crop grown in over 50 countries on a total area of some 7 million ha. In common with most crops, climate and soil are the two main determinants of yield per unit area. Both are largely outside the growers' control but next most important is the plant's nutrition, which can be manipulated to the advantage of producer and processor. An adequate supply of macro- and micronutrients is crucial and, not surprisingly, this has led to a plethora of research wherever the crop has become established.

Nutrition research was reviewed and collected together in one place for the first time in English in the early 1970s (Draycott, 1972). Until then results had been scattered through many published and unpublished reports, and in papers in numerous scientific journals not readily accessible to many of the people who could make most use of them. The object then was to review UK research and experimental work on sugar beet nutrition from other countries. Much material about residual effects of fertilizers on sugar beet was obtained from the results of classical and long-term experiments at Rothamsted, Woburn, Saxmundham and Broom's Barn. From many comments received from the UK and abroad, the exercise was a useful one and the book became the standard reference for researchers, teachers, advisers and farmers involved with growing the crop.

Thirty years later the earlier book is not only out of print and unavailable but also very much out of date. Research on sugar beet nutrition expanded rapidly during the 1970s, 1980s and 1990s, shedding fresh light on many subjects, some not even contemplated at the last review. For example, the environmental impact of fertilizers is at the forefront of our considerations now but was not even mentioned before. Soil and plant analysis to diagnose nutrient requirements was becoming accepted in the early 1970s but is fundamental now, resulting in nutrient applications being much more focused. Other topics such as organic farming, sustainability, satellite mapping and remote sensing in relation to sugar beet nutrition have all appeared in the past 30 years.

This new book, now with two authors, one from each side of the Atlantic, attempts to cover these new subjects and review work published over the past 30 years, setting it in the context of earlier knowledge. We have included work not only from the UK and USA but also from many other countries, internationalizing the treatment and hopefully the book's relevance and appeal. Where information on some topics was lacking for sugar beet, we have drawn on research with other crops, where relevant, to fill the gaps.

Sugar beet yields have risen rapidly during the period, while the amount of fertilizer used, particularly nitrogen, has fallen steeply. In economic and environmental terms this is a great success story, which may be credited to the development of research, covered in detail in subsequent pages. Simultaneously, atmospheric sulphur deposition has declined to a fraction of that in 1970 as a result of cleaner air and this is expected to present a new crop-nutrition challenge of diagnosis and treatment.

Those of us who are closely involved with sugar beet are fortunate that during the second half of the last century the crop did in fact become 'international'. Research and development know no country or continental boundaries, results being freely shared. This is due in no small measure to two bodies: in Europe the International Institute for Sugar Beet Research and in North America the American Society of Sugar Beet Technologists. We have drawn on their many meetings and publications, which have helped in the production of this volume. Likewise we have both benefited from our close association with the fertilizer industries on both sides of the Atlantic, and the seed companies and sugar processors. In addition, experience gained in our positions with Broom's Barn and Michigan State University has proved invaluable.

A. Philip Draycott, UK
Donald R. Christenson, USA

Acknowledgements

We thank our many friends and colleagues around the world for discussions and information – in particular, J. Pidgeon, J. Johnston, I. Perry, C. Sutton and A. Smucker for their comments on selected chapters, and T. Schwartz for suggesting that we should work together on a sugar beet book.

APD thanks British Sugar and UK sugar beet growers for support and encouragement over many years. R. Hull, G. Cooke and H. Penman were mentors at Broom's Barn and Rothamsted.

The following kindly provided information on the nutrition of sugar beet in their countries: M. Agbani (Morocco), J. Amaral (Portugal), M. Armstrong (UK), K. Burcky, R. Merkes, M. Burba and B. Märländer (Germany), I. Buzás (Hungary), V. Demis (Greece), R. Duval and M. Richard-Molard (France), M. Erjala (Finland), H. Elsherif (Egypt), D. Hindle, T. Schwartz, A. Cattanach, J. Eckhoff, C. Frate, J. Giles, D. Jacobs, S. Kaffka, J. Lamb, D. Melin, S. Sackman, R. Sams, C. Shock, D. Traveller (USA), S. Kuharic (Slovenia), G. Legrand (Belgium), J. Malec (Poland), C. Marcussen (Denmark), G. Markevich (Belorussia), T. Masutani (Japan), R. Morillo, M. Sanz, M. Bermejo and M. Bilbao (Spain), G. Montovani (Italy), M. Reyes-Lacalle (Chile), F. Ryan (Ireland), C. Sperlingsson (Sweden), P. Wilting and B. Smit (Netherlands), and S. Yücel (Turkey).

DRC thanks Monitor and Michigan Sugar Companies for financial support for over 30 years. Also the help and friendship of D. Sunderland, R. Braem, R. Zielke and the late P. Brimhall from Michigan Sugar; G. Nichol, R. Sanborn, R. Fogg, P. Pfenniger, R. Roslund and L. Hubbell from Monitor Sugar, along with the agriculturists from both companies is deeply appreciated. Thanks also go to S. Poindexter and H. Rouget of Michigan State University Extension for their leadership in research and extension programmes in the State of Michigan. Special mention of C. Bricker is made for assistance with research projects at Michigan State University.

As far as possible we have used our own pictures of nutrient-deficiency symptoms on leaves and plants where we knew their history and location. We are grateful for some from Broom's Barn. The picture of sulphur deficiency in the field was provided by G. Connors. The rest are reproduced, with the permission of the University of California, from Ulrich and Hills (1969).

Both authors thank Mrs J. Robertson for typing several drafts and for her attention to detail in assembling references and the index. K. Foster drew the diagrams and W.

Martindale helped with the literature search, for which we are grateful. Finally we thank our wives, Joyce and Carol, for patience.

We dedicate the book to our children and grandchildren, hoping they will continue writing – Melissa, Roger, Jorge, Karen, Melissa, Laurie, Tyler, Joshua and Beth.

APD
DRC

1

2

3

4

ıitrogen

ᵓlate 1. Leaf number 24 showing uniform yellowing, no green veining, small size on the left compared with leaf well ɩupplied with nitrogen.

ᵓlate 2. Leaf colour series, uniform yellow, decreasing severity of deficiency.

ᵓlate 3. Petiole and leaf blades growing horizontally.

ᵓlate 4. Long, narrow-shaped yellow leaves, horizontal habit of petioles.

5

6

7

8

Phosphorus

Plate 5. Seedling with yellow cotyledons and first pair of true leaves green.
Plate 6. Severely deficient field-grown plant at the eight-leaf stage.
Plate 7. Stunted greening of eight- to ten-leaf stage plant, brown veining of older leaves.
Plate 8. Leaves showing various degrees of deficiency.

9

10

11

12

Potassium

Plate 9. Marginal leaf scorching and smooth leaf surface.
Plate 10. Mature leaf with interveinal scorch and crinkled leaf surface.
Plate 11. Plant grown without sodium. Centre leaves green, often hooded similar to calcium deficiency, mature leaves with crinkled leaf surface and scorching.
Plate 12. Plant grown with sodium. Centre leaves green, mature leaves less severely scorched than without sodium.

14

13

15

16

Magnesium

Plate 13. Early stage of deficiency. Margins of middle-aged leaves yellow, edges dying and becoming brittle.
Plate 14. Later stage of deficiency showing chlorosis between the veins.
Plate 15. Final stages of deficiency showing black necrotic tissue between veins followed by brown necrosis.
Plate 16. Deficient beets in autumn showing 'washing out' of necrotic tissue.

17

18

19

20

Calcium

Plate 17. Tip burn on centre leaves, hooding of older leaves, younger leaves normal.
Plate 18. Centre leaves with severe tip burn, blades of older leaves crinkled, deformed, fail to expand.
Plate 19. Hooding of centre leaves.
Plate 20. Black-tipped deformed leaves.

21

22

23

Sulphur

Plate 21. Uniform yellowing, leaves light green, mature leaves very brittle.
Plate 22. Leaf age series, left to right, young, young-mature, mature and old leaves.
Plate 23. Sulphur-deficient beets growing in the field – a rare sight.

25

24

27

26

Boron

Plate 24. Initial stages of deficiency – death of the growing point.
Plate 25. Netting and cracking of mature leaves.
Plate 26. Deficiency in mature field-grown beet. Proliferation of small centre leaves, centre black necrosis, prostrate habit of petioles, cracked and corky areas on petioles.
Plate 27. Mature roots showing heart and crown rot.

28

29

30

31

Manganese

Plate 28. Early stages showing yellow speckling between veins.
Plate 29. Later stages showing speckling becoming translucent.
Plate 30. Yellow spots coalescing to necrotic areas and holes developing from tissue removal.
Plate 31. Field-grown sugar beet, leaves growing in typical upright posture.

32

33

34

35

Iron

Plate 32. Uniform yellowing of young leaves.
Plate 33. Uniform interveinal chlorosis, prominent green veins.
Plate 34. Field-grown mature beet showing vivid yellow mature leaves.
Plate 35. Leaf recovering from iron deficiency after treatment.

36

37

38

39

Zinc

Plate 36. Uniform yellowing of young centre leaves, interveinal necrosis and formation of large green veins on older leaves.
Plate 37. Initial pitting of mature leaf blade.
Plate 38. Coalescing of pitted areas, main veins remain green.
Plate 39. Yellowing followed by interveinal necrosis and pronounced veining.

40

41

42

43

Copper

Plate 40. Chlorosis of young centre leaves.
Plate 41. Pronounced veining.
Plate 42. Diffused light chlorosis and veining.
Plate 43. Later stages of deficiency showing start of leaf bleaching.

44

45

46

47

48

49

Chlorine

Plate 44. First signs of chlorosis of young centre leaves.

Plate 45. Normal leaf on left and interveinal chlorosis of chlorine deficiency on right.

Plate 46. Chlorine deficient on left and normal on right through transmitted light.

Molybdenum

Plate 47. Uniform yellowing of all leaves.

Plate 48. Uniform yellowing with veins standing out and start of necrotic spotting.

Plate 49. Veining and necrotic spotting viewed through transmitted light.

1
Introduction

Elements in Sugar Beet

Sugar beet is composed primarily of carbon (C), hydrogen (H) and oxygen (O), but other elements are necessary as components of structural tissues or as participants in biochemical reactions. Those known with certainty to be essential for this plant are nitrogen (N), phosphorus (P), potassium (K), calcium (Ca), magnesium (Mg), sulphur (S), sodium (Na), boron (B), chlorine (Cl), copper (Cu), iron (Fe), manganese (Mn), molybdenum (Mo) and zinc (Zn). Other elements where there is still doubt over their essentiality for sugar beet (but which may be needed for some other plant species) are cobalt (Co), nickel (Ni), selenium (Se) and silicon (Si).

The quantity of each of these nutrient elements needed by sugar beet to perform satisfactorily for sugar production varies widely from element to element. Some, such as nitrogen, make up several per cent of the dry matter of the plant. Others, such as molybdenum, make up only a few parts per million. This is because the former is a major constituent of proteins and nucleic acids, whereas the latter is essential only in enzymatic reactions. Those elements needed in large quantities (the major elements or macronutrients) by sugar beet are N, P, S, K, Ca, Mg and Na. Those needed in small amounts (the trace elements or micronutrients) are B, Cl, Cu, Fe, Mn, Mo and Zn.

Origins of Crop

Sugar (the common name for sucrose) is produced from only two crops, cane and beet. Cane sugar has been produced in large quantities in the tropical regions for several centuries and continues to dominate the world supply of sugar. In contrast, sugar beet is a relatively new crop, appearing in the 19th century in the temperate regions. At the beginning of the 21st century it provides about a quarter of the world's sugar and cane three-quarters.

What we know as sugar beet was first grown at least 2000 years ago as a garden vegetable. The sugar beet currently grown is far removed from the garden plant. The vegetable was probably selected from various *Beta* species growing round the shores of the Mediterranean. It was widely used for culinary purposes throughout Europe from the Middle Ages onwards (Winner, 1993).

During this period bee honey was prized for its sweetness, being the only such food readily available. Limited quantities of cane sugar were imported into Europe for the tables of the rich and the rest made do with fruits and vegetable juices for sweetening. Any plant whose juice was sweet was therefore highly valued.

Beet was grown on a field scale first in the 17th century but only as fodder for cattle. A range of different types of *Beta vulgaris* were grown for this purpose, some for their storage root, some for their leaf and always in a variety of colours. It was from a white root type of Silesian beet with high sugar concentration that sugar beet was developed early in the 18th century. A new and important crop was born.

Production

Several German chemists discovered how to extract sugar from beet roots and showed that the crystals were the same as those derived from cane. Various attempts were made to industrialize the process in primitive factories in the late 18th century, coupled with field-scale cultivation of the crop. None was very successful.

By the beginning of the 19th century cane sugar had become an important world commodity, being imported by many European countries from their colonies. This trade was then severely interrupted by the Napoleonic wars, which led to a reawakening of interest in sugar from beet. Over the past 200 years, after many false starts, the sugar beet industry became firmly established. Winner (1993) has detailed its progress worldwide and Francis (2002) in the UK.

Plant breeding has contributed most to improvements in the productivity of the plant and hence the crop. Sugar concentration has been increased from some 12% of the fresh root to current values of 17–20%. Improvements in yield and chemical properties of the root by plant breeding continue to increase the amount of white sugar extracted at the processing factories. In addition, resistance and/or tolerance to pests and diseases aid in the production of this crop. Perhaps the biggest breakthrough by plant breeders was the introduction of monogermity, allowing 'sowing to a stand'.

Alongside these improvements from selection, yield and quality have been greatly influenced by progress in plant nutrition. Over the past century, research in every country where the crop is grown has gone on apace. The work detailed in the following chapters recounts how research has focused on ensuring that the crop has sufficient macro- and micronutrients to perform to the limits imposed by the climate in each locality.

Where Grown

Once established in Europe, with field production and processing methods proved, sugar beet was taken to other regions of the world. Processing factories have been built in areas favourable to the crop in the Americas, Asia and North Africa. In 2000, nearly 7 Mha of sugar beet were grown worldwide.

Despite its widespread production, sugar beet is essentially a crop grown in temperate regions. Most is grown at latitudes between 30 and 60°N, as a summer crop in maritime, prairie, semi-continental and some semi-arid and arid climates and as a winter and/or summer crop in Mediterranean and other semi-arid and arid conditions (Draycott, 1972). The crop is now grown with supplemental irrigation in regions where low rainfall previously limited its production.

Besides being widely spread geographically, the crop is produced successfully on a range of soil types. On a textural classification, the crop is found on virtually all types – clays, silts, sands and organic soils. Production may be limited on soils with excessive wetness in spring and autumn. Delayed sowing or difficulties in harvest can limit its production on soils with a very high clay content. In many areas this may be overcome through the use of artificial drainage, such as mole drains, tile drains and other methods. Generally the crop is grown on fairly level land to aid mechanical field operations. Where soils are deep and in a good physical state, sugar beet thrives almost everywhere if the climate allows.

An essential aspect of all soils where the crop is grown is pH, which must be near the neutral point. Problems in this respect are easily rectified (Chapter 5). When this criterion is met and if the nutrients detailed herein are available, with good husbandry the crop will produce sugar commensurate with the limits set by the climate (Scott and Jaggard, 1993).

World Production of Sugar from Beet and Cane

Total production of sugar now approaches 150 Mt, having risen from 30 Mt over the course of the last 60 years (Fig. 1.1). It now increases at about 2 Mt per year. Sugar made from the two crops strives to satisfy world demand for sweet food, notwithstanding increasing production of syrups from other crops and of artificial sweeteners. A small proportion of sugar produced from cane and beet is fermented to alcohol and this may rise as use of biofuels increases worldwide.

In 1900 beet and cane each provided about half of the sugar produced and increases in each rose in parallel during the early part of the 20th century. During the second half of the century to the present day, sugar from cane rose rapidly whereas from beet it has risen more slowly (Fig. 1.1) and currently shows signs of levelling off or even decreasing slightly.

Between 1950 and 2000 the proportion of sugar produced from cane changed from about two-thirds to three-quarters of world sugar production. These changes are due to complex social, political and economic pressures, which will inevitably continue to beset the sugar industry in the new century.

Europe is a major producer of sugar from sugar beet, accounting for about 77% of the total beet sugar produced (Table 1.1). The USA and Asia make up 17% of the production with South America and Africa accounting for the remainder.

A comparison of sugar beet production between Europe and the USA is given in Table 1.2. Yield of sugar per hectare in the European Union (EU) and in the Far West of the USA are comparable. Shorter growing seasons and less certain moisture supplies contribute to the lower yields in the Great Lakes, Upper Midwest and Great Plains. In the Far West (Idaho and California), a longer season and irrigation contribute to the higher yields. Yield of both roots and sugar is high in the Imperial Valley of California and growers there hold the world record for sugar beet production, exceeding 93 and 12 t ha^{-1} of roots and sugar, respectively.

Table 1.1. Annual world sugar production from beet.

		Mt	Per cent
Europe	West	22	56.4
	East	8	20.5
Africa		1	2.6
America	North	4	10.2
	South	1	2.6
Asia		3	7.7
Total		39	

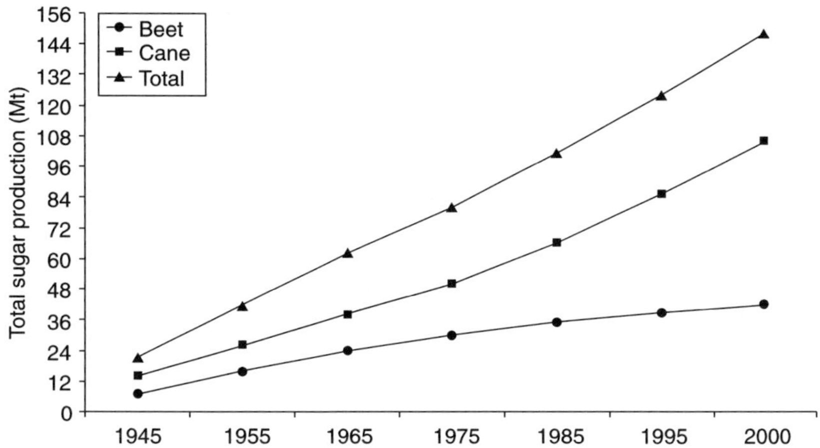

Fig. 1.1. World sugar production from cane and beet, 1940–2000.

Table 1.2. Comparison of area in sugar beet production, amount of beet processed, amount of sugar produced and yield of beet and sugar for Europe and the USA (from Licht, 2000; Bartens, 2001; Haley *et al.*, 2001).

	Area (Mha)	Beet processed (Mt)	Sugar produced (Mt)	Beet yield (t ha^{-1})	Sugar yield (t ha^{-1})
Europe					
West	2.6	137	22	53	8.5
East	3.0	65	8	22	2.7
Total	5.6	202	30	36	5.4
Included in West above					
European Union	2.0	116	19	58	9.5
USA					
Great Lakes	0.07	3.1	0.45	44.3	6.4
Upper Midwest	0.27	12.8	1.77	47.4	6.6
Great Plains	0.09	4.3	0.58	47.8	6.4
Far West	0.13	8.8	1.25	67.7	9.6
Total	0.56	29.0	4.05	51.8	7.2

Application of Nutrients in Various Countries

Table 1.3 shows the amounts of the three major nutrients being applied in fertilizers in the 1990s (excluding organic manures) in the EU. Thirty years ago (Draycott, 1972) the amounts per hectare for the EU countries (excluding Portugal which had no sugar beet industry at that time) were 136, 128 and 192 kg N, P_2O_5 and K_2O ha^{-1}, respectively. Thus there has been a significant reduction in quantities of all three nutrients. Causes, effects and benefits of this reduction are dealt with in later chapters.

Similar information for 11 countries in other parts of the world where recent information was readily available shows a wide variation, some amounts being extremely large (Table 1.4). The following chapters, which define the

Table 1.3. Area of land in sugar beet production and approximate amounts of NPK applied for sugar beet in the EU.

	Area (1000 ha)	Application (kg ha^{-1})		
		N	P_2O_5	K_2O
Austria	41	130	126	151
Belgium/Luxembourg	95	130	80	290
Denmark	60	110	68	115
Finland	33	142	95	65
France	414	130	110	240
Germany	452	100	70	140
Greece	42	139	111	127
Ireland	33	160	100	220
Italy	245	95	135	125
Netherlands	115	120	105	135
Portugal	9	165	155	90
Spain	140	190	140	40
Sweden	56	106	67	58
UK	150	105	65	125
Average		130	102	137

amounts needed by sugar beet, question these applications as possibly being excessive in some cases, and insufficient in a few.

Improvements in Sugar Beet Performance due to Nutrition and Other Factors

As mentioned earlier, conventional plant breeding and variety testing in many countries have gradually produced the highly efficient sugar beet grown today. This yields more weight per unit area and is of much better processing quality than its primitive ancestors. Annual improvements through selecting from the best continues to increase productivity and probably contributes most to the dramatic change seen for more than 70 years (Fig. 1.2). With novel methods of plant breeding researched during the past 20 years, perhaps even greater advances will be made. At the time of writing, however, sugar production from genetically modified beet is banned in most, if not all, countries.

Several authors have attempted to separate the effects of better nutrition from plant breeding and other factors (Watson, 1952; Scott and Jaggard, 2000). Draycott (1996) tried to put a value on the benefit of individual macro- and micronutrients. To estimate the consequences for yield of changes in nutrition alone in Fig. 1.2 is impossible due to many other simul-

Table 1.4. Area of land in sugar beet production and approximate amounts of NPK fertilizer applied for sugar beet in various countries in 2000.

	Area (1000 ha)	Application ($kg\ ha^{-1}$)		
		N	P_2O_5	K_2O
Belorussia	49	112	56	151
Chile	49	140	300	90
Egypt	60	190	37	0
Hungary	59	76	68	134
Japan	70	171	315	160
Morocco	66	240	120	250
Poland	315	118	75	135
Slovenia	11	100	80	250
Turkey	335	133	77	100
USA				
Upper Midwest	268	85	65	15
Great Lakes	68	160	65	220

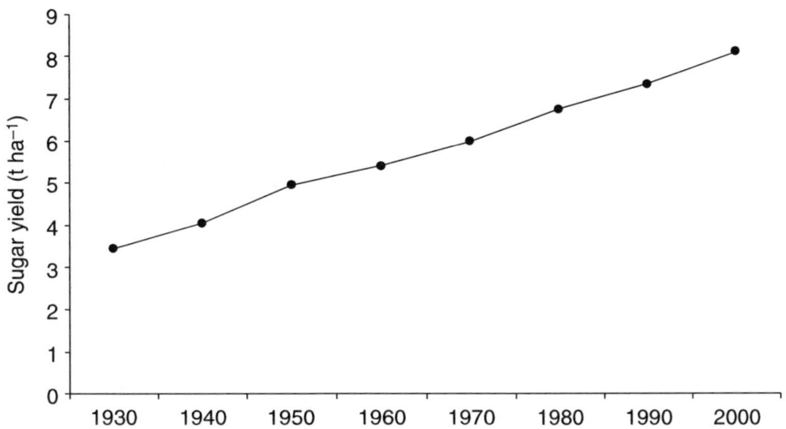

Fig. 1.2. Average improvement in sugar beet performance in the UK and USA, 1930–2000.

taneous changes. The following chapters will evaluate the effect of individual nutrients and identify the magnitude of their effect on yield and quality from strictly controlled experiments in many countries where the crop is grown. The results leave no doubt that on virtually all the 7 Mha of crop sown annually worldwide, nutrition plays a vital role.

Pidgeon *et al.* (2001) have recently sounded a warning over a likely slowing of the future rate of rise in yield per unit area. Their analysis suggests that the national average production in the most efficient countries (France and the UK are quoted) is now some 80% of estimated potential yield, as set by climate, soil and variety. They imply that nutrition must be near optimal (or possibly still excessive in the case of some nutrients) in these countries but suggest that there is a long way to go to optimize factors affecting growth of the crop (including nutrition) in other parts of the world.

Determining Optimum Amounts of Fertilizer Nutrients

Over the past 70 years, considerable strides have been made in determining optimum amount of nutrients for sugar beet. Soil-testing procedures have been developed to assist in this process. Field trials correlate and calibrate these procedures with response to applied fertilizer. Laboratory analyses have become fairly straightforward and rapid with advanced technologies. While not perfect, soil testing provides a tool for assessing the nutrient status of soil prior to sowing. When coupled with plant-tissue analyses, most, if not all, limitations due to nutrient supply can be eliminated.

Test values are often expressed in units such as mg kg^{-1} soil. In our view these should not be converted to kg ha^{-1}. Such conversions lead to attempts to utilize soil-test values as absolute amounts of nutrient available to the crop. At best they give a snapshot of the quantity available, e.g. for nitrate and sodium. However, for other less mobile nutrients, such as phosphorus, potassium and micronutrients, such treatment is not acceptable. Analytical procedures usually remove only a proportion of the available nutrient from the soil, which should be regarded as 'an index of availability'. This index will serve two purposes: (i) to provide an indication of the probability of response (low, medium or high); and (ii) to suggest the amount of nutrient to apply.

Numerous studies have shown that soil-test values do not change on a quantitative basis when compared with the amount removed or added. For example, in the case of phosphorus, Yerokun and Christenson (1990) showed a reduction in soil-test concentration of 0.13–0.31 mg P kg^{-1} soil for each mg phosphorus removed by plants. Havlin *et al.* (1984) reported a change of 0.17, Leamer (1963) 0.21 and Adepoju *et al.* (1982) 0.48 mg P kg^{-1}. White and Doll (1971) reported a similar range of values for increasing soil-test concentrations with added phosphorus fertilizer.

Summary

On a world scale sugar beet is an important source of sugar and the crop occupies nearly 7 Mha each year. While complex political and social tensions exist among nations, there will be a need for sugar production from sugar beet. The Napoleonic wars enhanced the establishment of the sugar beet industry in Europe because availability of cane sugar was restricted. Such a situation could occur again and this explains why many other countries have set up their own sugar beet industries.

Wherever sugar beet is grown in the world, climate and soil are the two major determinants of success. Diseases, weeds and other pests can usually be overcome. Except where irrigation is available, climatic factors cannot be changed. However, soil can be modified, to improve and then maintain optimum nutrient availability for sugar beet production.

The following chapters show how this can be achieved. A maximum economic yield of quality roots is essential to grower and processor. More importantly, nutrition of the crop should have a minimum impact on the environment, and much recent research on this topic is reviewed.

2
Nitrogen

In common with the majority of crop species, in sugar beet nutrition nitrogen is an important element. From the beginning of growing beet for animal and human food, it was immediately obvious from the pale yellowish-green leaves when nitrogen was in short supply. Growers quickly realized that organic manures (and, later, mineral fertilizers) corrected deficiencies, increased growth rate and greatly improved the yield of roots.

The simple reason is that nitrogen in a plant-available form is usually in short supply in soils under continuous cropping throughout the world. Only a few arable soils can regularly provide more than 100 kg N ha^{-1} during the growing season. Sugar beet takes up soil nitrogen more efficiently than most crops but needs double this amount in most circumstances for maximum production. How best to exploit the effect of additions of nitrogen in its various forms has been central to research programmes with sugar beet in every country where it is grown.

Until the second half of the 20th century, work concentrated almost entirely on the effect of the element on yield. Then attention turned to the negative effects of nitrogen on root quality and sugar extraction. This followed from overuse in many countries. Inexpensive fertilizer was at every farmers' disposal, thanks to the worldwide adoption of the Haber–Bosch process, which converts atmospheric nitrogen to plant-available forms. Research in the 1960s and 1970s not only looked at yield but judged the effect of nitrogen fertilizer on efficient sugar production too.

In the final period of the 20th century, work continued apace but much of it took on quite a different slant. Attention was drawn to the possible negative effects of excesses of nitrogen in soils on health and on the quality of the environment. Nitrate leaching from soil into water supplies led to limits being set. No review of nitrogen nutrition of sugar beet is now complete without considering not only the use of fertilizers and manures for the crop, but also broader issues of their effect on soil, water and the environment.

Nitrogen in the Atmosphere and Soil

Many chemical and biological pathways have an impact on the availability of nitrogen to growing plants and its effect on the environment, particularly nitrate moving into surface and subsurface water resources. The following discussion focuses on the parts of the nitrogen cycle affecting nitrogen availability to the sugar beet crop.

Nitrogen in the atmosphere

Approximately three-quarters of the air in the atmosphere covering the earth's surface is nitrogen gas (N_2), amounting to 77,000 t above each hectare (Foth and Ellis, 1997). Atmospheric nitrogen is available only to certain plants and plant associations capable of breaking the triple bond of the nitrogen molecule, of which sugar beet is not one. Industrial fixation requires expenditure of energy, generally from fossil-fuel resources. It was not until the Haber–Bosch process was developed that such fixed nitrogen was economically available for fertilizer in the ammonium (NH_4^+), nitrate (NO_3^-) and associated forms, e.g. urea ($CO(NH_2)_2$).

Nitrogen in soil

More than 90% of nitrogen in soil is bound in organic matter containing approximately 5% N. A hectare of soil (30 cm deep, bulk density of 1.333 t m^{-3}) contains 2000 kg of organic nitrogen for each per cent of organic matter. With an expected annual decomposition rate of about 1%, 20 kg of nitrogen would be mineralized for each per cent organic matter in soil. The amount of nitrogen mineralized is widely variable across soils because of different organic-matter contents and disparate mineralization rates. Mineralization also varies both within and among seasons and climates.

Other pools of the element in soil include available mineral nitrogen (NH_4^+ and NO_3^-) and NH_4^+ fixed in clay minerals. The latter accounts for less than 10% of total nitrogen while the former accounts for less than 1%. Addition of fertilizer increases soil mineral nitrogen, but the effect is relatively small and temporary. Long-term effects of fertilizers and manures are dealt with in Chapter 9.

Nitrogen Cycle

Amounts of nitrogen in a typical sugar beet field in an arable rotation are shown in Fig. 2.1. Several pools of nitrogen provide the crop requirement. The most important are usually organic matter when oxidation occurs as described above, fertilizer given specifically for the sugar beet and, in semi-arid regions, fertilizer applied to crops grown in rotation with sugar beet.

In rotations containing legumes, biological fixation is an important source. Field beans may fix 50–100 kg N ha^{-1} and lucerne up to 400 kg N ha^{-1}. Initially this nitrogen is incorporated in soil organic matter but it quickly provides a large quantity of mineral nitrogen for the following sugar beet.

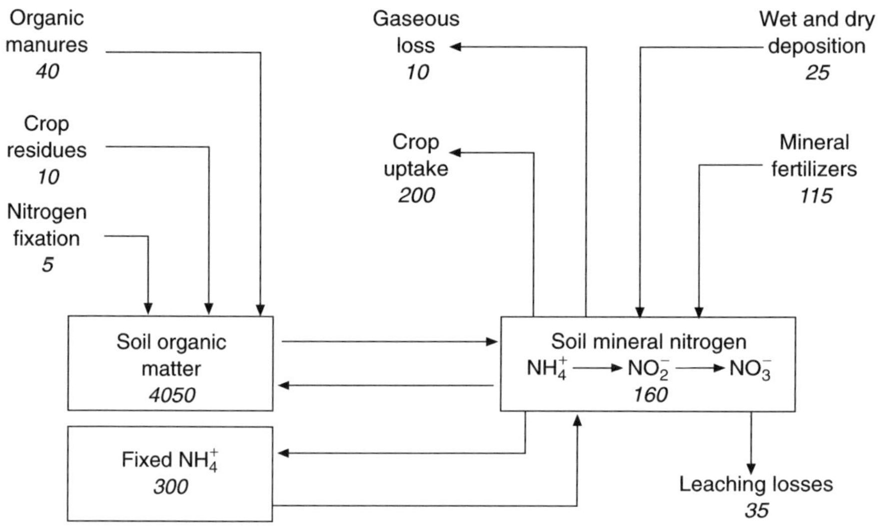

Fig. 2.1. Nitrogen dynamics of a typical sugar beet field (kg N ha^{-1}).

All forms of organic nitrogen (soil organic matter, organic manures, crop residues) are mineralized first to NH_4^+ at widely varying rates by many organisms present in the soil. This process is followed by nitrification, first from NH_4^+ to NO_2^- (by *Nitrosomonas*) and then from NO_2^- to NO_3^- (by *Nitrobacter*). All these reactions are temperature- and pH-dependent.

There may also be losses during the changes from organic to plant-available forms of nitrogen. The NH_4^+ may be lost as volatilized ammonia (NH_3) and NO_3^- by leaching. In some circumstances there are losses through denitrification where oxygen is lacking, such as when there is an excess of water in the soil. Denitrification converts nitrate into gaseous nitrogen (N_2) and oxides of nitrogen (NO and N_2O), which are lost to the atmosphere. In well-drained soils in good structural condition, both essential requirements for sugar beet (see Chapter 10), such losses are probably small. However, we have found no direct measurements of such losses made in sugar beet fields.

The C:N ratio of residues in soil affects both the immobilization and the mineralization of nitrogen. When the ratio is above 20:1, the net effect is for immobilization, while below that value mineralization is rapid. It has been suggested that additional nitrogen is therefore needed when a large quantity of cereal straw (C:N ratio 80:1) or other low-nitrogen residue is incorporated into soil.

Fixation of NH_4^+ in some clay minerals (vermiculite and smectite) may account for small amounts of nitrogen removal from the available pool. In turn, NH_4^+ may be released from the fixed form to become available to plants. However, this reaction is not of significant importance in most agricultural soils and prediction of the amounts is not possible with current technologies.

Nitrogen in the Plant

Nitrogen uptake

With any nutrient, the crop obtains part from applied fertilizer and part from soil reserves. In the case of nitrogen, the latter is mainly in the form of decaying organic matter or unused fertilizer given for previous crops. Table 2.1 shows the amount of nitrogen in the crop when it is harvested. The values (product of dry matter and total nitrogen) cover an extensive range of locations and soils and nitrogen-supply situations.

With low-yielding crops, as in much of the pre-1970 data, uptake by roots and tops is very small (50 and 125 kg ha^{-1}, respectively). Where soil provides or the crop receives large amounts of mineral nitrogen, uptake by roots and tops can be very large (140 and 215 kg ha^{-1}, respectively). Summarizing these and other data suggests that optimum values for uptake by roots, tops and total are 90, 110 and 200 kg ha^{-1}, respectively. There is scant evidence that even very high-yielding crops should contain greater than these.

Rising yields and decreasing nitrogen usage over the past 30 years in UK have dramatically changed the amount of nitrogen applied to produce a tonne of sugar beet. In 1970 about 6 kg N was being used for each tonne of roots produced at 16% sugar. By 2000 this had fallen to less than a third (1.7 kg N t^{-1}), as reported by Draycott and Martindale (2000a). It would appear that there is still scope for further improvement by optimizing uptake with minimum fertilizer application, as outlined below.

Concentration of nitrogen in dry matter of sugar beet at harvest

Nitrogen concentration in whole tops varies depending on nitrogen supply, but ranges from 1.0 to 3.5%. Concentration in roots is less variable, ranging from 0.5 to 0.8% (Table 2.2). The range for leaves (or laminae) is 2.2–3.5% and for stalks (petioles) 1.0–1.5%. Where crops have been growing for a time either in a deficient or luxury supply of available soil nitrogen, values outside these ranges may occasionally be found.

Process of nitrogen uptake

Unusually among common arable crops, the nitrogen uptake pattern of sugar beet has two desirable components. First, there must be

Table 2.1. Quantity of nitrogen in sugar beet at harvest.

Quantity of N taken up ($kg\ N\ ha^{-1}$)			Amount of N applied ($kg\ N\ ha^{-1}$)	
Roots	Tops	Total		Reference
Mean of pre-1970 published data				
50	125	175	113	Draycott, 1972
Range of soil types				
		164–215	125	Last and Draycott, 1975b
Optimum values				
85	115	200	125	Armstrong and Draycott, 1983
99	108	207	–	Hébert, 1987
Range of years and locations				
		105	0	Last *et al.*, 1983
		195	125	Last *et al.*, 1983
		245	207	Last *et al.*, 1983
		180–310	125	Armstrong *et al.*, 1986
18	59	137	0	Christenson *et al.*, 1993
95	71	166	45	Christenson *et al.*, 1993
100	81	181	80	Christenson *et al.*, 1993
111	92	203	110	Christenson *et al.*, 1993
126	106	232	145	Christenson *et al.*, 1993
Winter crop				
		134–249	0–240	López-Bellido *et al.*, 1994
Range of soils with large N residues				
110–130	160–200	270–330	0–200	Horn, 1994
Range of soil types, organic manure, etc.				
		134–193	0–100	Allison *et al.*, 1996a
		180	160	Märländer and Windt, 1996
80–140	75–215	155–355	0–240	Vereerstraeten *et al.*, 1997
		156–214	0–120	Mambelli *et al.*, 1997b
50–140	75–215	105–355	0–240	Range
90	110	200	–	Optimum values – see text

Table 2.2. Concentration of nitrogen in dry matter of sugar beet at harvest.

N concentration (% dry matter)			Fertilizer ($kg\ N\ ha^{-1}$)	
Roots	Petioles	Leaves		Reference
0.60	1.40	3.20	126	Draycott, 1972
0.70		2.00	100	Greenwood *et al.*, 1980
0.50	0.94	2.15	0	Christenson *et al.*, 1993
0.52	0.96	2.19	45	Christenson *et al.*, 1993
0.58	1.05	2.31	80	Christenson *et al.*, 1993
0.62	1.10	2.43	110	Christenson *et al.*, 1993
0.67	1.20	2.64	145	Christenson *et al.*, 1993
		2.4–3.5	120	Olsson and Bramstorp, 1994

sufficient to sustain growth; otherwise the crop will not produce optimum yield. For example, Armstrong *et al.* (1986) found that the rate of uptake could exceed 5 kg N ha^{-1} day^{-1} in summer. Secondly, when harvested, roots should contain the least possible amount of soluble nitrogen compounds to ensure maximum extraction of the end-product – sugar.

This balancing act has long intrigued sugar beet researchers and growers alike. The ideal for sugar beet is a plentiful but not excessive supply during much of the growing season and scarcity just before harvest. It contrasts sharply with a crop like wheat, where nitrogen is needed not only for growth but also to form protein in the end-product – grain.

With sugar beet the target is to achieve the required leaf canopy, as explained below, and to sustain it without excess uptake of nitrogen. Figure 2.2 illustrates the uptake pattern for beets with very high yields in north-west Europe. There is no reason to expect different results elsewhere. A small number of well-designed studies would be useful to test this result in other regions.

During the early stages of growth, the nitrogen demand of seedlings is relatively small but important for rapid early growth. Nitrate is taken up from the soil at an ever-increasing rate. Most of this demand is met by soil mineral nitrogen, supplemented by fertilizer.

In a year of average sowing date and weather, after 60 days the crop will have taken up only about 10 kg N ha^{-1}. However, as the longest day approaches and temperatures rise, there is always a phase of rapid growth. By August a good crop will have increased uptake from 10 to 150 kg N ha^{-1}. Thus 140 kg N ha^{-1} must pass from the soil into the crop in some 60 days at an average 2.5 kg N ha^{-1} day^{-1}. On warm, sunny days with moist soil, rates of uptake of nitrogen of nearly 6 kg N ha^{-1} day^{-1} have been reported.

After the middle of September days shorten and temperatures fall, leaves senesce and the uptake curve ceases to rise. During the autumn period less than 50 kg N ha^{-1} is taken up. At harvest (in November) with a large tonnage of roots (approaching 100 t ha^{-1}), high sugar percentage (18.5%) and low amino nitrogen (less than 100 mg 100 g^{-1} sugar), tops should contain no more than 100 kg ha^{-1} N and roots a similar amount (Draycott and Martindale, 2000a).

Effect of nitrogen on the growth and physiology of sugar beet

Of all plant nutrients, an application of nitrogen in most situations has the most spectacular effect on the appearance of the crop. Leaf colour changes from pale green or yellow to dark green. The area of leaves and the general vigour of the crop are visibly improved. More importantly, taproots increase in size and so do yield and profit.

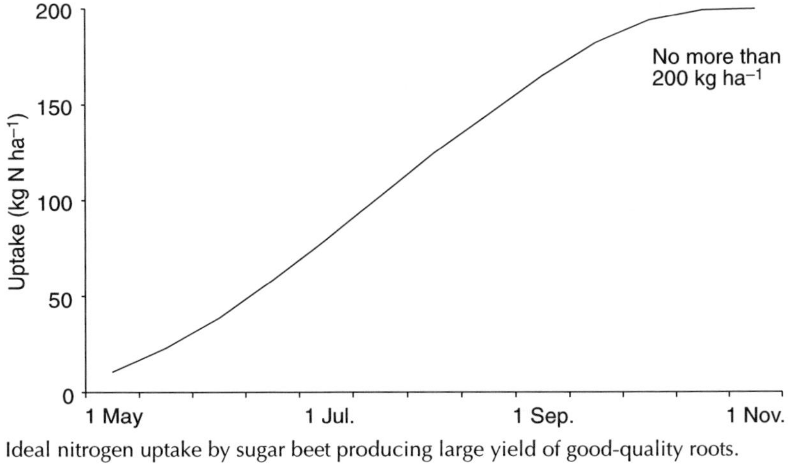

Fig. 2.2. Ideal nitrogen uptake by sugar beet producing large yield of good-quality roots.

The first detailed work on the causes of these changes is credited to Watson (1952) at Rothamsted, who investigated several crop species, including mangolds and sugar beet. He measured some of the effects of additional nitrogen (or a shortage) and tried to explain how nitrogen improved production. The procedures introduced for measuring cotton growth and yield (Balls and Holton, 1915) were modified for UK crops.

Watson (1947) introduced the concept of leaf-area index (LAI), which has proved to be extremely useful. He argued that, just as agricultural yield is expressed in terms of weight of crop per unit area of land, leaf area should be expressed in a similar way, and this measurement is the LAI. LAI is defined as the area of leaf on a plant divided by the area of land occupied by the plant. This concept has greatly helped the analysis of causes of variation in dry-matter yield.

For example, the measurement of LAI of sugar beet sown at different times and in several years helped explain why different yields resulted (Watson, 1952). Watson's group found that increased nitrogen supply increased LAI through increases in number of leaves and leaf size. They went on to show that leaves were larger because cells were larger than in the corresponding leaves of low-nitrogen plants. In contrast, the efficiency of the canopy of leaves was affected little by nitrogen, as determined by the 'net assimilation rate' (Watson, 1952).

Watson showed that for sugar beet the main function of nitrogen was to help the crop cover the ground with leaf so that energy from sunlight was not wasted by falling on bare soil. Although he did not prove it conclusively by experiment, he inferred that the crop needed an LAI of 3 as soon as possible. At LAI 3, the canopy captured nearly all the sunlight energy. In the USA, Stout (1961) stressed the importance of an early and sufficient supply of nitrogen for rapid leaf cover.

The next major step forward in understanding how factors (such as nitrogen) influence growth and yield through changes in leaf development was by Monteith (1978). He made measurements of sunlight energy being intercepted by different leaf canopies. Crop dry-matter yield increased linearly with the amount of radiation intercepted. This work led to the developments of Scott and Jaggard (1993), who applied the principles to the sugar beet crop. They measured radiation intercepted by crops given four amounts of nitrogen, from too little, through optimum to excess. Figure 2.3 shows the result. On plots where small amounts were

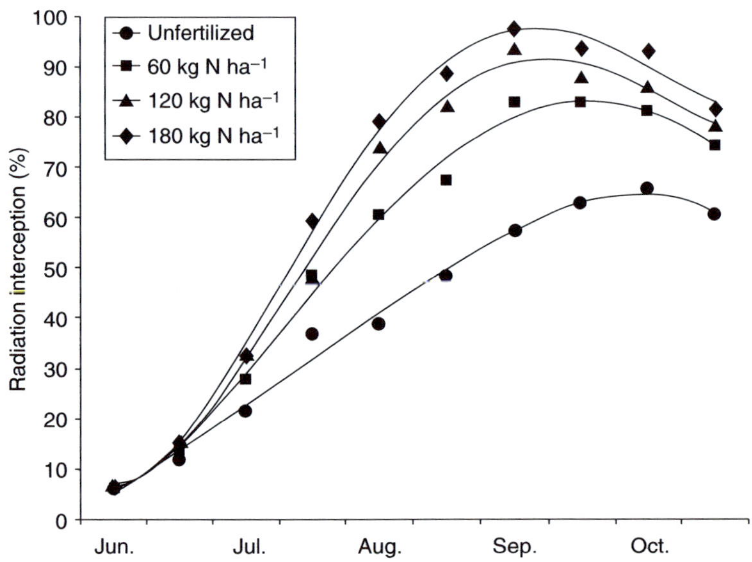

Fig. 2.3. Effect of nitrogen fertilizer on light interception by sugar beet (after Scott and Jaggard, 1993).

added, leaves grew slowly during June and July and throughout the summer much of the sun's energy fell on bare soil. Canopy growth was rapid where 120 kg N ha^{-1} was applied, and a leaf cover that intercepted more than 85% of the sunlight was produced by the end of July. Although fertilizer in excess of 120 kg N ha^{-1} resulted in more prolific leaf growth, it led to only trivial increases in radiation interception and failed to increase yield (Table 2.3), despite making the leaves appear darker green. These and other experiments have shown that this increase in chlorophyll concentration does not increase the efficiency of the leaf per unit area. Only severe deficiency of nitrogen associated with senescence or disease decreases this conversion efficiency (Armstrong *et al.*, 1983).

Table 2.3. Total dry-matter and sugar yields of the various patterns of light interception (Fig. 2.3) created by different nitrogen-fertilizer amounts at Broom's Barn in 1984.

Fertilizer nitrogen (kg ha^{-1})	Yield (t ha^{-1})	
	Dry matter	Sugar
0	10.9	5.9
60	14.0	7.9
120	18.3	9.5
180	18.3	9.5

Leaves are initiated and commence growth before they are visible. Nitrogen must be available during the initiation and growth stages before the leaves are visible in order to improve their subsequent size. Figure 2.4 shows this effect on leaf number 5 from initiation, appearance and unfolding to maximum length. Under ideal conditions it took at least 2 weeks for applied fertilizer to be effective. This finding has implications for the time of application of nitrogen.

Work at Rothamsted over the past 30 years (G.F.J. Milford, UK, 2002, personal communication) has done much to elucidate the underlying relationships between sugar beet leaf number per plant, individual leaf area, LAI and the effects of nitrogen supply. Each successive leaf produced by all sugar beet grows to a larger size than its predecessor up to a point in the leaf sequence when leaf size progressively decreases. Nitrogen increases individual leaf size (Watson, 1952) and Milford has since shown that nitrogen also causes the largest leaf to be produced later in the leaf sequence. The overall effect is to prolong top growth and increase leaf area per plant and maximum LAI (Armstrong and Milford, 1987).

Work at Broom's Barn by a group working on the growth of sugar beet under the direction of the late R.K. Scott during the

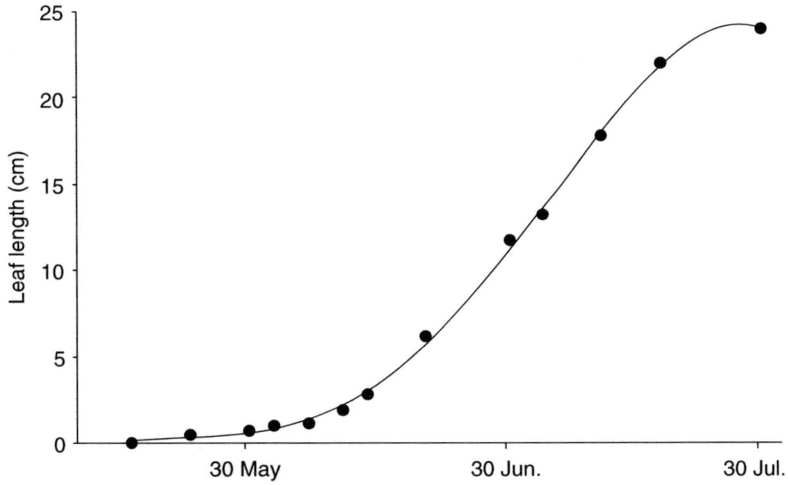

Fig. 2.4. Growth of leaf 5 from initiation to final size.

1980s and, as yet, not fully published (but see Scott and Jaggard, 1993, 2000) examined the effect of soil nitrogen supply on top growth and light interception. A wide range of soil nitrogen was ensured by fertilizer and soil type. With little nitrogen (no fertilizer, sandy soil) the largest leaf produced was the tenth. With optimum nitrogen (125 kg N ha^{-1}) the largest leaf was the 13th. With excessive nitrogen (from peat) it was the 20th. Research in progress (Malnou *et al.*, 2003) aims to improve understanding of leaf-canopy development throughout the growing season in relation to soil nitrogen.

Increasing nitrogen supply influences the partitioning of dry matter between plant parts, most notably increasing the proportion of leaf and petiole tissue. For example, Christenson *et al.* (1993) found in 13 experiments that applied nitrogen increased top growth 150%, but increased root growth only 120%. It appears that residual nitrogen creates the same effect. Shock *et al.* (2000) found that the root:total dry-weight ratio declined from 0.8 to 0.7 as the carry-over nitrogen from an onion crop increased.

Increasing the supply of nitrogen to sugar beet not only decreases the proportion of dry matter partitioned to tap roots but has profound effects on their physiology (Milford *et al.*, 1980) and processing quality (Pocock *et al.*, 1990). At the optimum, sugar yield is near its peak with minimum detrimental effect on quality, as discussed below and in Chapter 13. However, excessive supply of nitrogen increases the size of storage cells in the tap root and tissue hydration (i.e. the weight of water per unit dry matter). This is the main reason why the sugar percentage decreases with an increasing supply of nitrogen.

Effect on quality

The depressing effect of excess nitrogen (nitrogen supply above that for optimum sugar extraction) was reported very early in the 20th century (Headden, 1912). Gardner and Robertson (1942) showed that excessive nitrogen reduced sugar concentration and the purity of sugar beet. They found a near-linear relationship between nitrate nitrogen in the root at harvest and decrease in sugar percentage. Numerous reviews have been published concerning the effects of excess nitrogen on sugar beet quality (Rounds *et al.*, 1958; Haddock *et al.*, 1959; Stout, 1961; Hills and Ulrich, 1971; Draycott, 1972). A full account is given in Chapter 13, which deals with the effect of nutrition on root quality.

Partitioning and Redistribution of Nitrogen

Sampling plants from soon after germination at weekly or monthly intervals throughout the growing period to harvest gives much information on the need and pattern of uptake of nitrogen. Dividing samples into components (usually leaves (laminae), stalks (petioles) and roots) shows how nitrogen is partitioned. More detailed analysis also indicates that considerable quantities of the element are redistributed within the plant.

When sugar beet seed germinates, the radicle emerges first and becomes the primary root. Both are covered with root hairs. Usually when the primary root is 5–10 mm long, the hypocotyl emerges from the seed with yellow, folded cotyledons and progresses towards the soil surface. Within hours of the cotyledons appearing above the soil they turn green and start photosynthesizing, and growth in the true sense begins.

Information on nitrogen nutrition during the very earliest phase from germination to the beginning of photosynthesis is sparse. It has often been assumed that the seed itself contains enough nitrogen to supply this need. Work in France (Dürr and Mary, 1998) has shown that nitrate metabolism is important before cotyledons unfold. This is perhaps not surprising and the presence of so many roots suggests the uptake of nitrate in water from soil. Dürr and Boiffin (1995) reported that, 10 days after sowing, seedlings contained from 0.1 to 0.2 mg N. These quantities were roughly equal to the amount of total nitrogen in the seed initially and were directly related to seed size.

Mäck and Tischner (1990) studied the effect of endogenous and externally supplied ammonium and nitrate nitrogen on uptake by seedlings. They found that the two forms of nitrogen had large differences on amino nitrogen concentration in storage roots from a very early stage in the life of the plants. Lexander (1993) reported that lipids and proteins used during germination were stored in the cotyledons. Detailed measurements by Dürr and Mary (1998) in growth chambers showed that, out of a total of 0.1 mg N in a seedling at emergence, 0.05 mg was in the cotyledons, 0.03 mg in the hypocotyl and 0.02 mg in the roots. There is a need for this work to be continued in the field situation to define the optimum soil and plant nitrogen regime during the very earliest phase.

Measurements were made in UK on seedlings during the next stage of growth when average plant dry weight was about 1.5 g (Last *et al.*, 1983). Six amounts of nitrogen fertilizer were tested, from 0 to more than 200 kg N ha^{-1}. Seedling dry weight was increased by over 50% on average in 6 years of trials, showing the importance of fresh fertilizer in April/May. Seedlings weighed the most with 150 kg N ha^{-1} fertilizer. The ideal amount of mineral nitrogen in the top 0–30 cm of soil was approximately 40 mg kg^{-1} compared with 13.5 mg kg^{-1} when no new fertilizer was applied.

The spring-sown crop then enters the phase of most rapid growth. In north-west Europe this starts in May and continues through to October. Figure 2.5 shows how the nitrogen uptake rises rapidly as the crop attains full leaf cover and photosynthesizes at the maximum rate about the time of the longest day. During this period the amount of nitrogen uptake per day may exceed 5 kg N ha^{-1} day^{-1} (Armstrong *et al.*, 1986), the range reported being from 1.6 to 5.4 kg N ha^{-1} day^{-1}. Parallel studies in France (Jourdan *et al.*, 1992) on a high-yielding soil (final root yield, 100 t ha^{-1} at 16% sugar) showed that during June sugar beet tops were taking up 3 kg N ha^{-1} day^{-1}. When complete ground cover was reached, nitrogen began to be redistributed to the roots. Figure 2.5 is based on data from several European countries and is suggested as a target uptake and redistribution pattern for high yield.

Armstrong *et al.* (1986) measured the net uptake and amounts of nitrogen redistributed between the plant parts from August through harvest over 3 years on two contrasting sites. Autumn uptake ranged from no net uptake to 1 kg N ha^{-1} day^{-1}. Remobilized nitrogen represented as much as 80% of the net increase in taproot nitrogen.

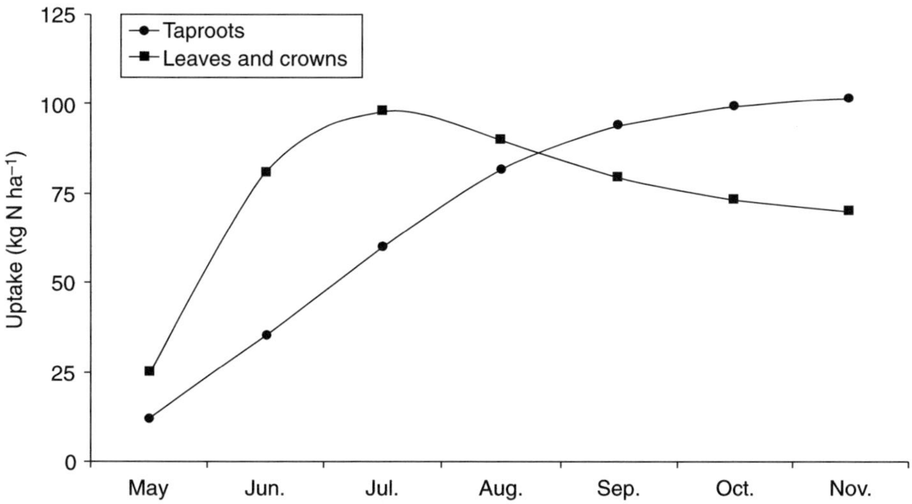

Fig. 2.5. Nitrogen uptake by leaves plus crowns and taproots of a high-yielding crop.

Relationship between nitrogen uptake at harvest and sugar yield

Graphs of total nitrogen uptake and sugar yield in field experiments made in the 1970s first indicated that there is probably a desirable upper limit at about 200 kg N ha^{-1} (Last *et al.*, 1983). Even in much higher-yielding crops subsequently grown in the last 20 years in the UK and other countries in the EU, there has been no clear case for more than 200 kg ha^{-1} uptake (Armstrong and Milford, 1985; Armstrong *et al.*, 1986; Pocock *et al.*, 1988; López-Bellido *et al.*, 1994; Wendenburg, 1996). Figure 2.6 summarizes the data presented by these and other workers. Studies are continuing to help substantiate and explain this suggested relationship.

Effect on yield of sugar

This has been measured in many field experiments in all parts of the world where sugar beet is grown. On most soils there is insufficient available nitrogen present to produce a full yield of sugar without nitrogen fertilizer. On all these fields, increasing increments of fertilizer at first rapidly increase the amount of sugar produced by the crop. Eventually, further increments increase yield little, if at all. In some cases there are reports of large amounts of fertilizer decreasing yield. There have been few investigations of the precise shape of the nitrogen/sugar-yield response.

Crowther and Yates (1941) assumed and used an exponential curve (Fig. 2.7a) to estimate the nitrogen requirement of sugar beet. The form adopted was similar to that used by Mitscherlich (1954), with no provision for decreases in yield with large dressings of fertilizer. However, Boyd (1961) proposed a parabolic curve (Fig. 2.7b), suggesting that there was a clearly defined optimum dressing and that, on average, more than 100 kg N ha^{-1} decreased sugar yield considerably.

During the following 10 years many experiments were made on farmers' fields in the UK testing 0–225 kg N ha^{-1}. These were partly for research purposes on the use of plant and soil analysis, and partly in a drive to decrease excess usage of nitrogen fertilizer by growers. Boyd *et al.* (1970) collated and analysed data from these experiments and found that on nearly all fields two straight lines (linear-plateau) best fitted the relationship between nitrogen application and sugar yield (Fig. 2.7c). Nitrogen fertilizer increases yield steeply along a straight line until a point is reached where more had little effect.

Plant population may be decreased by fertilizer, as described below. A linear plus exponential form of relationship then fits best (Fig. 2.7d). This has been described by Allison *et al.* (1996a).

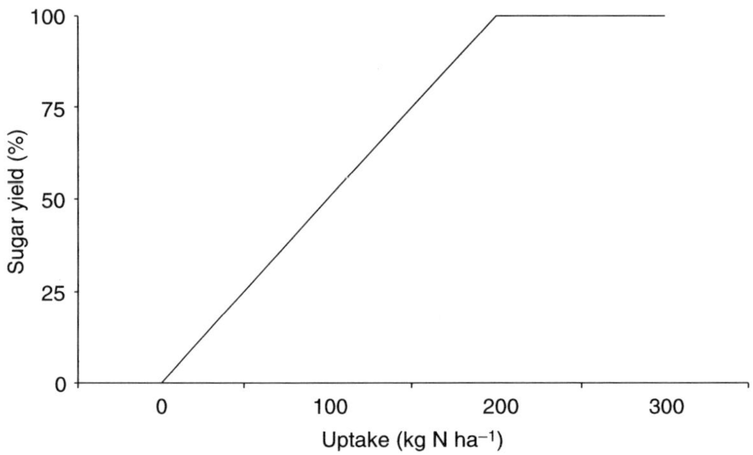

Fig. 2.6. Relationship between nitrogen uptake at harvest and sugar yield.

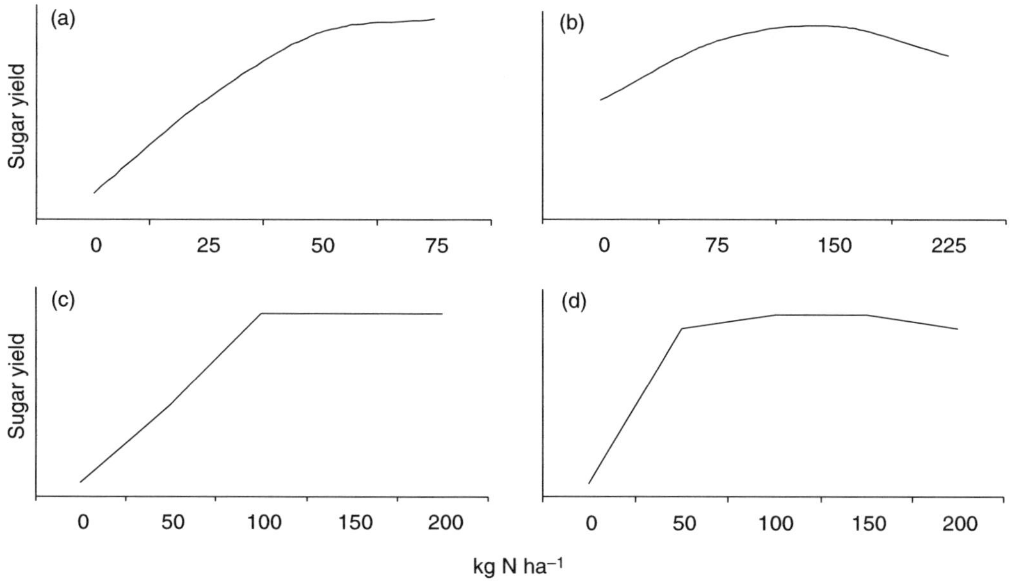

Fig. 2.7. Various proposed relationships between applied nitrogen and sugar yield: a, exponential; b, parabolic; c, two straight lines (linear-plateau); d, combination of straight line and quadratic.

Märländer and Ladewig (1995) compared the use of various forms of the relationship between nitrogen fertilizer and sugar yield in northern Germany. A parabolic fit of the type in Fig. 2.7b produced an optimum of 120–130 kg N ha^{-1}. Fitting two straight lines (Fig. 2.7c) produced an optimum of 70–80 kg N ha^{-1}. The authors found that the latter fitted the data best and was desirable because it would result in the use of less fertilizer and in better beet quality and have less effect on the environment.

Determining Nitrogen-fertilizer Need

Plant-tissue analysis

Over the past 50 years there has been considerable progress in deciding the nitrogen status of sugar beet plants by analysis of plant parts, usually the lamina or petiole. Ulrich (1948) began the work in California and his techniques have been adopted with varying success elsewhere. Laminae and petioles of healthy, well-nourished sugar beet plants contain a large amount of nitrogen in the nitrate form (Wright and Davidson, 1964; Murphy and Smith, 1967), and Ulrich and others (Ulrich, 1950; Ulrich and Hills, 1952; Ulrich *et al.*, 1959) set out the basis of using the values to improve nitrogen applications. Sørensen (1960, 1962) tested their usefulness in Denmark, and White (1959), Last and Tinker (1968), Last and Draycott (1975a,b) and Armstrong *et al.* (1986) in the UK.

A major problem with the technique is ensuring that the sample taken (whether laminae or petioles) produces analytical results that can be interpreted correctly. Two factors affect the result greatly: the position of the plant part taken and the age of the plant (see Draycott, 1972). At any stage of growth of a plant, the outer leaves contain the greatest concentration of nitrate nitrogen. Progressively towards the central leaves, nitrate drops off rapidly, often by a factor of 5–10 in dry matter. Thus it is important to take clearly defined laminae or petioles. To complicate matters further, nitrate in both plant parts decreases rapidly from spring through summer to autumn. In this case a fivefold change is common.

Brown (1943) and Ulrich *et al.* (1959) recommended using the youngest fully expanded lamina and/or its petiole; it is the only one that can be defined easily and most workers have since used it. To decide whether a crop contains sufficient nitrogen to yield fully, Ulrich (1961, 1964) introduced the concept of 'critical concentration' for sugar beet, below which a plant is considered nitrogen-deficient. This was 1000 p.p.m. of nitrate nitrogen in the dry matter (100 p.p.m. in fresh material) of petioles. He considered that this was the desirable state at 11–12 weeks before harvest, from when nitrate concentration should fall off to improve sugar accumulation.

The theory was put to the test in the UK first by Last and Tinker (1968). As the amount of nitrogen fertilizer needed by sugar beet for maximum sugar yield varies from field to field, they suggested that a small application of nitrogen might be made in the seedbed and the rest (based on plant analysis) given in June. The petiole nitrate concentration decreased rapidly with time, from about 1000 p.p.m. (in fresh tissue) in early June to less than 100 p.p.m. in early September. On average, petiole nitrate concentrations of about 800 p.p.m. in June were associated with the largest sugar yields, but Last and Tinker (1968) considered that the method was not accurate enough to predict the nitrogen top dressing requirement of sugar beet on individual fields. Although the nitrate concentration was greatly increased by nitrogen fertilizer and seemed to be a good indication of the immediate nitrogen status of the crop, the effect both of age of plants and of nitrogen applications differed between experiments. This, together with the possibility of a varying supply of soil nitrogen during the rest of the season, made it difficult to estimate with any accuracy the amount of additional nitrogen fertilizer needed.

The study was continued by Last and Draycott (1975a,b) on widely differing soil types. Measurements were made at monthly intervals from June to November on petioles of plants given 0, 125 and 250 kg N ha^{-1}. Soil mineral-nitrogen measurements showed that the petiole nitrate concentration did reflect the soil supply in spring and early summer but not in autumn. The experiments led to the definition of critical concentrations for UK conditions for the fresh petiole of the youngest fully expanded leaf. *See table below.*

These were greater than Ulrich's values but work by Armstrong *et al.* (1986) has confirmed the need for an early rapid uptake of nitrate, a shortage always leading to depressed sugar yield.

In Israel, Albasal *et al.* (1970) found a close linear relationship ($r = 0.81$) between yield of roots of autumn-sown sugar beet and petiole nitrate concentration. Nitrate concentration reached its maximum value in December, about 3 months after seedling emergence. Albasal *et al.* (1970) stressed that the crop should not be allowed to become short of nitrogen during this period; otherwise loss of yield results. Hale and Miller (1966) in California tested the technique of petiole analysis to decide whether it was suitable for predicting optimum harvest date. They found that nitrate concentration in the petiole was not sufficiently correlated with sugar yield for it to be useful in this way. Also in California and using the same technique, Hills *et al.* (1963) made experiments to determine how long sugar beet should be deficient in nitrogen (< 1000 mg kg^{-1} nitrate-N in the dry petiole) prior to harvest to obtain maximum sugar percentage. Analysis of petioles did not give the answer. Even with apparent mid-season nitrogen deficiency, there was little effect on rate of root growth and they concluded that more research was needed into the value of the test in commercial practice.

In a somewhat different approach Last *et al.* (1983) measured nitrate concentration in plants throughout the growing season. Rate of dry-matter production (kg ha^{-1} day^{-1}) up to the beginning of June was related to nitrate concentration. By September the relationship completely disappeared. Ulrich (1950) sug-

	July	August	September	October
Nitrate-N (mg kg^{-1})	400–500	200–300	100–150	100

gested that the concentration should be above 1000 mg kg^{-1} until mid-season. Concentration values below that for any appreciable length of time may result in lower root yield.

Other subsequent research in the USA explored the possibility of predicting petiole nitrate concentration, e.g. from soil nitrogen concentration in spring and amount of fertilizer applied (Carter *et al.*, 1971; Giles *et al.*, 1977; Gilbert *et al.*, 1981). In Colorado and Idaho where the work was conducted, the peak concentration occurred in late June or early July, following a mid- to late-April sowing. Since late-season nitrogen application may reduce sugar in the harvested crop, the information would need to be available prior to the latest date that nitrogen should be applied. This procedure would have merit in areas where sugar beet has a longer growing season such as where sowing is in early September and harvest in May or June.

Another use would be for prioritizing the order in which fields should be harvested. A delay in harvest could be scheduled for those with higher predicted nitrogen petiole nitrate, while those with lower predicted values could be harvested earlier. As farms get larger and harvest is spread over longer periods of time, this could be a factor in getting better-quality beets at all harvest times.

It would seem from this and other work that critical nitrate concentrations exist early in the season but much more research is needed to further define just when and how they could be useful in commercial practice.

Taproot analysis for amino N

Many tare house laboratories throughout the world routinely test incoming lorry-loads of sugar beet at harvest for amino N concentration. Often this forms the basis of payment for quality and sugar extraction (Chapter 13). Several researchers have suggested that these measurements might be used retrospectively to advise on the amount of nitrogen fertilizer needed by the crop.

Marcussen (1985) in Denmark produced encouraging evidence. He found that 1 mg amino N per 100 g sugar in roots corresponded to 2 kg N ha^{-1} fertilizer. For an optimum total uptake of 220–260 kg N ha^{-1} the theoretical need was for roots to contain 110–130 mg 100 g^{-1} sugar of amino N. By advising growers of their data, Marcussen reports a large decrease in overuse of nitrogen fertilizer. Over a 10-year period it fell from 190 to 140 kg N ha^{-1}. Similar findings have been reported in the UK (Draycott *et al.*, 1997), discussed further in Chapter 13.

Chlorophyll meters

At the beginning of the new millennium, new technologies making full use of electronic sensing are just beginning to be applied to sugar beet. After some 50 years of work with chemical analysis of soil and plant, research still aims at finding a reliable, accurate and simple method of fine-tuning nitrogen-management decisions during the growing season so that they relate to weather and crop conditions. Recently, hand-held chlorophyll meters that read leaf chlorophyll concentrations (which are correlated superficially with leaf nitrogen) have become readily available for testing on sugar beet and other crops. In theory the need for additional nitrogen could thus be rapidly assessed on site (Guérif *et al.*, 1995).

Workers in Italy (Mambelli *et al.*, 1997a, b; Tugnoli and Bettini, 2000) looked at sugar beet crops given 0–180 kg N ha^{-1} at various stages after emergence. They compared leaf total nitrogen concentration with petiole nitrate and chlorophyll-meter readings. The first two were very closely correlated ($r < 0.9$). Relationships between them and the chlorophyll meter were good in young plants but decreased with time. After the 10–12-leaf stage the relationships were not significant. Further work is needed to determine whether the instrument could be used to predict an early shortage of nitrogen, so that it could be rectified with more fertilizer.

Remote sensing

Other non-destructive testing methods of soil and plants for nitrogen (and other nutrients) are being developed and some are at the start of commercial application. They fall into three main types: those which use sensors in the crop, those which employ positioning satellites to produce field maps of soil or plant characteristics, and those which employ aerial or satellite instruments sensitive to leaf characteristics.

Steven (2001) and Ferguson (2001) reviewed these various approaches while Jaggard *et al.* (1997a) tested some of them with sugar beet. Chlorophyll absorbs red light while leaf structures strongly reflect light in the near infrared. Conversely, soils show a quite different effect. Thus the spectrum measured over a growing crop of sugar beet changes from that of soil to a spectrum characteristic of a leaf cover. Scott and Jaggard (1993) described great progress with understanding the basis of yield in sugar beet through such measurements and linked them to nitrogen supply.

Accounting for the nitrogen returned to the soil from sugar beet tops is important in the management of nitrogen for subsequent crops. In dry areas large quantities of nitrate may accumulate in the soil between sugar beet crops. Moraghan *et al.* (2000) utilized aerial colour photographs of a late-season sugar beet canopy to evaluate the amount of nitrogen contained in tops. This in turn was an excellent predictor of the response of a following wheat crop to applied nitrogen. Reduction of nitrogen use on crops grown between sugar beet crops in a rotation have the potential to reduce residual nitrogen before sugar beet, particularly in areas where there is less than 500 mm of annual precipitation.

Site-specific fertilizer management

Site-specific fertilizer management aims to utilize the information described above. Measurement of variability in the field may include remote sensing, soil sampling and yield mapping. The ability to respond to variability has been enhanced by development of equipment that can apply varying amounts of material as the spraying or spreading machine traverses the field. The common term used for this is 'variable rate application'.

Work directed to the better management of nitrogen has centred on the determination of how intensively fields should be sampled. Sampling every 0.5 ha on a grid gives sufficient sensitivity for the evaluation of variability encountered across the field. However, this sampling intensity has proved too expensive for practical application (Franzen, 1999a). Zone sampling appears to be a superior method. Zones delineated by soil type, topography, remote-sensing imagery of previous crops or a combination of these factors appear to be more practical. The data collected are used to construct a map that serves as the basis for nitrogen-fertilizer application. A computer utilizing the map generated controls the variable-rate applicator.

Results from North Dakota and Minnesota are inconsistent in regard to the economic benefit of site-specific use for sugar beet production. One problem associated with this work is the lack of replication. Large tracts of land are required to test adequately variable rates, making replicated research difficult. Smith (1997) conducted a well-designed study for 3 years. Grid soil sampling and variable-rate nitrogen fertilization were compared with conventional soil sampling and a single nitrogen application rate under three distinctly different residual-nitrogen situations. Average net increase to variable-rate nitrogen application ranged from 7 to 10% of gross revenue. However, Lamb and Rehm (1999) reported that, in the southern Minnesota growing region, variable-rate nitrogen application based on grid sampling did not cause root yield and quality to be any different from those with the use of a constant rate of application.

Additional research and experience are needed to evaluate this technology. The science supporting the tools at hand is still being developed. A series of fact-sheets (Franzen, 1999a,b,c,d) provides some discussion of these new tools.

Soil Analysis to Predict Nitrogen Requirement

The uptake of nitrogen by a sugar beet crop producing a large yield of roots needs to be about 200 kg N ha^{-1}. In nearly all situations a large proportion of this comes from the soil – usually much more than half – the remainder being from fresh fertilizer. To be in a position to predict the portion provided by the soil is highly desirable.

Work began in earnest in the 1970s in several countries, and continues, measuring ammonium- and nitrate-nitrogen concentrations in soils before sugar beet. Measurements are often made to about a metre depth while with other major nutrients it is conventional to measure only the surface 0–30 cm. In addition, many researchers also studied the amount of nitrogen mineralized during the growing season because it plays a significant part in the nitrogen taken up by the crop.

In Europe

The total quantity of nitrogen in soils where sugar beet is grown in Europe varies widely but most of the crop is grown on mineral soils that have been tilled for hundreds of years. These old arable soils contain little organic matter, usually no more than 1–2.5%; consequently the total nitrogen concentration is also small, often as little as 0.1%. Most of this nitrogen is tightly bound in the organic matter, with only a small portion released during the decomposition or 'mineralization' of organic matter. Together with residues of unused nitrogen fertilizer given for previous crops, this represents the whole of the soil nitrogen that is available for the sugar beet.

Mineralization of organic nitrogen takes place very slowly in the northern latitudes of the sugar beet-producing areas of the world because of the low temperatures experienced during late autumn, winter and early spring. These conditions are often accompanied by loss of nitrate by leaching when soils drain after rainfall or by denitrification where soils do not drain freely. When soil temperatures rise in spring, a flush of mineralization takes place that provides some of the nitrogen needed by the crop.

The first occasion when the value of soil mineral-nitrogen analyses for sugar beet was aired thoroughly on a European stage was in Brussels in 1978 (Proceedings of 41st Institut International de Recherches Betteravières Winter Congress). As would be expected in north-west Europe, much attention was given to the leaching of nitrate in winter and spring. Kolenbrander (1978) reported that the 'water surplus' or 'through drainage' (the amount by which precipitation exceeds evapotranspiration) ranges from 0 to 500 mm $year^{-1}$, with an average of 250 mm $year^{-1}$. In The Netherlands, September–March was a period of leaching but, during the growing season, April–September losses were nil, except in extremely wet conditions.

Müller (1978) reported 10 years of mineral-nitrogen measurements in West Germany. During winter there was movement out of the plough layer but not complete loss from the profile. Consequently sugar beet took up 100–200 kg N ha^{-1} from soil alone. Useful comparisons were made between sugar beet plots and those remaining as clean fallow. This showed that in March both contained 50 kg N ha^{-1} in the mineral form. By late August fallow plots contained 160 kg ha^{-1} (110 kg ha^{-1} of mineralization). In contrast, sugar beet plots contained almost none, the crop having taken up the spring 50 kg ha^{-1} plus the mineralized 110 kg ha^{-1}.

Günter (1978) reported parallel studies in southern Germany. Widely differing soil types were compared, from deep loess clay to sandy loam. In the former, there was 120 kg N ha^{-1} present in spring and 260 kg ha^{-1} mineralized by June to 120 cm depth, giving 380 kg ha^{-1} available to the crop. In the latter, there was 40 kg ha^{-1} in spring and 200 kg ha^{-1} mineralized by June. Given no fertilizer, the sugar beet crops took up 220 and 200 kg ha^{-1}, respectively, suggesting little, if any, need for fresh fertilizer.

Draycott and Last (1978) also summarized 10 years' work in the UK, during which they attempted to relate measurements of soil mineral nitrogen in autumn, winter and spring and at monthly intervals from before sowing until harvest to the nitrogen requirements of the crop. They also tested methods of predicting the amount of nitrogen that would mineralize during the growing season by laboratory incubation of soil taken before sowing. All fields studied had been in cereals prior to sugar beet and nitrogen residues were small. To a sample depth of 90 cm, soils contained 110 kg N ha^{-1} in autumn, 75 kg ha^{-1} after winter and 100 kg ha^{-1} when the crop was sown. After growing sugar beet, 40 kg ha^{-1} remained the following November. About 60 kg ha^{-1} was released by mineralization, the amounts being weakly related ($P < 0.05$) to the optimum nitrogen dressing and the magnitude of the response to nitrogen ($P < 0.01$). Christmann (1978) reported similar average values of mineral nitrogen in French fields. At the end of the winter, 50–100 kg ha^{-1} was present to 1 m. Mineralization produced 100–200 kg ha^{-1} during the growing season. Long-term studies suggested that this was 1.5% of the total nitrogen in the soil.

Neeteson and Smilde (1983) compared methods of recommending nitrogen fertilizer based on soil analysis in Belgium, The Netherlands and Germany. In all three countries a highly significant relationship had been found between the amount of mineral nitrogen present just before sowing and the optimum dose of fertilizer. They thought that sampling to 60 cm depth was sufficient, after comparisons with values to 100 cm. Analogides (1983) came to the same conclusion after detailed experiments in Greece. Pfleiderer (1983) examined nitrate-nitrogen values in 1000 fields in the Lower Saxony area of Germany to decide whether it was necessary to sample the 60–90 cm layer. Measurements on the 0–30 and 30–60 cm layers were sufficient if a weighting factor was included.

Working in Belgium, Boon and Vanstallen (1983) suggested that accuracy could be improved by including a factor for soil organic matter. In France, Machet and Hebert (1983) found that a third of fields performed best with no fresh nitrogen fertilizer. These soils provided (ideally) 250 kg N ha^{-1} in good growing conditions. Similarly, in Switzerland, Walther (1983) found that for optimum production the soil profile should contain 180–200 kg N ha^{-1}.

In the USA

In areas with less than 500 mm of annual precipitation NO_3-N may accumulate in the soil while in those areas with greater than 500 mm nitrate will leach or denitrify over winter. Erickson and Ellis (1971) examined concentrations of nutrients in water emitted by tile-drainage lines in sugar beet-growing region of Michigan. They found that the outflow carried less than 10 kg N ha^{-1} $year^{-1}$. This did not vary appreciably across different management systems. Christenson *et al.* (1993) sampled soils in both spring and autumn. These were poorly drained, but had had tile drainage. High concentrations of NO_3-N present in the autumn had disappeared when sampled in the spring. Based on the previous report, it is assumed that much of the nitrogen present in sugar beet soils in autumn is lost by denitrification.

In the Red River Valley, nitrate accumulation in the soil profile was reducing the quality of the sugar beet grown with less than 450 mm annual precipitation. Hilde *et al.* (1987) reported on a programme to reduce the amount of residual nitrogen in these soils. Sugar percentage had declined from 16.5 to 15.6 over a 30-year period. A change in management of the nitrogen supply to crops was put in place over a period of 6 years. Residual nitrogen in fields was reduced from 300 kg to 190 kg N ha^{-1}. This was accompanied by an increase in sugar recovery of 16%.

Christenson *et al.* (1993) examined results from over 50 experiments across a range of soil types where spring soil samples were taken to a depth of 60 cm. They concluded that there was a very low probability of response to applied nitrogen when the surface 60 cm of soil contained more than 50 kg N ha^{-1}.

Electro-ultrafiltration (EUF)

When used for measuring available nitrogen in soil samples, this technique not only determines mineral nitrogen present but also estimates the amount that may be mineralized during the growing season. It is used in several countries in Europe, notably Austria, Germany and Ireland. Soils from 0–30, 30–60 and 60–90 cm are sampled, dried, ground and analysed.

Mengel (1994) reported on a working-group study of EUF. The method extracts inorganic and organic ions out of a soil suspension by means of electrostatic forces. Nitrogen compounds extracted were nitrate, ammonium and some amino-nitrogen compounds. Pot and field experiments led to the conclusion that EUF values were highly significant for assessing optimum fertilizer amounts. In an early study in Austria and Germany, Wiklicky *et al.* (1983) found highly significant correlations between EUF-N and the amount of nitrogen fertilizer needed ($r = 0.86$, $P < 0.001$). In a later report, Wiedemann (1994) said that EUF had been adopted as the basis for recommendations for nitrogen in Austria as early as 1974 and Germany in 1982. It was used on 180,000 ha of various crops each year. This has resulted in improved yields and quality. Britton (1994) also reported that the EUF method had been in use in Ireland for 10 years with great success.

Other researchers have not been so enthusiastic. Andkjaer *et al.* (1994) had grave doubts about the value of the EUF system in Denmark, because of large fluctuations in the amount of mineralization during the growing season from year to year. Similarly workers in France (Machet *et al.*, 1994) compared EUF with soil mineral nitrogen concentrations and found the EUF forecast to be less reliable than conventional mineral nitrogen. Likewise, Allison and Armstrong (1995) and Allison *et al.* (1996a) compared various methods of analysing soil in the UK, including the EUF system. They reported in their tests that EUF was less reliable than an index system already in place in the UK. With EUF, only 40% of fields would have received within 30 kg N ha^{-1} of the correct amount. Forty per cent would have received more than 60 kg N ha^{-1} too much or too little. They concluded that the method was both expensive and not very accurate in UK conditions. Houba *et al.* (1994) concurred and suggested that a neutral salt, such as $CaCl_2$, be used.

The difference in the conclusions between countries remains to be resolved. It would seem that, where the amounts of nitrogen present are large and vary over a wide range (Austria, Germany, Ireland), EUF is useful. Where amounts and range are often small (the UK, France, Denmark, the Great Plains in the USA), its value is doubtful. More comparative work is needed to make the most of EUF.

Potentially available nitrogen soil tests

The first large-scale study in the UK (Last and Draycott, 1971), on 65 fields in 1961–1965, attempted to predict the amount of nitrogen fertilizer which sugar beet would need from analysis of the soil from each field. Both aerobic and anaerobic incubation of fresh and air-dried soil was tested to determine the amount of nitrogen released in the laboratory. Topsoils and subsoils to 60 cm were sampled in autumn and spring. It was concluded that autumn sampling was not useful but spring values were related ($P < 0.01$) to optimum nitrogen application.

Working in Austria, Bronner (1983) reported that the amount of nitrogen released during incubation provided an index of the capability of soils to supply nitrogen to the growing crop. In addition, measurements on subsoil samples improved the forecasts. Lindén and Nouno (1983) made measurements of mineral nitrogen in spring in Sweden in soil from 0–90 cm and found they varied from 25 to 200 kg N ha^{-1}. There was a loose relationship with optimum dressing due to a large amount of mineralization during the growing season. They considered that a measure of mineral nitrogen in spring coupled with an incubation test would be needed.

Nitrogen recommendations should account for nitrogen mineralized from organic matter in the soil. Excellent reviews of procedures that measure organic nitrogen in the soil are available in the literature (Keeney, 1982; Campbell *et al.*, 1993; Mulvaney, 1996; Stevenson, 1996). These procedures extract organic nitrogen from the soil with the goal of providing indices of availability that may be used in fertilizer recommendations. Several workers have reported that mineralizable nitrogen is important in the nutrition of sugar beet (Roberts *et al.*, 1972; Carter *et al.*, 1974; Westerman and Crothers, 1980).

Crop residues have an influence on the mineralization potential of soils. Christenson and Butt (1997) showed that nitrogen mineralization potential (Campbell *et al.*, 1993) was closely related to the estimated amount of crop residue returned to the soil in a long-term cropping-systems study. Crop residues returned over a 20-year period ranged from 75 to 145 t ha^{-1}, with an associated increase in soil mineral nitrogen from 70.7 to 94.3 mg kg^{-1}. It required 0.33 t crop residues ha^{-1} to increase the soil mineral nitrogen by 1 mg kg^{-1}.

Carter *et al.* (1974) suggested that mineralization in soils of southern Idaho was relatively constant from year to year and therefore it was not necessary to measure mineralizable nitrogen each year. Relationships between mineralizable nitrogen and sugar beet-yield response or uptake have not shown much promise. Bronner and Bachler (1980) found poor correlations between hot-water-extractable organic nitrogen and nitrogen uptake, yield and fertilization. Varsa (1970) also found poor correlations between nitrogen released on incubation or by autoclaving and yield of roots or sugar.

Christenson and Butt (1998) investigated the use of mineralization potential for the major soil series in the sugar beet production area of Michigan. It was thought that a specific index of organic nitrogen availability could be applied to each series, aiding in nitrogen-fertilizer recommendations. They reported as much variability within as between soil series, rejecting the original hypothesis.

Conclusions on the value of soil analysis for predicting the need for nitrogen fertilizer

Over the past 30 years a large amount of research has been done in nearly every country where sugar beet is grown. Initially the work set out to aid the grower. Concerns about overuse of nitrogen and resultant poor processing quality of roots increased the interest in soil tests. Work is still in progress to extend and refine the work described in the reports above. It is clear that introducing soil tests leads to reductions in fertilizer use.

Measurements from soils from 0–30 and 30–60 cm are necessary on soils with more than 500 mm of annual precipitation. Samples to at least a metre in depth are needed when the annual precipitation is less than 500 mm. Mineral nitrogen measured before sowing appears to be the best time to take samples for the more moist conditions, but autumn sampling is satisfactory in drier conditions. Good correlations can then be expected between soil nitrogen summated from 0–60 cm with the optimum amount of fertilizer, provided there is a wide range of values of soil nitrogen. When fields all follow a run of crops leaving little residue, correlations are poor (Last *et al.*, 1994).

Several countries have adopted EUF to extract nitrogen from soil samples, with considerable success. Again the method works well when profiles vary greatly in amounts of nitrogen present. Laboratory incubation methods also hold out hope for predicting what is still the most difficult question to answer: how much nitrogen will become available during the growing season?

Real and apparent recovery of nitrogen fertilizer

Most sugar beet crops contain much more nitrogen at harvest than was given in fertilizer, double in many cases. When available nitrogen measured in soil in spring is added to the amount mineralized and taken up by the crop during the growing season, much of the fertilizer appears to be accounted for by the total amount in the crop at harvest. Thus sugar beet has acquired a good reputation as an efficient scavenger for nitrogen. Even when huge amounts of fertilizer are given, the crop takes much up, often to the detriment of root quality.

Studies using fertilizer enriched with ^{15}N have provided an insight into just how much of the fertilizer given for sugar beet is used in the year of application. Powlson (1994) reported on a 2-year study by A.J. Macdonald (unpublished) where 122 kg N ha^{-1} containing ^{15}N was applied for sugar beet. Results for the 2 years were very similar. The average recovery of applied fertilizer was 60% (tops 33%, roots 27%). Analysis of the soil showed that 24% had entered the organic matter, only 1% remained as mineral nitrogen and losses averaged 15%.

Nitrogen Recommendations

The huge amount of information now available on the nitrogen requirement of sugar beet summarized in this chapter so far should put the grower in a strong position to decide exactly how to supply the crop. Regrettably, despite over 100 years of research, it is still difficult to make specific recommendations for a field. Many experiments worldwide, even as recently as the past decade, prove how hard it still is to forecast the nitrogen requirement of sugar beet accurately. Not surprisingly, research continues in most countries to improve on current recommendations.

Nitrogen-fertilizer advice in Europe

During the early years of sugar beet growing in Europe and through to the middle of the 20th century, average crop requirements based on field experiments were published (Crowther and Yates, 1941). In many countries 75 kg N ha^{-1} sufficed for this crop. Then it was realized that taking account of the soil type could improve the recommendation (Boyd *et al.*, 1957), as could residues of previous crops and the use of organic manure (Adams, 1962).

The next step was to analyse soil and plant tissue to further refine recommendations. These attempts have been partially fruitful, as already described above, and are being used in some countries. Now techniques such as chlorophyll testing, satellite mapping and computerized crop modelling are being widely tried. Perhaps in the future a more reliable method of prediction will be available than at present but it is more likely that a combination of information will be needed.

Recommendations for nitrogen in the UK

The main source of independent advice is the Ministry of Agriculture, Fisheries and Food (MAFF), now the Department of the Environment, Food and Rural Affairs (DEFRA). The first comprehensive set of recommendations for the major nutrients for all the important crops was published in 1973. This has been updated at intervals and the seventh edition was published recently (MAFF, 2000). In this current edition a new field-classification system is introduced, called the soil nitrogen supply (SNS) index. This is defined as the amount of nitrogen (kg ha^{-1}) in soil that becomes available for uptake by the crop from establishment to the end of the growing season. In the case of sugar beet it is soil mineral nitrogen at sowing plus that mineralized during the growing season.

Provision is made for use of either: (i) soil sampling and measurements to 90 cm depth to determine the amount before sowing; or (ii) no sampling but an estimate based on field-specific information for previous cropping, fertilizer and manure, soil type and winter rainfall. Tables then allow the SNS index to be assessed from rainfall, soil and previous crop.

Excess winter rain is grouped into three categories of 50–150 mm, 150–250 mm and over 250 mm excess winter rain. Soil is categorized into six main groups, from light sand through to peat. The 12 main crops are identified dependent upon the residual nitrogen they are known to leave. For example, cereals leave very little, while vegetables receiving a large amount of fertilizer leave considerable amounts.

The SNS index thus established allows the total amount of nitrogen fertilizer to be read off as in the example in Table 2.4. The recommendations in Table 2.4 are decreased by the amount of available nitrogen in organic manures (see Chapter 8 for amounts). It is also regarded as good practice to apply the nitrogen fertilizer in two doses: 30–40 kg N ha^{-1} in or on the seedbed (to prevent establishment problems and avoid the possibility of leaching in very wet springs) and the remainder after full emergence.

Table 2.4. Nitrogen recommendations (kg ha^{-1}) for sugar beet in areas of moderate rainfall (150–250 mm excess winter rain) (after MAFF, 2000).

	Soil nitrogen supply (SNS) index					
Without soil sampling	0	1	2	3	4	5
With soil sampling	SNS = Soil mineral nitrogen to 90 cm + estimate of mineralizable N					
	< 60	61–80	81–100	101–120	121–160	161–240
All mineral soils, except deep silty soils	120	100	60	30	30	0
Deep fertile silty soils	–	60	40	30	30	0
Organic soils	–	–	–	30	30	0
Peaty soils	–	–	–	–	–	–

A computer model is being tested at Rothamsted to determine if recommendations could be improved. The project is called SUNDIAL (simulation of nitrogen dynamics in arable land). To date, SUNDIAL has been used to interpret results of field experiments on soil type and different weather patterns but it is not yet in commercial use for sugar beet.

Recommendations in the USA

The primary consideration is the extent of nitrate accumulation in the soil profile. Generally in the production areas west of the Mississippi River, evapotranspiration exceeds rainfall and leaching potential is very low. In such conditions nitrate accumulates in the rooting depth of sugar beet. East of the Mississippi River (Great Lakes region) nitrate is leached away or lost by denitrification in the wetter winters.

In areas with low leaching potential, nitrogen recommendations are adjusted to account for nitrate accumulated in the soil profile to a depth of 1 m or more. Since winter percolation is small and little denitrification occurs, autumn testing can be utilized to adjust nitrogen recommendations for the following year. Autumn sampling is advantageous, since sugar beet is usually sown as soon as the soils are sufficiently dry for field operations, leaving little time for adequate sampling in spring.

Sugar beet production in the Great Lakes region is primarily on medium- and fine-textured soils. For the most part, these soils are somewhat poorly drained. Winter losses of nitrate prohibit autumn testing for adjustment of nitrogen recommendations for the following year. While early spring soil sampling could provide useful information, a concerted programme has not been put in place.

However, there has been some success in adjusting the amount of nitrogen needed by the pre-side-dressed nitrate test of Christenson *et al.* (1993). This is patterned after the Magdoff *et al.* (1984) test developed for maize production, where nitrogen is applied as a side-dressed application in early to mid-June. The basis of this approach is that a significant amount of the mineral nitrogen produced from mineralization has been produced by this time and is therefore taken into account.

This approach could be helpful for adjusting nitrogen amounts for sugar beet since excess fertilizer costs 2.4 times as much per kg of nitrogen as underfertilization in the Great Lakes region (Christenson *et al.*, 1993). There needs to be a short 'turnaround time' for the results to be useful. Rapid transit of samples to laboratory and electronic transmission of the results to the user will aid in this process.

Independent fertilizer recommendations in the USA are provided by land-grant universities. Such recommendations for several states are given in Table 2.5. In every case, recommendations are based on expected yield multiplied by a factor. This factor ranges from 3.9 to 4.3 kg N t^{-1} in all regions except the Upper Midwest, where 7.0 is used. As discussed earlier, nitrogen recommendations are adjusted for nitrate concentrations in the surface 0.5–1.8 m where conditions merit the approach. Adjustments are made for crop-management practices, such as ploughing down legumes, cereal

Table 2.5. Nitrogen-fertilizer recommendations for sugar beet in various growing regions of the USA as given by universities[a] involved in fertility research.

Region	Basic recommendation	Adjustments soil N	Adjustments cropping
Great Lakes	4.3 × Expected yield	No adjustment	Legumes (−)
Upper Midwest	7.0 × Expected yield	NO_3-$N_{(0–60\ cm)}$	Legumes (−)
Great Plains	4.0 × Expected yield	NO_3-$N_{(0–180\ cm)}$	Legumes and livestock manure (−)
Far West	3.9 × Expected yield	NO_3-$N_{(0–120\ cm)}$	Cereal straw (+)

[a]References: California – Hills *et al.*, 1982; Colorado – Mortvedt *et al.*, 1996; Michigan – Christenson *et al.*, 1992; Montana – Lichthardt and Jacobsen, 1992; Nebraska – Binford *et al.*, 2000; North Dakota and Minnesota – Franzen and Cihacek, 1996; Wyoming – Blaylock *et al.*, 1996.

straw and animal manure, and soil management, such as methods of tillage, irrigation and leaving land fallow.

Nitrogen Fertilizers

The primary means for chemically fixing atmospheric nitrogen is by the Haber–Bosch process. Atmospheric N_2 is combined with H_2, forming NH_3. Large quantities of energy, (usually fossil fuel) are needed for this process. Methane gas is usually used for the energy as well as a source of H_2. The reaction is shown as follows:

$$3\,H_2 + N_2 \xrightarrow[\text{High pressure}]{\text{Catalyst, 1200°C}} 2\,NH_3$$

This ammonia is then used in the preparation of other nitrogen fertilizers such as urea, ammonium nitrate or sulphate and mono- or di-ammonium phosphates. Some selected properties of the various fertilizers are given in Table 2.6.

Generally all sources of nitrogen, applied at equivalent rates and in a proper manner, give similar yield results when used for sugar beet production. Further discussion of sources of fertilizer, methods of application and quantities of nutrients is in Chapter 11.

Nitrification Inhibitors

Compounds toxic to nitrifying organisms have potential as nitrification inhibitors in crop production. Such material should inter-

Table 2.6. Nitrogen concentration and selected characteristics of nitrogen fertilizers.

Nitrogen source	Per cent N	Physical state	Main advantage	Main disadvantage
Anhydrous ammonia	82	Gas	Low cost	Needs to be injected
Ammonium sulphate	21	Solid	Supplies sulphur	Low concentration
Ammonium nitrate	34.5	Solid	Immediate availability	Used in explosives
Urea ($CO(NH_2)_2$)	46	Solid	Low cost	Volatilization losses Toxicity to seeds
Nitrogen solutions	28–32	Liquid	Ease of handling Compatible with pesticides Applied in irrigation systems	Volatilization losses
Mono-ammonium phosphate ($NH_4H_2PO_4$)	11–13	Solid	Low toxicity to seed Primarily used as a source of phosphorus	Low concentration
Di-ammonium phosphate ($(NH_4)_2HPO_4$)	16–18	Solid	Primarily used as a source of phosphorus	Low concentration Toxicity to seed

rupt the conversion of NH_4^+ to NO_2^- by specifically retarding *Nitrosomonas* activity without interfering with formation of NO_3^- by *Nitrobacter*. Nitrapyrin, dicyandiamide (DCD) and Etridiazol are effective in retarding nitrification of applied ammonium-nitrogen fertilizer. All three compounds are compatible with urea, anhydrous ammonia and urea–ammonium nitrate (UAN). Nearly 2 Mha (primarily maize) are treated annually in the USA.

Hoeft (1984) reported that the stability of nitrapyrin declined with increasing temperature. The half-life when soil temperature was less than 5°C was in excess of 100 days and above 21°C 11 days. Walters and Malzer (1990a, b) working with maize suggested that the greatest effect on inhibition of nitrification would occur on sandy soils where leaching potential is high.

Limited research has been conducted with nitrification inhibitors in sugar beet production. Hills *et al.* (1981) measured the effect of nitrapyrin on the uptake of nitrogen by sugar beet grown with irrigation in California, USA. They showed a slight suppression of nitrogen concentration in petioles up to 10 weeks after application of nitrapyrin with ammonium sulphate, compared with nitrogen fertilizer alone. When nitrapyrin was used, yield was improved by about 1 t ha^{-1} and sugar percentage decreased 0.3%. They suggested that nitrogen fertilizer might be reduced by about 11 kg N ha^{-1} with the use of the inhibitor. There were no further studies found evaluating the use of nitrification inhibitors for sugar beet production. It is possible that the slight gain in nitrogen efficiency discouraged further work.

Urease Inhibitors

The first step in the formation of available nitrogen from urea is hydrolysis to the ammonium ion. Paulson and Kurtz (1969) have shown that urea hydrolysis is carried out by an exo-enzyme attached to soil colloids. Urea hydrolyses rapidly in soils (Gasser, 1964). Schlegel *et al.* (1986) concluded that urease inhibitors would be most beneficial when added to urea applied to the soil surface. Christenson (2000) summarized recommendations concerning the use of urease inhibitors on certain crops but there are no reports concerning urease inhibitors for sugar beet production.

Nitrogen-fertilization Practices

There are two main considerations that affect the way nitrogen is applied in practice, once the amount needed has been decided. First, nitrogen fertilizers are phytotoxic when concentrated in quantity near the seed. Secondly, in nitrate form, nitrogen fertilizers are prone to leaching by rainfall in late winter and early spring. Thus application must be timed to match the crop's need as closely as possible.

In Europe

Thielebein (1960) in Germany made the first study of the phytotoxic effects of nitrogen fertilizer. Fertilizer caused some seeds to fail to germinate, other seeds to produce seedlings which died before emergence and yet others to produce plants which were initially retarded. Now that most sugar beet crops are produced from monogerm seeds sown to a stand, these negative effects cannot be tolerated.

Last *et al.* (1983) made one of the first detailed studies of the effect of nitrogen fertilizer on germination and emergence of monogerm sugar beet sown to a stand. Ammonium nitrate was broadcast as solid prills at 0–207 kg N ha^{-1} in the seedbed 2 weeks before sowing. Figure 2.8 shows that in the three dry springs (1973, 1974 and 1976) there were serious losses of plants. In moist springs (1975, 1977 and 1978) ammonium nitrate had little effect, both when the overall emergence of the early monogerm crops was good (1978) and when it was poor (1975 and 1977).

Nitrogen fertilizers differ in their phytotoxicity, depending on chemical form. Ammonium nitrate is probably the least damaging of those commonly used and ammonium sulphate the most damaging (Adams, 1961a). Urea is less damaging than the sulphate (Sawahata and Takase, 1966) but is not without problems, as shown by Armstrong

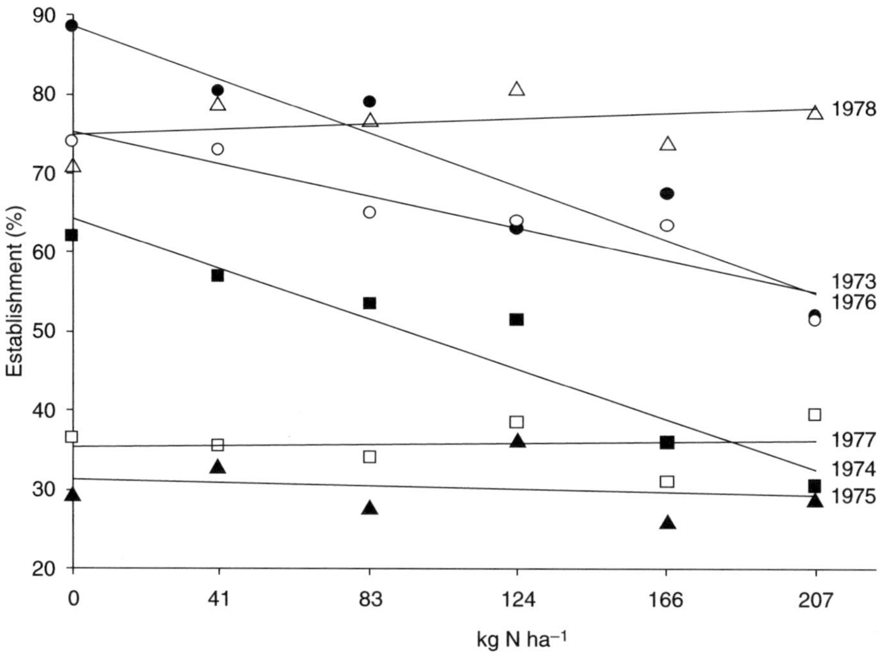

Fig. 2.8. Effect of nitrogen fertilizer on plant establishment in 6 years at Broom's Barn.

and Stillingfleet (1993). They found that 120 kg N ha^{-1} as solid urea reduced plant population by 30,000 plants ha^{-1} compared with split dressings of other forms of nitrogen fertilizer, with a consequent large loss in sugar yield of 1.25 t ha^{-1}. They thought this was a result of high pH causing ammonia gas to be produced, coupled with spring soil temperatures.

To avoid damage from one large dose, nitrogen applications are now separated from the germination and early growth stage by time and/or space. In the UK, it is normal practice to give a little nitrogen fertilizer just before, at or just after sowing and the remainder after full emergence. This decreases the likelihood of leaching, as well as removing the possibility of damage. In other countries, nitrogen is separated by placing fertilizer in a band near the seed but sufficiently distant to prevent damage (Dunham, 1991a; Marcussen, 1991; Vandergeten and Vanstallen, 1991; Allison, 1992). Placement has also been shown by several of these workers to increase the efficiency of uptake of fertilizer and to decrease the amount needed – see Chapter 11.

In the USA

Like the practice in Europe, application of large quantities of nitrogen in the seedbed is avoided due to reduction in the sugar beet stand. Blumenthal (2001) reported that broadcasting of 100 kg N ha^{-1} as urea prior to sowing in Nebraska reduced the stand by 15,000 plants ha^{-1}. A small amount of nitrogen may be placed in the seedbed near sowing, but the major portion is applied in autumn prior to sugar beet, side-dressed between rows when beets are in the two- to six-leaf stage or applied with irrigation water during the season.

Even with the risk of stand reduction, growers in the Great Lakes region may apply nitrogen before planting, followed by tillage for the seedbed. The risk is markedly reduced if there is at least 20 mm of precipitation within 10 days of sowing. Yerokun and Christenson (1989) showed that this amount of precipitation eliminated any effects of up to 40 kg N ha^{-1} applied in direct contact with maize seed.

Effect of Organic Residues on Need for Nitrogen Fertilizer

All forms of organic manure, crop residues and soil organic matter provide widely varying amounts of nitrogen for sugar beet. It is present in the form of readily available nitrogen (ammonium-N and uric acid-N) equivalent to fertilizer nitrogen. This may account for half the total nitrogen in some manure, such as poultry litter and slurries, or as low as 10% of the total nitrogen in strawy farmyard manure. The rest of the nitrogen is in the organic matter. Most of this is not available in the season of application. It is released slowly and unpredictably over a period of years.

Chapter 8 contains details of how best to utilize organic manures in relation to their properties and the needs of sugar beet. Most have a profound effect on nitrogen nutrition. It is necessary to take account of the amounts available in the tables in Chapter 8, which also covers green manuring. The longer-term effects of soil organic matter, crop residues in rotations and sugar beet tops are in Chapter 9.

Agronomic and Nutrient Interactions

Fuehring and Finkner (1973) conducted an elaborately designed experiment to evaluate the interrelationships of applied nitrogen, zinc, plant population and frequency of irrigation in regard to the yield and quality of sugar beet. While sugar beet has shown little response to applied zinc (Chapter 6), Fuehring and Finkner (1973) showed an interaction between the two nutrients when zinc was in marginal supply from the soil. With optimum nitrogen and water, 5–9 kg Zn ha^{-1} increased extractable sugar significantly. It was only under these ideal conditions that there was an increase due to zinc.

Voth and Christenson (1980a) investigated the interaction of nitrogen and manganese in regard to yield and sucrose production. The impetus for the work was from field observations where banded or side-dressed nitrogen increased the manganese concentration of sugar beet leaves. Similarly, manganese-deficient leaves contained above-average nitrate concentrations. In this study, increasing nitrogen application depressed manganese concentration in leaf tissue. Even though there was a significant interaction between nitrogen and manganese in the greenhouse (Table 2.7), there was no significant interaction between nitrogen and manganese in regard to sugar percentage or recoverable sugar in field studies.

Chapter 12 describes interactions between nitrogen and other major plant nutrients and between nitrogen and agronomic factors. Increasing nitrogen application also affects the concentration of nutrients in sugar beet. Bravo *et al.* (1989) reported that increasing nitrogen fertilizer increased sulphur, sodium and magnesium, but decreased calcium and potassium concentration in sugar beet plants (average of blades, petioles, crowns and roots). In the same study, Bravo *et al.* (1992),

Table 2.7. Effect of nitrogen and manganese in nutrient-culture solution on manganese concentration of sugar beet leaves (from Voth and Christenson, 1980a).

	Nitrogen concentration (mg l^{-1})		
Manganese concentration (mg l^{-1})	70	140	210
	Manganese in leaves (mg kg^{-1})		
0.00	10	10	9
0.13	114	85	94
0.25	198	149	125
0.50	267	236	204
1.00	336	264	210

showed that increasing nitrogen increased zinc and boron in blades, but decreased manganese concentration in all plant parts.

Environmental Considerations and Nitrogen Fertilization

When nitrogen fertilization of sugar beet was reviewed from its origins up to the early 1970s (Draycott, 1972), virtually every piece of work dealt with utilizing nitrogen solely as an aid to crop production. During the period of the late 19th century through to that date, researchers aimed at ensuring that growers and processors reaped maximum benefit. Early fertilizer recommendations (Boyd *et al.*, 1957) first based quantities of nitrogen on the yield of roots produced and a little later put more emphasis on the yield of sugar (Boyd *et al.*, 1970).

During the 1970s it was increasingly recognized that not only was the nitrogen given in fertilizers and manures improving crop production, but it was also having an environmental impact. The interest shown by researchers and the general public led to a very rapid change in emphasis during the 1980s and 1990s, which continues today. Not only is the effect of the nitrogen on the crop measured but, more importantly, its effect on the soil, water and aerial environment. In the latest recommendations for all crops grown in the UK (MAFF, 2000), great emphasis is now laid on minimizing losses of applied nitrogen.

The main concern has centred on the failure of soil to retain nitrogen, both natural and applied. As the nitrate ion (NO_3^-) nitrogen is very mobile but in the ammonium form (NH_4^+) it is not leached. There are three main pathways of nitrogen loss: first, leaching into groundwaters and hence into drinking-water, secondly, volatilization of ammonia and, thirdly, through denitrification.

In most sugar beet soils the major pathway, especially where soils tend to be sandy and free-draining, is by leaching. In most temperate regions this is usually in winter, particularly after dry summers when previous crop uptake has been restricted (Draycott *et al.*, 1997). Another time when leaching sometimes takes place is in wet springs. When heavy rain follows nitrogen-fertilizer application, the movement of nitrate from the rooting zone is significant. Last and Draycott (1975a,b) found that 40 kg N ha^{-1} was lost in wet years, whereas in normal weather (about 50 mm rain per month) losses were negligible.

The other two pathways of loss (volatilization of ammonia and denitrification) have not been measured much on sugar beet fields. Undoubtedly, when nitrogen is applied in forms such as urea and organic manures, there will be some volatilization of ammonia. When fertilizer is applied, sugar beet soils are warming up; often the soil is bare and some are of high pH and low cation-exchange capacity. Parallels with other crops suggest that these are all conditions that favour losses, especially when applied to the soil surface and not incorporated in the soil. Whitehead (2000) has suggested that for grassland up to 10–25% of the nitrogen from urea may be lost, 1–3% from ammonium nitrate and up to 25% from diammonium phosphate. Clearly, measurements are needed on sugar beet fields.

In contrast, these conditions favouring volatilization are different from those which encourage denitrification, the process by which nitrate is converted to NO, N_2O and N_2. These gases diffuse into the atmosphere. Denitrification is usually associated with anaerobic conditions caused by excess water, organic matter and near-neutral pH. There is a paucity of information on such losses from sugar beet fields but measurements on other crops suggest that losses could be significant where large amounts of organic manure are used, where soils are clayey and/or in poor structural condition and where drainage is inadequate.

Nitrate concentration in drinking-water

In the EU the admissible level is 50 mg nitrate-N l^{-1} of water (11.3 mg N as nitrate per litre). In the USA comparable levels are 45 and 10 mg l^{-1}. There is some doubt as to the need for such restrictions in regard to human health and the scientific basis of the choice of these concentrations has been questioned (Croll, 1990; Leifert and Golden, 2000). However, a much more widespread problem with water quality associated with nitrogen is the degradation of aquatic ecosystems, such as in the Baltic Sea, where extensive algal blooms have been seen in recent years. Together with phosphates, nitrates undoubtedly contribute to eutrophication.

Johnston (1989) reviewed the implications for sugar beet growing by drawing on information from long-term experiments at Rothamsted. In that context, for every 100 mm of through drainage, 10 kg N ha^{-1} as nitrate raised the nitrate concentration of the water by 44 mg nitrate l^{-1}. In an average winter in the UK much of the sugar beet growing area has through drainage of 100 mm $year^{-1}$ or more, so water-quality limits would in theory be exceeded. In practice, drainage from agricultural land is diluted with that from other sources within each aquifer.

Powlson (1998b) has recently reviewed the full nitrogen balance for Broadbalk, where wheat has been grown at Rothamsted for 154 years. He studied plot 8, given 144 kg N ha^{-1} every year. Atmospheric deposition was about 45 kg N ha^{-1} $year^{-1}$ and no organic manure was given. He found that, on average over the years, atmospheric loss as ammonia was 7 kg N ha^{-1}, denitrification 32 kg N ha^{-1}, leaching 25 kg N ha^{-1} and crop uptake 125 kg N ha^{-1}.

Measurements were made over a shorter period of nutrient concentrations in water draining from under a field in a sugar beet/cereal rotation at Broom's Barn (Draycott *et al.*, 1997). Fertilizer nitrogen was used to produce optimum yields of crops without excess. Water percolating through the soil and leaving the drainage system had nitrate concentrations well below the EU guideline in most years but occasionally exceeded it slightly. Similarly Neeteson and Ehlert (1989) made a detailed study in Dutch conditions of the environmental aspects of applying inorganic fertilizers to sugar beet. Provided sugar beet tops were removed, fertilizer had little impact. Where they were ploughed in, there was a potential hazard for leaching losses. In a lysimeter study, no nitrate was lost in spring with normal rainfall (230 mm April–June). At harvest, measurements on 27 sugar beet fields of nitrate present to 60 cm showed that there was, on average, only 3 kg N ha^{-1} more on plots with optimum fertilizer applied than on plots given none. They examined Broom's Barn data from several experiments and showed that the apparent recovery of fertilizer was over 75%.

Swennson (2002) has recently reported on a comprehensive study on 283 dairy farms in southern Sweden, looking at how much nitrogen goes in and out of the farm gate. Nitrogen efficiency was greatly improved by including sugar beet in the rotation of crops grown on these farms.

In general terms, if 125 kg N ha^{-1} is given for sugar beet, tops and roots will remove 200 kg N ha^{-1} from the field. Where roots only are removed, tops will contain some 100 kg N ha^{-1}, with the potential for leaching. In attempts to 'save' this nitrogen, several studies have been made with cover crops, which take up nitrogen and release it for a following crop. These are reported in Chapter 9.

Summary

When sugar beet was introduced for sugar production, nitrogen nutrition was one of the first topics investigated because it has a large effect on profitability. Research was devoted to increasing the yield of roots. In the next phase, effects on root quality and sugar production were emphasized. In the past two decades, attention has been paid to the deleterious effects of excess nitrogen on water quality. Research has focused on leaching losses and the effect on groundwater, since nitrate in drinking-water may pose a health hazard. Even though the standard of approximately 50 mg l^{-1} NO_3-N may be called into question, agricultural practices need to be adjusted to address the problem.

In spite of all the research, there are still difficulties in predicting the amount of nitrogen needed on a field basis. This is due to the many factors affecting the nitrogen cycle. Nitrogen released from organic matter is an important component of sugar beet nutrition and measurements relate to field response to fertilizer. Some success is reported with measuring mineral nitrogen within the soil profile to adjust the nitrogen recommendation and Table 2.8 summarizes expected amounts in Europe.

In the first phase of growth, sugar beet needs sufficient nitrogen to promote rapid leaf development. The goal should be to develop complete leaf cover quickly. There is scant evidence that nitrogen supply greater than needed for an LAI of 3 will increase either photosynthetic efficiency or sugar production. We have found no evidence that sugar beet should take up more than 200 kg N ha^{-1} by harvest. Ideally the crop should probably have about 90 kg N ha^{-1} in the roots and 110 kg N ha^{-1} in the tops. Sugar beet takes up nitrogen at a rapid rate during the 60-day period from the emergence stage to the point of an LAI of 3. Studies indicate that during this 60-day period the average rate of uptake is 2.5 kg ha^{-1} day^{-1}, with peak rates as high as 6 kg.

Models can be used to predict amounts of mineral nitrogen in soil in spring based on soil characteristics, previous cropping and weather, particularly through drainage. Still one of the most difficult areas is prediction of the amount becoming available during the growing season. Laboratory analyses and techniques such as soil incubation have met with limited success. More research in this area is needed because studies of nitrogen uptake by sugar beet show that this is a major source of the element in most countries. Only with this information will growers be able to fine-tune applications of nitrogen to suit crop need and avoid excess, with all the implications for fertilizer costs, root quality and the environment.

Table 2.8. Summary of average amounts of mineral (nitrate + ammonium) nitrogen in soil profiles to a metre depth in Europe where sugar beet is being grown (kg N ha^{-1}).

In autumn before sowing sugar beet in the absence of organic manures or large residues from previous crop	100–150	
At the end of the winter	50–100	
Mineralized during the growing season	60–200	(about 1.5% of the total soil N)
Taken up by the crop	100–250	
Remaining in soil after growing sugar beet	40	

3
Phosphorus and Sulphur

These elements are major plant nutrients, needed in similar quantities by sugar beet and both taken up in the anionic form. In soils of near-neutral pH where sugar beet is grown, phosphorus enters the plant as $H_2PO_4^-$ and HPO_4^{2-}, and sulphur as SO_4^{2-}. A major difference between the two elements is that the soil complex holds phosphate ions tightly and there is no appreciable leaching, whereas sulphate is not retained, being readily leached from the plough layer. Sulphate may accumulate in subsoils when there is an accumulation of iron and aluminium oxides and hydroxides.

During the 20th century and before, attention was directed mainly towards phosphorus nutrition of this and other crops. It was usually in very short supply, particularly in soils not previously cultivated. In contrast, sufficient sulphur was being deposited in rain to satisfy crop requirements. The main source of the sulphur at the time was fossil fuel being burnt and the gases emitted to the atmosphere in the industrialized regions. The relative position of the two elements is changing for three main reasons.

First, on most fields the crop now gives only small increases (or none at all) in yield from phosphorus fertilizer. This is because regular applications have increased soil supplies of available phosphorus to adequate concentrations for maximum yield without fresh additions each year. In the future phosphorus will only be needed periodically on such fields to maintain a satisfactory concentration of available phosphorus in soil.

Secondly, atmospheric pollution has been greatly reduced during the latter part of the 20th century as a result of pressures to clean the air. As an example, in sugar beet areas of the UK, deposition has decreased from 70 kg S ha^{-1} $year^{-1}$ in 1970 to 7 kg S ha^{-1} $year^{-1}$ in 2000. Estimated deposition in the USA ranges from less than 3 kg ha^{-1} west of the Mississippi River to approximately 10 kg ha^{-1} in the Great Lakes region. These values reflect a reduction of 25% and 50%, respectively, from 1985 to 2000.

Thirdly, some crops' needs were supplied by sulphur contained in fertilizers such as $(NH_4)_2SO_4$, now largely replaced by urea and NH_4NO_3 and the replacement of single by triple superphosphate. Not surprisingly, deficiencies of sulphur are appearing in a number of crop species. So far, there have been few reports of responses to sulphur by sugar beet but it seems likely that there is an increasing risk of shortage.

PHOSPHORUS

Phosphorus in Soil

Total

The amount of total phosphorus present in soil generally reflects the concentration of the element in the minerals of the parent material. Much of the phosphorus in rocks is in the form of apatite ($Ca_{10}(PO_4)_6X_2$, where X may be OH^-, Cl^- or F^-). During weathering, phosphorus is released as the soluble ions $H_2PO_4^-$ and HPO_4^{2-}, to be either taken up by roots or rendered insoluble again through precipitation or adsorption. Few soils contain much phosphorus in their natural state, which is why the nutrient is so important in fertilizers for crop production.

In surveys of phosphorus concentrations in soils of the temperate climatic regions, authors have reported a range of 0.04 to 0.4% P (400–4000 mg P kg^{-1}). For some 150 years, many soils in crop production have now received regular additions of phosphorus fertilizer, which has greatly increased the reserve of total phosphorus in soil. At Rothamsted, where plots have received a moderate amount annually for 100 years, total concentration has increased threefold, nearly all of the change being in the surface 25 cm of soil.

Additional inputs in intensive cropping are from farmyard and other organic manures and sewage sludges (see Chapter 8). In the case of phosphorus, atmospheric deposition is negligible. Rainfall and dry deposition provide only 0.2–1.5 kg P ha^{-1} $year^{-1}$, probably originating from dust transferred to the atmosphere from wind erosion of the soil.

Available

Phosphorus dissolved in soil solution is the controlling point of availability to sugar beet. Concentration in solution is usually very low – < 0.1 mg P kg^{-1} – but, as the crop removes phosphorus from solution, more dissolves from the solid phase. The rate of replenishment to the solution is critical during the rapid growth of the crop. 'Labile phosphorus' is a term used to describe near readily available forms of the element, distinguishing this portion from the slowly dissolving forms of the total phosphorus present.

Allison and Chapman (1995) showed that in a typical sugar beet topsoil (0–30 cm), less than 2 kg P_2O_5 ha^{-1} was in the immediately plant-available form (HPO_4^{2-} or $H_2PO_4^-$). Labile phosphorus amounted to 50 kg P_2O_5 ha^{-1} or about 1% of the total in the topsoil. Unavailable fractions predominated, with that bound in the organic matter accounting for most (3.5 t P_2O_5 ha^{-1} or 70%) and iron, calcium and aluminium phosphates accounting for 1.5 t P_2O_5 ha^{-1}, or about 30% of the total.

Phosphorus applied in commercial fertilizers or organic manures add directly to the available fraction and to labile and non-labile forms. In calcareous soils, added phosphorus reverts through a series of calcium phosphates, ultimately becoming apatite. The intermediate forms include dicalcium phosphate, octacalcium phosphate and tricalcium phosphate, formed in that order, followed by apatite. Each of these compounds becomes less soluble and takes longer to form. For example, dicalcium phosphate will form in days, octa- and tricalcium phosphate in months, while apatite will take years. A major reason why sugar beet grown on calcareous soils is often very productive is that these intermediate forms contribute to the labile fraction and hence to the phosphorus in the soil solution.

In slightly acid to neutral soils (pH 6.1–7.3), the phosphates listed above may exist, but additional minerals will form with the presence of K^+ and NH_4^+ from fertilizers or manures. Numerous compounds similar to taranakites ($Al_5(X_3)H_8(PO_4)_6$, where X may be K^+ and/or NH_4^+) are formed in these soils. In addition, aluminium phosphates form in this pH range and these and other compounds contribute to the labile pool.

Absence of mycorrhizas on sugar beet roots

The roots of most soil-grown plants have associations with mycorrhizas, which are microorganisms that can be mutualistic, neutral or parasitic. Usually with crop plants it is a mutually useful relationship, often termed mycorrhizal symbiosis. In the case of phosphorus and other slow-moving nutrients in soil, uptake by plants is aided by vesicular arbuscular mycorrhizas. External hyphae absorb and translocate nutrients to the plant from outside the root depletion zone of non-mycorrhizal roots (Tinker *et al.*, 1992). The rate of uptake per unit root length can be increased two- or threefold.

In the case of sugar beet, mycorrhizas are completely absent, as with all members of the two important crop families *Chenopodiaceae* and *Cruciferae*. Non-mycorrhizal plants are usually found in habitats where soils are either very dry or saline (which would apply to members of the *Beta* genus), waterlogged or of high or low fertility.

The practical implications of the absence of mycorrhizas on sugar beet roots have had little attention. From work on other species, it would appear that soil concentrations of nutrients such as phosphorus in the water-soluble phase need to be greater to achieve a given rate and amount of uptake. Correlations between crop response to phosphorus fertilizer and soil analysis may be closer than with mycorrhizal plants, due to the absence of variability introduced by differing amounts of root colonization.

Physiological Role and Uptake

In common with many other plants, phosphorus is taken up by sugar beet roots against a steep concentration gradient. It is an active process, with adenosine triphosphate (ATP) supplying the energy for uptake. Once taken up, the phosphate ions are mobile in the plant and rapidly incorporated into organic compounds. One important function of the phosphorus is in the formation of ATP itself. ATP is then used to transfer energy produced by photosynthesis into energy stored as sugar. Thus, throughout the life of the crop until the day of harvest, an adequate supply of phosphorus is essential to maintain effective production. Kirkby *et al.* (1987) reviewed the physiological role of phosphorus in the biochemical processes in sugar formation, and Terry and Ulrich (1973) described the effects of phosphorus deficiency on the photosynthesis and respiration of sugar beet leaves (see also Chapter 7 on nutrient deficiencies).

Uptake at harvest

The quantity of phosphorus present in the crop (tops and roots) at harvest is reviewed in Table 3.1. Historical data showed that this could be as little as 5 kg P_2O_5 ha^{-1} where yields and phosphorus concentration were

Table 3.1. Quantity of phosphorus in sugar beet at harvest.

Quantity (kg P_2O_5 ha^{-1})			Fertilizer	
Roots	Tops	Total	(kg P_2O_5 ha^{-1})	Reference
23	30	53	50	Draycott, 1972 (mean of pre-1970 published data)
30	35	65	–	Draycott *et al.*, 1972b
–	–	71	70	Jansson, 1987
50	26	76	50	Kirkby *et al.*, 1987
–	–	52–112	0–300	Siegenthaler, 1987
42	40	82	60	Vanstallen and Vandergeten, 1987
70	40	110	100	Analogides, 1987a
71	39	110	46–180	Neeteson and Ehlert, 1989
40	25	65	135	Jourdan *et al.*, 1992
45	35	80	75	Anticipated amounts in typical crops

both small due to long-term cropping without fertilizer (Warren *et al.*, 1962). With adequate fertilizer and average yields, this rises to about 50 kg P_2O_5 ha^{-1} (Draycott, 1972). With increasing yields in most countries uptake has risen and, with high dry matter, yields of roots and tops at harvest, uptake may now exceed 100 kg P_2O_5 ha^{-1} (Siegenthaler, 1987).

In a typical Mediterranean climate, Analogides (1987a) made a thorough study of phosphorus concentration in and uptake of phosphorus by various parts of sugar beet. At harvest his crops contained a total of 130 kg P_2O_5 ha^{-1}, 90 kg of which was in the roots and 40 kg in the tops (Table 3.1). In North Dakota, USA, Etchevers and Moraghan (1983) showed that the relative amount of phosphorus in roots increased from about 40% to about 50% with increasing application on a responsive field. The increase was primarily due to increased root growth in response to increasing supply. At harvest, they found that, at the optimum application (110 kg P_2O_5 ha^{-1}), the tops contained 0.6 and roots 0.8 kg P_2O_5 t^{-1} of fresh plant tissue.

In the UK, Hollies (1997) has recently reviewed the uptake of phosphorus by all common crops (Appendix A) and the values have been adopted by the MAFF (2000). In terms of kg P_2O_5 t^{-1} of fresh material, sugar beet roots contain 0.8 and roots plus tops 1.9 kg P_2O_5 t^{-1}. Thus, for an average crop, about 50 kg P_2O_5 ha^{-1} would be present in roots and nearly as much in tops, or about 100 kg P_2O_5 ha^{-1} in total. Of the crops reviewed by Hollies, this was one of the largest amounts of phosphorus taken up.

Concentration of phosphorus in the crop at harvest

Table 3.2 summarizes data from various countries over the past 30 years. In terms of concentration in dried roots and tops in an average crop given a moderate amount of fertilizer, roots can be expected to contain 0.15% P and tops 0.35% P (0.35 and 0.8% P_2O_5, respectively).

North Dakota data (Etchevers and Moraghan, 1983) suggested lower values, 0.068 and 0.40% P for roots and tops, respectively. Reasons for these smaller concentrations are not readily explained; however, the available water during the growing season was low, resulting in relatively low yields (30–36 t ha^{-1}).

Concentration and uptake during growth

In common with several other nutrients, concentration in both roots and tops declines over the course of the season. Healthy plants given adequate phosphorus fertilizer contain 0.5% P in roots and 0.75% P in tops (1.15 and 1.73% P_2O_5, respectively) in the early season,

Table 3.2. Concentration of phosphorus in dry matter of sugar beet at harvest.

P concentration (% dry matter)				
Roots	Tops		Fertilizer (kg P_2O_5 ha^{-1})	Reference
0.15	0.34		–	Draycott *et al.*, 1972b
0.14	0.29		0	Draycott *et al.*, 1977
0.16	0.32		100	
0.13	0.29		0	Last *et al.*, 1985
0.14	0.31		50	
0.15	0.34		100	
0.15	0.30	(blades)	100	Analogides, 1987a
	0.25	(petioles)		
				Expected concentrations
0.13 or less	0.29 or less		–	Deficient
0.15 or more	0.34 or more		–	Adequate

falling slowly through to harvest to the values shown in the previous section.

The pattern of uptake has been described by Jourdan *et al.* (1992) in France for a large crop yielding 100 t ha^{-1} roots at 16% sugar. Total final uptake at harvest of phosphorus on the chalky soil of the Aube *département* was a modest 65 kg P_2O_5 ha^{-1}. The dry-matter yield of roots was nearly 20 t ha^{-1} and tops 5 t ha^{-1}. Uptake by roots rose throughout the growing period to 40 kg P_2O_5 ha^{-1} in November. Uptake by tops increased rapidly until August to 25 kg P_2O_5 ha^{-1} and remained the same to November. The maximum rate of uptake was in June/July, which for roots plus tops averaged 0.8 kg P_2O_5 ha^{-1} day^{-1}. The authors reported that there appeared to be little redistribution from one part of the plant to the other.

In UK conditions, Draycott and Martindale (2000b) reported on monthly measurements of phosphorus uptake from April until November. Figure 3.1 summarizes the results from some high-yielding sites. When the maximum leaf-area index was achieved in July/August, uptake by tops reached a maximum of just over 20 kg P_2O_5 ha^{-1} (cf. 25 kg on the very-high-yield site in France) and then declined slightly to harvest. Phosphorus in roots continued to rise rapidly till harvest, reaching 40 kg P_2O_5 ha^{-1}, the same as in France.

In summary the evidence suggests that the minimum total-uptake target is about 80 kg P_2O_5 ha^{-1}. The values of over 100 kg P_2O_5 ha^{-1} reported above in Table 3.1 from some countries look large. A possible explanation would be that a very large yield of tops at harvest brought about excessive growth associated with oversupply of nitrogen from organic matter or soil reserves. Certainly, French and UK data show that yields of 100 t roots ha^{-1} are obtainable with a much smaller total uptake.

Sources of Phosphorus Fertilizer

Apatite or rock phosphate is the main mineral from which phosphorus fertilizers are made. It is found in sedimentary, igneous, residual, phosphatized rock and guano deposits. Most of that used in the fertilizer industry comes from marine (sedimentary) deposits (Cathcart, 1980). Total concentration in rock phosphate ranges from 11.5 to 17.5% P (27–41% P_2O_5). Phosphorus availability from apatite is limited since water solubility is near zero and only 5–17% of the total is citrate-soluble. A further limitation affecting its use is the cost of handling, storage and transportation due to the low concentration of available phosphorus.

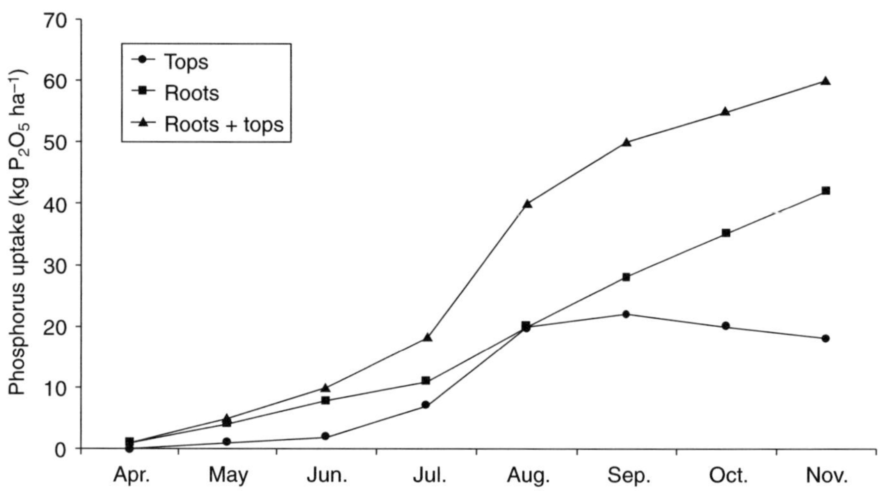

Fig. 3.1. Ideal uptake pattern of phosphorus by high-yielding sugar beet measured in France and the UK.

Finely ground, it may be effective on acid soils in warm climates and moist conditions and may meet the need of long-season crops and perennials under these conditions. This material is largely ineffective in the pH range where sugar beet is produced, though there have been some reports (Saive, 1987) that microbial activity in the rhizosphere may improve availability in some soils.

Phosphoric acid

Green or wet-process acid is manufactured by treating rock phosphate with concentrated sulphuric acid, resulting in an acid containing 48–53% available phosphate. This may be injected into soil or added to irrigation water applied to the growing crop. Phosphoric acid is not used for direct application due to the obvious handling difficulties. Nearly all wet acid produced is used to manufacture other fertilizers.

White acid is produced in an electric-furnace process and is used in the food industry. The cost of this form makes it prohibitive for use as a fertilizer.

Ammonium polyphosphate (APP)

Although APP has been used as a fertilizer for over 30 years, it is the most recent development in the phosphorus fertilizer industry. Two or more orthophosphate ions combine together by the loss of one H_2O molecule for each pair of $H_2PO_4^-$ ions. APP is manufactured by reacting pyrophosphoric acid ($H_4P_2O_7$) with NH_3. APP is a liquid containing 75 and 25% of the phosphorus present as polyphosphate and orthophosphate, respectively. The most common analysis is 10–34–0 ($N–P_2O_5–K_2O$).

Granulation of APP results in a product with an analysis of 11–55–0, which may be applied directly to the soil or blended with other fertilizer materials. Liquid 10–34–0 may be mixed with other liquid materials for direct application to the soils. APP has a chelating/sequestering action with metal ions. Silverberg *et al.* (1972) showed that APP may sequester 3% (w/w) Zn, 0.7–1.5 % Cu, 1% Fe, 0.9% B, 0.5% Mo and 0.2% Mn.

Other sources

Common sources are listed in Table 3.3. The important property of all of these sources is that nearly 100% of the phosphorus present is water-soluble. Early work with phosphorus showed that effectiveness is controlled to a large degree by water solubility.

Effect of Phosphorus Fertilizer on Yield and Soil Phosphorus

When sugar beet became established in various parts of the world, as in Europe in the 19th and early 20th century, it was soon recognized

Table 3.3. Sources of phosphorus used for crop production.

Sources	Available P_2O_5 (%)	
Superphosphate ('Single') $Ca(H_2PO_4)_2 + CaSO_4.2H_2O$	18	Largely replaced now by the more concentrated forms below
Concentrated superphosphate ('Triple') $Ca(H_2PO_4)_2$	44–52	Most common analysis 0–46–0; may be blended in some fertilizer formulations
Mono-ammonium phosphate $NH_4H_2PO_4$	48–55	Most common analysis 11–52–0; used in blended fertilizer formulations
Di-ammonium phosphate $(NH_4)_2HPO_4$	46–53	Most common analysis 18–46–0; used in blended fertilizer formulations
Ammonium polyphosphate $(NH_4)_3HP_2O_7$	34–37	Most common analysis for liquid fertilizer 10–34–0; granulated, 11–55–0
Potassium phosphates KH_2PO_4 and K_2HPO_4	30–60	Low salt index; contains 30–50% K_2O; economics control future production

that phosphorus was an essential element for successful production. In deficient soils it not only improves yield greatly but also increases plant establishment and early vigour. This was the case in the UK, where the crop started to be grown on a wide scale in 1920s.

Over the following 50 years average application of phosphorus increased gradually because advisory services and subsidies encouraged its use. By the late 1960s, the amount peaked at about 112 kg P_2O_5 ha^{-1} (Draycott, 1983). Application has since steadily decreased to about half this amount over the 30-year period 1970–2000 (Fig. 3.2).

There are several reasons for the decrease, the main one being the introduction and adoption of soil analysis to decide the optimum dressing. Soil analysis is widely accepted and used by UK growers. Soil concentrations of available phosphorus have also risen and this has reduced usage. About a quarter of fields now receive no phosphorus whereas 30 years ago every field received some (Draycott and Martindale, 2000b). Another, probably less significant, factor is the increased cost of fertilizer over the 30 years. In future, the use of the element will be more closely balanced against crop offtakes and approached on a rotational basis (see Chapter 9).

Draycott *et al.* (1971a) re-examined responses to phosphorus fertilizer in earlier experiments by extracting phosphorus from stored soil samples taken before the experiments started. Sodium bicarbonate (Olsen's P), anion resin, ammonium acetate/acetic acid and calcium chloride solutions were compared. Sodium bicarbonate solution was most closely related to response and, in common with many countries, has become the standard method in the UK for analysing soils for available phosphorus. Table 3.4 summarizes the results.

Draycott and Durrant (1976a) added information by making further experiments on fields chosen for their small concentration of soil phosphorus. This was to define more accurately the magnitude of the response to fertilizer and the economic optimum application. In the group of fields containing least available phosphorus (0–9 mg P l^{-1}), responses were up to 2 t sugar ha^{-1}. In the next group (10–15 mg P l^{-1}), there were responses of over 1 t of sugar and, in the third group (16–25 mg), of nearly 0.5 t. In higher groups, responses were negligible. These experiments showed however, that on economic grounds smaller quantities of fertilizer were needed than expected (Table 3.5). No detailed experiments of this type have since been made in the UK and these still provide the basis of current recommendations. The increasing cost of phosphorus fertilizer relative to the value of sugar has, however, led to a reduction in the recommended optimum for the group of soils containing least phosphorus from 150 to 100 kg P_2O_5 ha^{-1} (see Table 3.5).

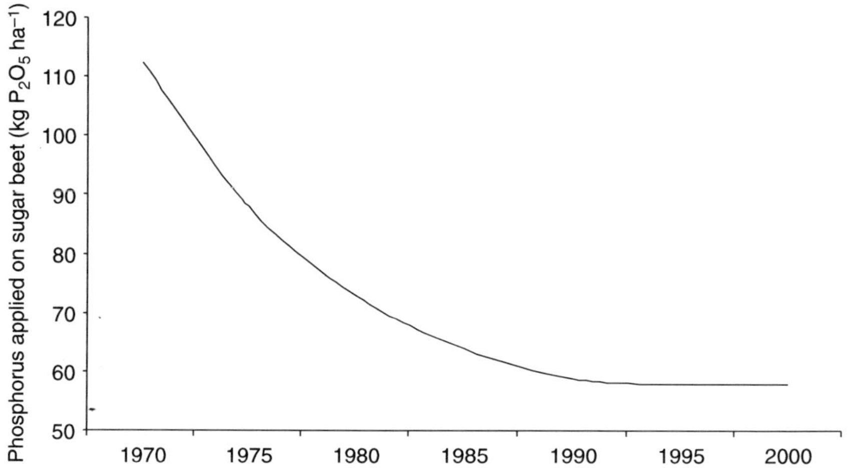

Fig. 3.2. Amount of phosphorus applied before sugar beet in UK, 1970–2000.

Table 3.4. Response to phosphorus fertilizer in relation to soil phosphorus extracted by sodium bicarbonate and optimum fertilizer application.

	mg P kg^{-1} soil				
	0–10	11–15	16–25	26–45	> 45
Increase in sugar yield (t ha^{-1})	+1.1	+0.3	+0.2	+0.1	−0.1
Number of fields	4	9	19	26	12
Optimum amount of fertilizer (kg P_2O_5 ha^{-1})	180	120	60	30	0

Table 3.5. Economic optimum application.

mg P l^{-1} soil	kg P_2O_5 ha^{-1}
0–9	100
10–15	75
16–25	50
>25	0

In a similar manner to the treatment of the UK data, Analogides (1987b) re-examined results from 73 phosphorus-fertilizing field experiments in Greece made over 12 years (1970–1981). Olsen's soil-phosphorus values varied widely, from 2.5 to over 60 mg P kg^{-1} soil. Responses to the largest amount of fertilizer tested (120 or 160 kg P_2O_5 ha^{-1}) were related loosely to soil phosphorus and he proposed a three-way classification: 0–10 mg P kg^{-1} (where all the large responses occurred), 10–20 mg P kg^{-1} (where the crop gave small average response) and above 20 mg P kg^{-1} (where there was never a significant response). Plotted graphically the results closely resemble those from the UK.

Sanz Saez (1985) examined responses to phosphorus fertilizer in both the central and the northern regions of Spain, where the crop is spring-sown. He also found all the large responses where soil contained less than 10 mg P kg^{-1}. Responses were moderate at 11–18 mg P kg^{-1} and non-significant with 19–29 mg P kg^{-1}.

Combining data from the countries reported above shows how Olsen's extraction method is able to predict the response to phosphorus fertilizer (Fig. 3.3).

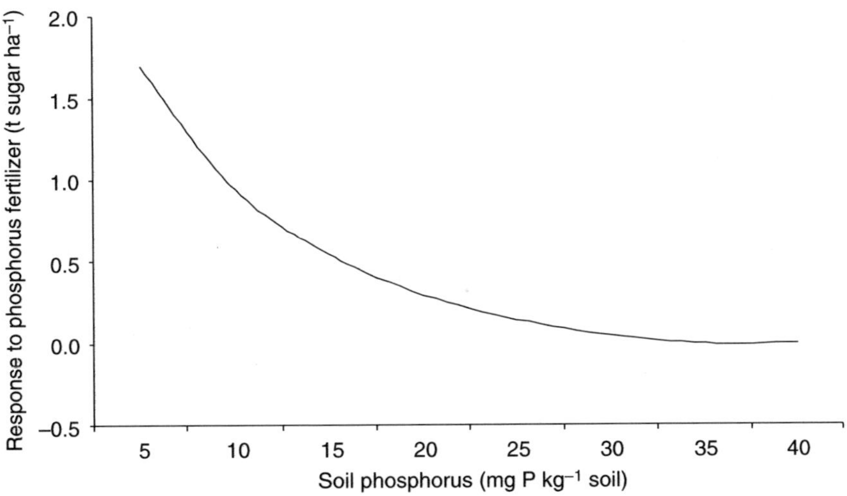

Fig. 3.3. Relationship between sodium bicarbonate-extractable soil phosphorus and sugar-yield response to phosphorus fertilizer in several European countries.

Long-term experiments on phosphorus fertilizer and soil analysis

Johnston *et al.* (1976) established plots of widely differing sodium bicarbonate-extractable soil phosphorus by applications of fertilizer to previously impoverished loamy sand at Woburn. In a long-term study, they were then able to measure the yield of sugar beet at a range of 12–150 mg P kg^{-1} soil both with and without fresh fertilizer.

Yields increased as the amount of soil phosphorus increased up to 32 mg P kg^{-1}. Yields were also increased by fresh fertilizer on soils containing up to 21 mg P kg^{-1} but not beyond. On the light soil with little organic matter, results suggest that sugar beet needs about 25 mg P kg^{-1} soil to yield near the optimum.

Johnston *et al.* (1986) reported on a similar study but on a very different soil – a heavy, notoriously difficult-to-manage, sandy clay loam at Saxmundham. During 1899–1968 a range of concentrations of soil phosphorus were established, surprisingly from as low as 3 to nearly 70 mg P kg^{-1}. The work confirmed the Woburn result that for sugar beet to yield fully, a background value of 25 mg P kg^{-1} was needed. Yield responses to fresh fertilizer were 1.1, 0.3 and 0.0 t sugar ha^{-1} at 9, 15 and 25 mg P kg^{-1} soil. On the very impoverished soils (less than 10 mg P kg^{-1}) a large fresh application of fertilizer did not give full yield (see also Chapter 9 for the long-term effects of phosphorus).

Allison and Chapman (1995) reported on an experiment continued for 30 years comparing, among other treatments, the yield of sugar beet with and without phosphorus fertilizer. The soil started off at 36 mg P l^{-1} soil extracted with sodium bicarbonate, which was considered more than adequate for a maximum yield of all the crops in a sugar beet/cereal rotation. It was nearly 20 years before plots that had received no phosphorus fertilizer gave significantly less sugar yield than those given fertilizer every year. This shows that the crop has a remarkable ability to obtain the element from soil reserves. However, for the remaining period of the experiment, the response to fresh fertilizer rapidly increased, showing that reserves had been seriously depleted.

Recovery in crop

In contrast to some other mobile elements, phosphorus fertilizer is taken up by plants relatively slowly because it is not mobile. Roots on vigorously growing plants need to explore fresh soil to meet requirements. Phosphorus moves to root surfaces over short distances by diffusion. In addition, there is no mycorrhizal symbiosis with sugar beet to aid uptake.

Chapter 2 showed that more than half the fresh nitrogen fertilizer applied for sugar beet is taken up by the crop. In the case of phosphorus the apparent recovery measured in the crop at harvest was as shown in Table 3.6 in two separate studies (Draycott *et al.*, 1977; Etchevers and Moraghan, 1983). In both cases recovery was far below 50% and declined with increasing amounts of applied phosphorus.

Table 3.6. Recovery in the crop measured at harvest.

Draycott *et al.* (1977)		Etchevers and Moraghan (1983)	
Applied P_2O_5 ($kg\ ha^{-1}$)	Recovery (%)	Applied P_2O_5 ($kg\ ha^{-1}$)	Recovery (%)
30	15.3	25	27
60	9.5	50	20
125	8.9	100	21
250	4.6	200	13

Efficiency

Draycott and Martindale (2000b) calculated the amount of phosphorus fertilizer applied to produce a tonne of roots from the UK average-usage data and national yields over the past 30 years. In 1970, 3.3 kg P_2O_5 ha^{-1} was given to produce each tonne of roots ha^{-1}. By 1999 this had declined threefold to 1.1 kg P_2O_5 ha^{-1}. This remarkable result reflected both a decline in excessive usage (Fig. 3.2) and increasing yield over the period (Fig. 1.2).

Placement

In contrast to some other crops, such as grass and cereals, sugar beet is nearly always grown in wide-spaced rows. Looking for increased efficiency of uptake of a relatively immobile nutrient in soil, many workers for over half a century have investigated the value of placing phosphorus fertilizer near the seed. Usually these placement methods have been compared with broadcast fertilizer before or after ploughing or on the seedbed.

Some early reports, probably when soils were relatively less fertile than today, showed benefits of placement a few centimetres below and/or to the side of the seed (Haddock, 1952; Davis *et al.*, 1961; Moraghan and Etchevers, 1981). Using labelled phosphorus (^{32}P), Anderson and Peterson (1978) showed that placement greatly improved recovery.

Many other reports record no better results from placed fertilizer over broadcast equivalent (Romsdal and Schmehl, 1963; Christenson *et al.*, 1975; Dunham, 1991a). Murphy and Walsh (1972), however, found an enhanced efficiency of fertilizer manganese and zinc and possibly copper and iron when banded with acidic phosphate. Further details of placement studies can be found in Chapter 11.

Effect of Phosphorus on Plant Establishment

Where sugar beet is grown in soil containing little readily available phosphorus, seedlings lack vigour, grow slowly and produce small plants. The result is that they are subject to physical damage and some do not survive attack by pests and diseases. A supply of phosphorus fertilizer round the roots has a visible effect on vigour, size and plant number (Davis *et al.*, 1962; Romsdal and Schmehl, 1963; Sipitanos and Ulrich, 1969). Sims and Smith (2001) showed that a reduction in yield due to insufficient phosphorus is initiated very early and is maintained throughout the growing season. Even though the above-ground sugar beet growth appears to return to near normal as the growing season progresses, root yield potential may have already been reduced.

Currently most soils under cultivation for sugar beet contain sufficient phosphorus to ensure full plant establishment. Plant counts done on plots with and without fresh phosphorus fertilizer for 12 years (Draycott *et al.*, 1972b, 1978a) showed no effect. No other examples have been found in recent literature of phosphorus increasing plant number. In contrast, there are reports of visible response in plant size to fresh phosphorus fertilizer. This is particularly where it is placed or banded near the seed.

Interactions with Other Nutrients

Many experiments have been made testing interactions between phosphorus and other nutrients. Generally interactions are negligibly small. Reviewing all the pre-war published information from Great Britain and Western Europe, Crowther and Yates (1941) found a small positive interaction between nitrogen and phosphorus but none between potassium and phosphorus. Boyd *et al.* (1957) examined the response to 125 kg P_2O_5 ha^{-1} in over 300 factorial experiments in Great Britain from 1934 to 1949. On mineral soils it increased sugar yields by 0.2 t sugar ha^{-1}, the $N \times P$ interaction was +0.04 t sugar ha^{-1}, the $P \times K$ interaction +0.03 and the $N \times P \times K$ +0.01.

Gallagher (1967) in Eire found that the interaction between phosphorus and potassium was positive and statistically significant (confirmed by Fuehring *et al.* (1969) in

Lebanon) although not as important as the N × K interaction. Jønsson (1969) in Sweden investigated the N × P interaction and found that it did not affect yield significantly.

There is some evidence that where phosphorus is in very short supply interactions are important. Trist and Boyd (1966) in the Rotation I Experiment at Saxmundham showed that, where soil phosphorus had been depleted, nitrogen fertilizer alone had little effect on yield but, when phosphorus fertilizer was given, nitrogen increased yield greatly. Also Hills *et al.* (1970) showed that, on phosphorus-deficient soil and in culture solution, a shortage of phosphorus decreased the absorption of nitrate nitrogen. In Russia, Gurevich and Boronina (1964) found that nitrogen given to sugar beet only increased the percentage of nitrogen greatly when phosphorus was also given.

Suggestions that giving large amounts of phosphorus fertilizer might improve the quality of sugar beet given large amounts of nitrogen were ill-founded. Peterson *et al.* (1966) found that additional phosphorus fertilizer did not improve sugar percentage but neither did it do any harm. Ogden *et al.* (1958) also found that large amounts of phosphorus did not offset the detrimental effects of excess nitrogen fertilizer.

Phosphorus Supplied by Organic Manures

All forms of organic manure and animal and plant remains provide variable amounts of phosphorus for sugar beet and other crops. Poultry manures and slurries from intensively managed animals are particularly rich sources; thus sugar beet receiving these need little, if any, fertilizer, particularly if the practices have been continued for many years. Other sources that may or may not require supplement fertilizer are sewage sludges and traditional strawy farmyard manures. Many short- and long-term experiments have been made to investigate sugar beet requirements for phosphorus against a background of these organic materials. These are reviewed in full in Chapter 8.

Future Fertilizer Practice for Phosphorus Based on Soil Analysis

UK

Practices have been reviewed and updated recently by the state sector and recommendations published (MAFF, 2000). These are based on a measurement of sodium bicarbonate-extractable soil phosphorus (Olsen's method). The amounts of fertilizer are calculated for yields of 60 t roots ha^{-1}, slightly higher than the national average at the turn of the millennium. Account is taken of cost of fertilizer, value of crop, the need to improve soils with a very low phosphorus value and a deduction for any organic manure (see Chapter 8).

Table 3.7 summarizes the recommendations. A maintenance application of 50 kg P_2O_5 ha^{-1} is identified, that being slightly more than the removal in roots of the 60 t ha^{-1} crop. For the first time, target soil phosphorus values are defined for rotations of: (i) sugar beet and cereals; and (ii) sugar beet and vegetables. For (i) the target range is 16–25 mg P l^{-1} and for (ii) 26–45 mg P l^{-1}. Guidance is given in Chapter 9 on how to build up or run down soil concentrations and then maintain them in the desirable ranges.

Table 3.7. Amounts of phosphorus recommended in the UK based on bicarbonate-extractable soil phosphorus (MAFF, 2000).

	Soil phosphorus (mg l^{-1})			
	0–9	10–15	16–25	26 and above
Fertilizer phosphorus (kg P_2O_5 ha^{-1})	100	75	50	0

Ireland

Power and Herlihy (1988) reviewed the use of phosphorus in Eire and published recommendations based on phosphorus extracted from the soil by Morgan's solution. The use of phosphorus on sugar beet had fallen rapidly during the 1980s from 215 to 154 kg P_2O_5 ha^{-1} as a result of a large rise in fertilizer price and better use of soil analysis. Herlihy (1986) proposed using the Morgan test for soil phosphorus which he thought suited their soils best. The recommendations are shown in Table 3.8. Herlihy also distinguished between the amount needed for optimum compared with maximum yield, the recommendations being a good compromise between the two, containing a small improvement component for deficient soils.

USA

Phosphorus fertilization practices in the USA also show a reduction in the amount applied for sugar beet. In Michigan between 1970 and 1995, the amount applied decreased from 165 to 80 kg P_2O_5 ha^{-1}. Soil testing plays a key role in determining amounts for sugar beet. Concerns over the environment and economic pressures continue to promote programmes for carefully monitoring reserves and adjusting fertilizer application accordingly.

In the sugar beet-growing regions of the USA, two methods for extracting available phosphorus are commonly used. The Bray–Kurtz P_1 method (Bray and Kurtz, 1945) was calibrated for a large number of crops including sugar beet in the 1930s and 1940s and is still widely used on non-calcareous soils (Frank *et al.*, 1998). Another method calibrated later (Olsen *et al.*, 1954) is also used in the USA and in other countries, as detailed above.

Bray-Kurtz P_1 extractant is 0.025 M HCl plus 0.03 M NH_4F, with a soil-to-solution ratio of 1 to 10. Olsen extractant is 0.5 M $NaHCO_3$, with a soil-to-solution ratio of 1 to 20. While the former is probably more suitable for slightly acid soils and the latter for calcareous soils, through careful calibration with crop response to applied phosphorus, both are very successful methods for sugar beet.

The minimum soil concentration for optimum yield of sugar beet by the Bray–Kurtz P_1 method is generally agreed to be 20 mg P kg^{-1} soil (Bray and Kurtz, 1945; Christenson *et al.*, 1981; Franzen and Cihacek, 1996; Binford *et al.*, 2000). For the Olsen method opinions differ in the USA, ranging from 10 to 18 mg P kg^{-1} soil (Table 3.9). It is interesting to compare these values with those mentioned above for the UK, where the target range by the same method is 16–25 mg P kg^{-1} soil.

Phosphorus Fertilizer and the Environment

The decreased application and increased efficiency of phosphorus use by sugar beet has environmental benefits. Phosphorus loss from agricultural land by erosion and in extreme cases, leaching can promote plant growth in waterways. Topsoil erosion is the usual route of phosphorus loss from sugar beet fields and careless management of manures will make the problem of sheet erosion and run-off worse. There is, however, an increased chance of phosphorus leaching when soil concentrations exceed 50 mg P l^{-1} (Olsen's P), as suggested by Blake and Johnston (1999), or 75 mg P kg^{-1} extracted by Bray–Kurtz P_1 (Jacobs,

Table 3.8. Amounts of phosphorus recommended in Eire based on Morgan's soil-phosphorus test (from Herlihy, 1986).

	Soil phosphorus (mg l^{-1})				
	0.5	1.0	3.0	10.0	30.0
Fertilizer phosphorus (kg P_2O_5 ha^{-1})	160	130	90	45	0

Table 3.9. Minimum soil concentration for optimum yield of sugar beet by the Olsen method.

	mg P kg^{-1} soil	
Washington and Idaho	10	James *et al.*, 1967; Westerman *et al.*, 1977
Utah	12.5	Haddock, 1959
North Dakota and Minnesota	15	Franzen and Cihacek, 1996
Montana	18	Lichthardt and Jacobsen, 1992
Wyoming	15	Blaylock *et al.*, 1996

1995b). Both these types of phosphorus loss can potentially increase the natural process of eutrophication in water catchments. This promotes the growth of algae and water plants, resulting in degradation of water-supplies for drinking and amenity uses. Reducing the chance of excessive concentrations of phosphorus on agricultural land is an important environmental indicator for reducing water pollution.

Neeteson and Ehlert (1989) examined the inputs of phosphorus for sugar beet in The Netherlands in relation to awareness of the effect of residues on the environment. When both roots and tops were removed from the field there was no danger of residual phosphorus; in fact a gradual decrease in soil phosphorus was expected. However, where tops were left on the field, large quantities were left behind. In The Netherlands most of the sugar beet is grown on heavier soils with a high phosphorus-absorbing capacity, so leaching of the element is unlikely. There was a slight worry that on the lighter soils some movement could occur. Balance-sheets of inputs in relation to soil phosphorus and offtakes on the two main soil types illustrate this.

SULPHUR

Physiological Role

Sulphur is essential for the formation of proteins in sugar beet, as in all plants, and the ratio of nitrogen to sulphur is usually about 15 to 1 by weight in healthy plants. Where the ratio is greater, a shortage of sulphur might be the cause. There may also be unusual concentrations of certain amino acids (e.g. arginine, as suggested by Hocking (1995)) when the balance between nitrogen and sulphur is upset. When sulphur is deficient, the concentration of the sulphur-containing amino acids is decreased, but other amino acids may be greatly increased (e.g. glutamine, as found by Sexton (1996)). McGrath (2000) found that the ratio of malate to sulphate was affected in sulphur-deficient plants and proposed that the ratio could be used as an indicator of deficiency. A detailed account of the functions of sulphur in higher plants can be found in Marschner (1995).

Amount of Sulphur in the Crop

Concentration and uptake

There is a paucity of information on the range of concentrations expected in healthy and deficient sugar beet crops and the quantity taken up by roots and tops. Table 3.10 suggests that either: (i) sugar beet concentrations vary widely in both top and root dry matter; or (ii) there are discrepancies in the analytical techniques employed. Clearly, more work is needed to determine accurately the expected concentration and its range in healthy and deficient crops.

Similar discrepancies appear for the sugar beet crop uptake of sulphur (Table 3.11). Syers *et al.* (1987) reported uptakes by low-yielding crops (40 t fresh root and 20 t top ha^{-1}) of 7–23 and 8–16 kg S ha^{-1} by roots and tops, respectively. In contrast, Armstrong (1985), who made measurements on field grown crops in 1978–1983, reported an uptake by roots plus tops of 50–70 kg S ha^{-1}. For some of his very high-yielding crops, the uptake was 100 kg S ha^{-1} and removal in roots 20–50 kg S ha^{-1}. These appear to be unusually large values, particularly alongside detailed measure-

Table 3.10. Concentration of sulphur in healthy and deficient sugar beet plant parts.

Portion of plant	% S in dry matter: Healthy	% S in dry matter: Deficient	Reference
Leaf blades	0.05–1.4	0.005–0.02	Ulrich and Hills, 1969
Leaf blades	0.4–0.7	–	Bravo *et al.*, 1989
Petioles	0.1–0.3	–	Bravo *et al.*, 1989
Crowns	0.08–0.28	–	Bravo *et al.*, 1989
Roots	0.01–0.15	–	Bravo *et al.*, 1989
Whole leaves (solution culture)	0.4–0.5	0.1–0.2	Hocking, 1995
Tops	0.8–1.1	0.6	Sexton, 1996
Roots (field-grown)	0.2	0.1	Sexton, 1996
Recent fully expanded leaves (field-grown)	> 0.3	< 0.3	Connors, 2000
Tops	0.30–0.35	–	F. Zhao, UK, 2002, personal communication
Roots	0.03–0.04	–	F. Zhao, UK, 2002, personal communication

ments on a high-yielding crop reported in France (96 t roots ha^{-1}) where uptake was only 6 kg S ha^{-1} by roots and 10 kg S ha^{-1} by tops (Jourdan *et al.*, 1992). Clearly more work is urgently needed to clear up these discrepancies.

Sulphur in Soil

Total sulphur

Sulphur is present in organic and inorganic forms. In leached sandy soils there may be less than 200 mg S kg^{-1}, whereas fine-textured soils with little leaching and some saline soils may contain more than 3 g S kg^{-1} (Syers *et al.*, 1987). Organic forms of sulphur are present in dead plant material, such as roots and crop residues, and in organic manures, microorganisms and humified organic matter. Inorganic forms of sulphur derive from the soil parent material, atmospheric deposition (especially in industrialized regions of the world) and oxidation or reduction of soil organic sulphur. Scherer (2001) in a worldwide review of sulphur in crop production found that, whilst plant-available sulphur is in the sulphate form, 95% of soil sulphur is organically bound. Measurements of leaching losses of sulphur were also reviewed and amounts of sulphur lost were in the range 30–80 kg S ha^{-1} $year^{-1}$ in north-west Europe.

When soil conditions favour aerobic reactions sulphur compounds oxidize slowly to sulphates. Anaerobic conditions reverse the reactions and sulphates and other sulphur compounds are reduced to sulphides. These latter account in part for the 'bad egg' smell (H_2S) often encountered when organic matter is decomposing in waterlogged soil.

Table 3.11. Uptake of sulphur by field-grown sugar beet.

Tops	Roots (kg S ha^{-1})	Total	Yields	Reference
7–23	8–16	15–39	Low	Syers *et al.*, 1987
		50–70	Average	Armstrong, 1985
8	5	13	Average	Bravo *et al.*, 1989
	20–50	100	High	Armstrong, 1985
10	6	16	Very high	Jourdan *et al.*, 1992

Available sulphur

Thus differences in parent material, deposition, extent of leaching and amount of organic matter cause the wide variation in total sulphur concentration in soils. What determines plant availability of that sulphur is the amount present as sulphate ions. These are weakly held in most soils and swept to roots by mass flow in the transpiration stream of water. In this mechanism the transport contrasts sharply with phosphate ions. These are strongly adsorbed and only move to roots slowly by diffusion. This is also why sulphate ions are readily leached, whereas phosphate ions are not moved out of the root zone.

There is little evidence in the literature relating to soil analysis for sulphate and response by sugar beet. Sulphate concentration can be measured in extractant solutions such as those containing OH^-, $H_2PO_4^-$ or HCO_3^-, which displace sulphate. For other crops it has been found that sodium bicarbonate solution produced results that correlated with crop uptake. Several studies have also shown that phosphate solutions can be used with success to predict soils that are sulphur-deficient (Syers *et al.*, 1987) but more work is needed on sugar beet similar to that begun by Armstrong (1985). He measured concentrations of available sulphur in sugar beet soils in the UK. Concentrations ranged from 8 to 41 mg SO_4^{2-}-S l^{-1} soil.

In studies at Rothamsted, McGrath *et al.* (1996) proposed plant analysis as a better guide than soil analysis to sulphur status. Rather than a direct measure of sulphur present, McGrath (2000) now uses the malate-to-sulphate ratio. Malate is an organic acid present in all plants and is produced where sulphate is deficient. Malate and sulphate can be extracted simultaneously and measured using ion chromatography. Results for sugar beet are awaited with interest.

Effects of a Shortage of Sulphur

Appearance of the crop in the field

There are very few reports of clear evidence of shortage of sulphur in sugar beet anywhere in the world. The earliest was that of Ulrich and Hills (1969), who illustrated the response to sulphur fertilizer in California. Armstrong (1985) noticed slight paling of leaves in some of his experiments where sugar beet was grown on soils containing a small concentration of available sulphur, but there was no response.

More recently, Connors (2000) commented on the onset of sulphur shortage in field-grown sugar beet. He warned that a slow crop-growth response to applied nitrogen should arouse suspicions of sulphur deficiency and prompt further investigation by tissue analysis. Sulphur deficiency is more pronounced with large nitrogen inputs and is accompanied by a large nitrogen to sulphur ratio ($> 17:1$) in young tissue.

Chapter 7 contains details of sulphur-deficiency symptoms, photographs and plant analyses.

Nutrient-culture studies

Hocking (1995) grew sugar beet in a hydroponic culture system to study the effect of different concentrations of sulphur in the nutrient solution on growth and nitrogen-containing impurities. A concentration of 1.5 mM sulphur was estimated to provide 'adequate' amounts of the element under the conditions of his experiment. This was varied down to zero and to one-tenth and up to double the 'adequate' supply.

Giving no sulphur caused severe stunting, yellowing of leaves and decline in chlorophyll concentration. Plants quickly recovered when supplied with sulphur. Sulphur concentration in leaves was 0.1% S in dry matter with zero application, 0.2% S with slight deficiency and 0.35–0.45% S without deficiency. These measurements were made when plants were about 3 months old. Root weight, as expected, was maximal with the 'adequate' supply. This was decreased to as little as 20% at zero sulphur supply and about 80% at one-tenth adequate. The double dose depressed yield slightly.

Withholding sulphur also had a dramatic effect on amino acid concentration in the plants. In particular, arginine was greatly increased by sulphur deficiency. Hocking (1995) suggested that the switch in nitrogen metabolism was initiated either by a critical concentration of sulphur in tissue or by a specific change in the N : S ratio, or both.

Hocking proposed that arginine concentration could be used as an indication of impending sulphur deficiency and that there may be implications for sugar extraction if root impurities are affected by sulphur shortage.

Pot-experiment studies

Hoffmann in Germany (work reported by Beckers, 1999) demonstrated the effect of sulphur deficiency on sugar beet by growing plants in sand culture. With less than 4 mg S kg^{-1} sand, dry-matter production was decreased considerably. In soils below this concentration, leaves contained less than 0.3% S in dry matter. For full yield, leaves needed 0.35% S or more in this pot experiment. Field trials were also done, as described below, and on all sites leaves contained 0.32–0.56%, indicating little, if any, shortage of sulphur.

Field studies

Europe

One of the first field studies was that of Armstrong (1985). In 1984 the effect of soil-applied gypsum or elemental sulphur at 60 kg S ha^{-1} just after sowing was tested on six farms. Soil sulphur concentrations before application ranged from 8 to 41 mg SO_4-S l^{-1}. Analysis of growth throughout the season and detailed measurements at harvest showed no effect of the sulphur additions on any site in growth, yield or root quality. Armstrong concluded that soil reserves plus deposition (up to 40 kg S ha^{-1}) satisfied the crops' needs and no application was justified for sugar beet in the mid-1980s.

Following the nutrient-culture study of Hocking (1995), a field survey was made by Sexton (1996) in the UK in 1995 to assess the sulphur status of sugar beet grown in areas of different deposition. Total deposition (wet plus dry) was measured and mapped in 1990, and areas were chosen varying from less than 10 kg S ha^{-1} to above 40 kg S ha^{-1}. Plant samples were taken in July and September.

Sulphur concentrations in leaves varied from 0.6 to 1.1% and in roots from 0.1 to 0.2% in July. In September results were similar but more variable, probably due to use of sulphur fungicides in August. Generally, where deposition was below 15 kg S ha^{-1}, plants contained least sulphur (leaves 0.6–0.7% S) and, where deposition was greatest (30 kg S ha^{-1}), plants contained most (1.0% S).

A range of amino acids were measured in the plant samples. These concentrations were used to locate potentially sulphur-deficient fields, as suggested by Hocking (1995), a large value for arginine being taken as a sign of sulphur shortage. No evidence is presented to show that supplying these crops with additional sulphur improved yield or root quality.

In Germany work by Hoffman (reported by Beckers, 1999) at 21 locations all over the beet-growing area showed available soil sulphur 0–90 cm deep varied but exceeded 20 kg S ha^{-1}, which appeared to be sufficient. Not only was leaf concentration greater than 0.35% S (the critical value established in the pot experiments described above) but yield was never increased by sulphur fertilizer supplying 40 kg S ha^{-1}. Total uptake was 24 kg S ha^{-1}, of which 8–10 kg S ha^{-1} was in the tap root.

USA

As noted previously, Ulrich and Hills (1969) described sulphur deficiencies in California. Few reports have been published since that time concerning sulphur needs for sugar beet in the USA. This could indicate the lack of apparent need for supplemental sulphur.

Lamb (1989) presented data from a 3-year study in North Dakota showing no response. In Michigan, Robertson *et al.* (1976a) reported no difference in yield or per cent sugar between KCl and K_2SO_4 in a 2-year study on a sandy clay-loam soil. D.R. Christenson (unpublished data) summarized a three-site study comparing several amounts applied for sugar beet. There was no response on these medium- and fine-textured soils. Consequently there is no recognition of sulphur fertilizer needs for sugar beet in fertilizer-recommendation bulletins in the USA.

Even though there has been a reduction in the amount deposited across the USA between 1985 and 2000, no reports of yield response to applied sulphur have been reported in sugar beet. The lower deposition reported in the western USA is due in part to less industry as well as lower annual precipitation.

Future Need for Sulphur Applications on Sugar Beet Worldwide

Although there is a paucity of direct evidence that fertilizer is needed, it is clear from the above that sugar beet crops must take up some 50 kg S ha^{-1}. That atmospheric deposition is now only a fraction of what it was 25 years ago is also in no doubt. In the industrialized areas it was common for annual deposition to equate to or exceed crop uptake (50 kg S ha^{-1}). Now deposition may be a tenth of this amount. In the past this plus sulphur released from decaying organic matter more than satisfied sugar beet, with its deep, spreading root system.

When sulphate is leached it may be adsorbed on sesquioxides already accumulated as a result of soil weathering processes. The role of this form in sugar beet nutrition is largely unknown. It has been previously pointed out that sugar beet is a 'good scavenger' of nutrients from the whole soil profile. Further research is needed to evaluate this source in sugar beet nutrition.

In future, fertilizer sulphur may well be needed. Fortunately there is a range of inexpensive materials that could be used – gypsum, 15–18% S, kieserite, 20% S, single superphosphate, 10–12% S, and ammonium sulphate, 24% S – all of which suit sugar beet production. An alternative, being used on other crops, is also elemental sulphur. New experiments are needed now to establish the need for sulphur.

SUMMARY OF PHOSPHORUS AND SULPHUR

The reports described above show the great depth of knowledge that has built up from 150 years of detailed research on phosphorus. In contrast, the amount of work on sulphur is very limited. However, from both the extensive work on phosphorus and the small amount on sulphur, it would appear that currently, at least, a shortage of neither nutrient is depressing yield by much. In the case of sulphur this could change quickly and more extensive work is justified in many countries, if only to check that yield is not being depressed by a shortage. We think it unlikely but possible in the short term but quite likely in the long term.

When plant-nutrition research started in earnest on sugar beet, soils contained little available phosphorus because very few soils in their natural state contain much of this nutrient. Thus responses were large – establishment of the crop was improved, root growth was stimulated, plants had more vigour and the crop yielded several times more when given phosphorus fertilizer. In common with other crops, the element became one of the three main nutrients, with nitrogen and potassium.

It has been shown above that phosphorus is (almost) immune to leaching so reserves have now built up to such a degree that few crops respond. In future we believe that research, development and extension work should be devoted to monitoring phosphorus in soil. Then available concentrations can be maintained at the optimum of about 25 mg P kg^{-1} soil described for sugar beet and companion crops in rotation.

Until quite recently little attention has been given to sulphur as a plant nutrient. In the past the few tests conducted showed a sufficiency in the soil and no crop response to sulphur added as fertilizer. Supply from organic matter and the atmosphere provided sufficient for even the most demanding crops. Now the position is changing rapidly because sulphur pollution in the air has been greatly reduced and some crops with a large demand (e.g. oil-seed rape and winter wheat) are responding to sulphur fertilizer. More research on all aspects of the sulphur nutrition of sugar beet is urgently needed.

4
Potassium and Sodium

Potassium and sodium are two very important monovalent cations in sugar beet nutrition. Both are macronutrients, being taken up and utilized in large quantities by sugar beet crops producing optimum yield. This large uptake of sodium is unusual among crop plants, most having a sodium exclusion mechanism in their roots. Sugar beet is a halophyte due to its original shoreline habitat, absorbing and assimilating sodium, which partly replaces potassium.

The two elements have similar effects on the growth and productivity of sugar beet and are therefore considered here together. Many experiments dating back to the 19th century have shown how potassium greatly improves early vigour and growth and in turn sugar yield. More recently, it has been found that sodium plays similar roles and for peak performance a crop must take up a sufficiency of both elements.

In arable production, all crops, such as sugar beet, in a rotation remove potassium in the part sold. Such fields need additions of potassium in fertilizers and/or manures because few soils release sufficient to maintain yield. In contrast only sugar beet removes sodium and many soils already contain a great deal. It is only in humid areas, such as north-west Europe, where winter rains leach away sodium that the element needs to be applied in fertilizer. Indeed, in many arid areas, sodium accumulates in soils in concentrations harmful to plants. Of course, sugar beet under such conditions gives no response to fertilizer and many workers mistakenly believe sodium is not an essential macronutrient. The evidence is presented below that both potassium and sodium need to be supplied where soil cannot provide sufficient.

POTASSIUM

Potassium in Soil

The various forms of potassium present in soil exist in four main pools (or categories). That which is immediately available to plants is in the soil solution. Also readily available is that in the exchangeable form. There is also a pool of slowly available potassium, which is in the slowly exchangeable state. Finally, there is the potassium present in the clay lattice and in minerals, which is only available to plants as these weather or break down.

Potassium in the four pools is in reversible dynamic equilibrium, i.e. as plant roots take up potassium from soil solution, it is replenished by exchangeable and slowly exchangeable potassium. When potassium is added to soil in fertilizers and manures, it initially goes into soil solution and is then distributed between exchangeable and slowly exchangeable pools.

Total

Many minerals forming the parent material of soil when subject to weathering release potassium. In humid conditions some is gradually leached out of soil over long periods, though much is retained by incorporation into secondary minerals or held by cation exchange on clay minerals and organic matter. In arid conditions potassium may be retained through being reprecipitated in salts such as sulphates and chlorides.

Potassium concentrations in soil from the UK and USA as percentage of dry weight range from 0.3 to 2.5 (Whitehead, 2000). Land is commonly ploughed for sugar beet to 25 cm depth; with average bulk density this weighs 3000 t ha^{-1}. Thus the range of total potassium present is 9–75 t ha^{-1} (about 11–90 t K_2O ha^{-1}). An earlier estimate of the total amount of potassium present in the average furrow slice of sugar beet soils in the UK (Draycott, 1993) was 30 t K_2O ha^{-1}.

Much of the total potassium in soil is present as unavailable or slowly available forms. The former accounts for over 90% of the potassium in soils and is primarily contained in feldspars, but some is also present in micas. Slowly available potassium is held in clay minerals, such as mica and vermiculite, and represents less than 10% of the total potassium in soil. The amount is often related to past application of fertilizers and manures.

When plants remove potassium from the exchangeable pool, slowly exchangeable potassium begins to replace it immediately. Even though this replacement may be rapid, the rate of replacement is often inadequate to maintain maximum yields (McLean and Watson, 1985). Attempts have been made to measure the slowly available fraction in the laboratory, but results have not been found useful for making fertilizer recommendations. Indirectly, consideration is given to the supply from slowly exchangeable sources when routine laboratory procedures are tested in potassium fertilization trials in the field.

Available

What really matters is water-soluble and exchangeable potassium, comprising less than 2% of total potassium, which is readily accessible to sugar beet during its growth. Then any shortfall can be made up by additions in fertilizers and manures. The potassium dynamics of a typical sugar beet field is shown in Fig. 4.1. No slowly exchangeable potassium is shown because it is not readily determined.

Many soil scientists over the past 100 years have extracted that fraction of potassium loosely described as 'available' to sugar beet. A variety of extractants and procedures have been tested, from water to strong acids. Currently, the extract is also used for determination of other cations (particularly sodium, magnesium and calcium).

The first comprehensive work (Warren and Cooke, 1962) in the UK described 11 years of field experiments to compare methods of analysing soils for available potassium. The experiments were divided by soil analysis into groups of equal numbers of fields and average crop responses were used to value the analytical methods. Citric acid used to extract available potassium separated the soils into groups for differential applications, which was more profitable than giving a uniform amount to all fields. Acetic acid was less effective than citric acid; using dilute hydrochloric acid (HCl) to group the soils gave no more profit than a uniform application. Water-soluble potassium measurements were of even less value than acid-soluble values.

Davis *et al.* (1959) also used dilute HCl to analyse soils for available potassium in Michigan, USA. Extraction was with 0.135 M HCl and a soil to solution ratio of 1 : 4. The analyses were made on soil samples taken in August from plots of sugar beet where between none and 450 kg K_2O ha^{-1} had been applied. There was a close relationship between amount of potassium applied and amount extracted.

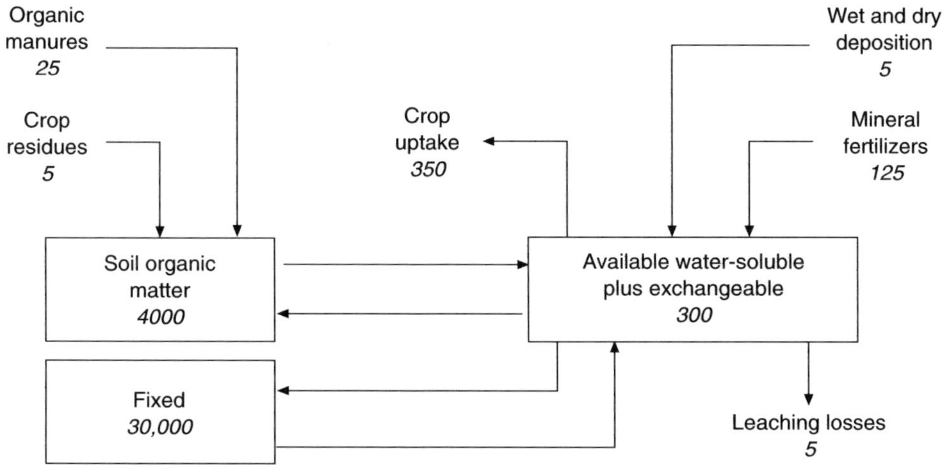

Fig. 4.1. Potassium dynamics of a typical sugar beet soil and crop (kg K_2O ha^{-1}).

Early work at Rothamsted tended to favour ammonium acetate (NH_4OAc) solution as an extractant for cations. This was termed 'exchangeable' potassium though later terminology refers to 'readily soluble' potassium. Warren and Johnston (1962a) analysed soils from the Barnfield experiment and found that the amount of potassium soluble in NH_4OAc was related to the amount of potassium in the roots of mangolds. They found that, with large amounts of fertilizer, the potassium applied exceeded the amount taken up by crops. Not all the extra potassium from repeated dressings accumulated in the surface soil: part moved down into the subsoil and some was lost in drainage. They suggest a 'saturation value' for soil. When more than this amount is applied, then leaching is rapid. On Barnfield, the saturation value was of the order of 300 mg kg^{-1} soil of readily soluble potassium.

It became clear at this time that the fraction of potassium that was immediately available for plant uptake was that in soil solution. However, for sugar beet with a long growing season and a large demand for potassium the most important fraction is the exchangeable potassium. This provides the source from which the water-soluble (or immediately available) fraction is replenished (Fig. 4.1).

Adams (1961c) took over the Rothamsted work on potassium for sugar beet and started to sort out its interrelationship with sodium. Available potassium was extracted by leaching soil samples with ammonium nitrate (NH_4NO_3) solution. At about this time NH_4NO_3 replaced NH_4OAc because of ease of analysis in flame photometers. The amounts of potassium extracted are closely correlated as they are with other methods which use an excess of NH_4^+ to exchange with K^+.

The soils were representative of the sugar beet-growing areas. Despite over 80% of fields being in the 'low' or 'very low' category, he was interested that the crop rarely showed obvious signs of potassium deficiency. He then realized that sodium application was making up, in part, for a poor supply of potassium. Adams pointed out that, when Boyd *et al.* (1957) and Warren and Cooke (1962) examined the relationship between soil potassium and response, they had considered only the responses where sodium was not given. Later work has confirmed the importance of examining response to potassium and sodium in the presence and absence of the other and of measuring both exchangeable potassium and sodium in soil.

Draycott and Durrant (1976b) evaluated the ammonium nitrate/shaking technique for potassium. Figure 4.2 shows response to potassium, with and without sodium, over a

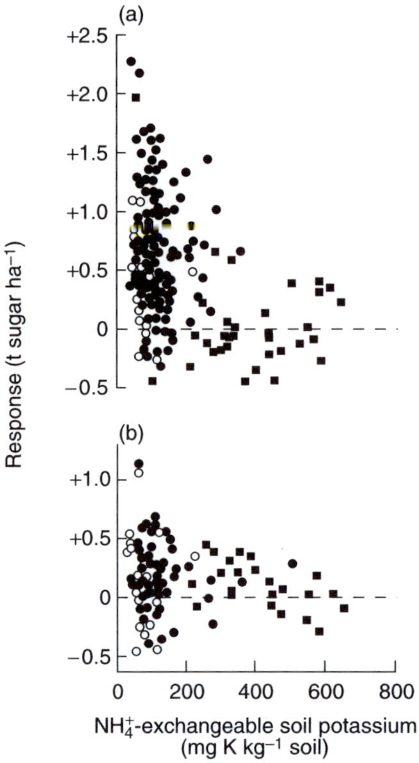

Fig. 4.2. Sugar-yield response to 200 kg K_2O ha^{-1} and exchangeable soil potassium: (a) no sodium fertilizer; (b) with sodium fertilizer. ● Mineral and ■ organic soils (1957–1960); ○ mineral soils (1970–1974).

wide range of soil values. Soil potassium extracted in this way was able to predict which fields would not respond but was not successful in predicting the magnitude of response on a field. All the large responses were, however, on fields where soil contained less than 200 mg K l^{-1}. Table 4.1 summarizes the increase in sugar yield from potassium fertilizer alone and underlines the value of the method. The magnitude of the average response on groups of fields decreased as soil potassium increased. This is a similar result to that of Beringer (1987) in Germany and elsewhere. Ammonium nitrate is now almost universally used in the UK to extract cations by shaking soil with an aqueous molar solution followed by filtration or centrifuging (Durrant *et al.*, 1974a).

The standard extraction procedure previously used of leaching the soil with NH_4OAc at pH 7.0 has been modified and is widely used in the USA for routine extraction of exchangeable cations. In general, modifications have included shaking in place of leaching, utilizing a narrow soil : solution ratio, and shorter extraction time (Doll and Lucas, 1973). The common soil : solution ratio is 1 : 10 with a shaking time of 5 min on a reciprocating shaker at 200 excursions min^{-1} (Warncke and Brown, 1998).

There is a paucity of work with potassium fertilizer for sugar beet in the USA. Most soils have rather high concentrations of potassium and the probability of response is low. In general the exchangeable potassium measured with NH_4OAc relates to uptake better than yield response.

Some soils, however, need potassium, and Moraghan and Cole (1978) in the Red River Valley reported potassium-deficiency symptoms and yield response to applied potassium when exchangeable potassium was less than 70 mg kg^{-1}, but not above. James *et al.* (1968) in Washington State reported yield response to applied potassium where soil contained up to 120 mg K kg^{-1} soil. Gascho *et al.* (1969) found no response to potassium supplied from four different forms of fertilizer on soils testing above 100 mg K kg^{-1} soil on different soils in Michigan. Similarly no response to applied potassium was found in Colorado on soils testing greater than 100 mg kg^{-1} (Ludwick *et al.*, 1980).

Table 4.1. Response to potassium fertilizer alone on UK fields in the 1970s grouped by soil analysis.

	Soil potassium concentration in ammonium nitrate extract (mg K l^{-1} of soil)			
	0–60	61–120	121–240	241–400
Increase in sugar yield (t ha^{-1})	+1.13	+0.70	+0.60	+0.40

Potassium in Sugar Beet

Concentration

Measurements throughout the growth of a crop show that the concentration of potassium in leaf and root dry matter is normally about 7% and 6%, respectively, in April, falling rapidly to 3% and 1%, respectively, in August. The average concentration of potassium in top dry matter at harvest is about 3%, whereas the concentration in root dry matter is about 0.8%, although values frequently range from 2.0 to 3.5% for tops and 0.6 to 1.0% for roots (Table 4.2).

Uptake

Of the cations taken up by sugar beet, potassium is taken up in the largest amount. All cations move to roots by a combination of mass flow and diffusion, potassium reaching roots mainly by diffusion. The element enters the root as K^+. There is some competition between the four major cations (K^+, Na^+, Ca^{2+}, Mg^{2+}) and experiments have shown negative interactions between potassium and magnesium. Bolton and Penny (1968) presented data showing the effect of potassium on magnesium uptake by sugar beet tops, ryegrass and kale leaves (Table 4.3). Increasing potassium supply depressed magnesium uptake to a far greater degree than magnesium suppressed potassium uptake. In reviewing potassium interaction with other nutrients, Dibb and Thompson (1985) concluded that increasing potassium supply has a fairly consistent effect on lowering tissue concentrations of calcium and magnesium. Conversely, there is little evidence to suggest that increasing calcium and magnesium supply suppresses potassium uptake by plants though unsubstantiated claims have been made on soil derived from magnesium carbonate. There is some evidence with potatoes grown on soils with very high exchangeable magnesium that uptake of potassium is depressed, possibly because Mg^{2+} is not easily replaced on the clay complex by K^+.

Potassium uptake during growth has been measured at intervals from May to November in the UK (Draycott, 1995), France

Table 4.2. Concentration of potassium in sugar beet dry matter at harvest.

% K			
Tops		Roots	References
3.0		0.77	Draycott, 1972
3.5		–	Moraghan and Ananth, 1985
3.6	Laminae	0.98	Analogides, 1987a
5.0	Petioles		
2.6		0.79	Bravo *et al.*, 1989

Table 4.3. Effect of potassium and magnesium fertilizers on magnesium concentration in crop dry matter (from Bolton and Penny, 1968).

	Ryegrass first cut	Sugar beet tops	Kale leaves
Treatment	g Mg kg^{-1} dry matter		
K0Mg0	1.10	3.05	1.30
K0Mg2	2.40	8.04	4.84
K2Mg0	0.92	2.91	0.96
K2Mg2	1.73	4.79	2.98

(Jourdan *et al.*, 1992) and the USA (Bravo *et al.*, 1989). Uptake was slower initially in the UK, probably due to slower spring growth, but later in the season total uptake was similar in the UK and France at about 400–500 kg K_2O ha^{-1} (Fig. 4.3). Potassium was lost more rapidly from tops, starting as early as August in France. Senescence and fall in quantity of potassium in tops started in September in the UK. In contrast, the overall pattern of uptake by roots in the UK and France was remarkably similar and ended in November at about 100 kg K_2O ha^{-1} in the UK and 150 in France. During the early period of growth when total uptake was most rapid, the rate averaged 8.5 kg K_2O ha^{-1} day^{-1}, with a maximum of 15 kg K_2O ha^{-1} day^{-1}. These are very large values among both crops and nutrients, confirmed in a recent report from Hungary, where final the uptake was 480 kg K_2O ha^{-1} and the maximum uptake rate was 11 kg K_2O ha^{-1} day^{-1} (Buzás and Johnston, 1999).

Work from the USA contrasts with that in Europe. The initial uptake in Colorado was slow, again probably due to slower growth in cool springs. During the peak uptake, the rate was a modest 2.7 kg ha^{-1} day^{-1}, the lower value related to the averaging of data over two planting dates (22 April and 27 May). A maximum uptake of 190 kg K_2O ha^{-1} occurred on 23 August. From that date to the middle of October, the amount in roots increased but was offset by the amount lost from the tops.

Offtake in roots at harvest

Early work summarized in Table 4.4 found average offtake by the relatively poor-yielding crops of the time was 112 kg K_2O ha^{-1}, but with a wide range of 30–340 kg K_2O ha^{-1}. To update this for current crops, Hollies *et al.* (2001) made measurements on 72 fields in 1998 of yield and potassium in roots towards the end of the growing season. With a root yield of 60 t roots ha^{-1}, offtake was about 100 kg K_2O ha^{-1}. This rose linearly with yield so that, at 100 t roots ha^{-1}, offtake was about 170 kg ha^{-1} (see 1998 data in Fig. 4.4).

Hollies (2000) therefore proposed that, for each tonne of roots harvested, 1.7 kg K_2O would be removed. This can be compared with an earlier estimate (Hollies, 1997), based on a literature survey, of 2.1 kg K_2O t^{-1}. The recent (Hollies, 2000) value has been adopted in commercial practice (MAFF, 2000).

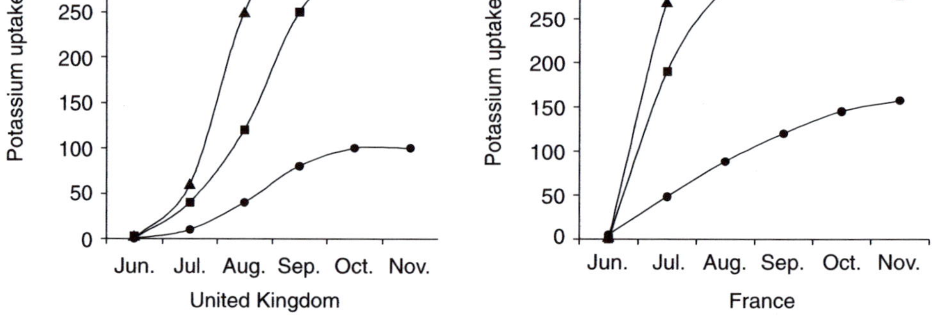

Fig. 4.3. Ideal uptake pattern of potassium by high-yielding sugar beet in the UK compared with France.

Table 4.4. Quantity of potassium in sugar beet at harvest.

Tops	Roots	Total	Root yield	References
	kg K_2O ha^{-1}		(t ha^{-1})	
204	132	336	50	Durrant and Draycott, 1971
144	112	256	42	Draycott, 1972
		306		Beringer, 1987
	134		63	Frankinet *et al.*, 1987
		308	50	Jansson, 1987
		660	81	Kirkby *et al.*, 1987
		350	94	Kirkby *et al.*, 1987
140				Vanstallen and Vandergeten, 1987
290	190	480	86	Analogides, 1987a
74	94	168	48	Bravo *et al.*, 1989
	70–170		70	Armstrong *et al.*, 1999
	25–230		20–100	Milford *et al.*, 2000
	70–240		60–105	Hollies *et al.*, 2001

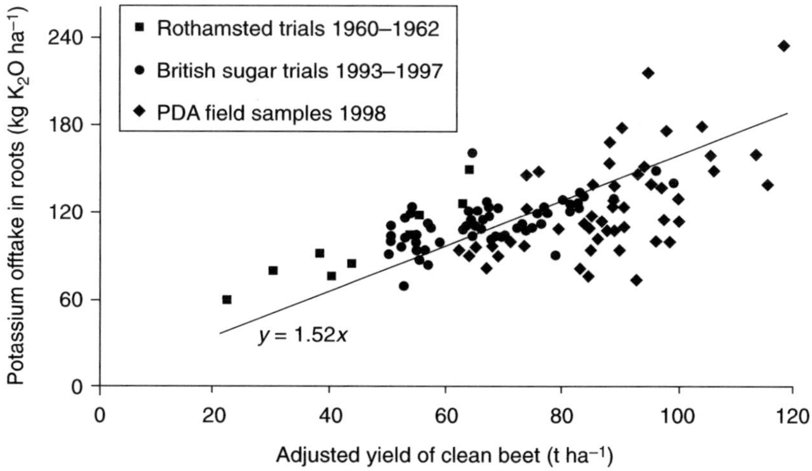

Fig. 4.4. Relationship between sugar beet yield and potassium offtake. PDA, Potash Development Association.

Using a combination of field trials data and factory tarehouse results, Armstrong *et al.* (1998) suggested that potassium offtake did not increase at all with increasing sugar yield over a wide range of sugar yields (5–14 t ha^{-1}). This statement was soon rejected (Armstrong *et al.*, 1999; Milford *et al.*, 2000) and it was then thought that potassium offtake probably rose asymptotically over a range of root yields from 20 t ha^{-1} to 120 t ha^{-1}. This comprised a linear phase of 20–70 t ha^{-1} and a less steep rise 70–120 t ha^{-1}. In a later paper (Armstrong *et al.*, 2000), several investigations were combined and, with a range of yields of 20–120 t ha^{-1}, a linear relationship was fitted to the data (Fig. 4.4). This interesting controversy over the exact nature of the yield/uptake relationship continues in the UK.

Physiological Role

In common with other cations, potassium helps maintain the osmotic potential of cells. Where potassium is deficient, the crop is more susceptible to drought. Cations also neutralize organic acids and help stabilize pH within the plant. Potassium has a special role in most plants in the opening and closing of stomata (Kirkby *et al.*, 1987). It is also very mobile in plant tissues and is found throughout the plant. It is important for photosynthesis, and transport of sugar produced in the leaves relies on potassium for movement to the storage root. At harvest, plants given potassium (and sodium) have a significantly greater sugar percentage than those given none. This has important financial implications because, for a given weight of sugar produced, growers are often paid commensurately more for roots with a high sugar percentage. In addition, costs are decreased because, for a given weight of sugar, less weight of roots has to be harvested and transported.

Potassium also improves performance by increasing leaf area in April–August. This allows the crop to intercept more radiation (particularly in the spring, when a large proportion falls on bare soil), giving proportional increases in sugar yield. Sugar beet given increasing amounts of potassium and sodium fertilizer with and without each other were analysed at monthly intervals from May–November (Farley and Draycott, 1974). Soils contained low to medium amounts of both elements prior to the experiments, which were repeated over three seasons on different fields. The two elements had similar physiological benefits every year, sodium being slightly quicker in action and superior overall. Both increased leaf area greatly early in the season and improved sugar yield consistently, partly as a result of increased light interception. They also acted by improving the amount of total photosynthate partitioned to the tap root in autumn and by increasing sugar percentage. Although each element increased its own concentration in the root juice, it proportionately decreased harmful nitrogen concentration. Thus there was no overall effect on juice quality.

Sources of Potassium and Sodium Fertilizers

Table 4.5 shows the nutrient concentrations of commonly available potassium and sodium fertilizers. All are of similar plant availability because they are almost entirely in the form of KCl and NaCl, which are water-soluble. Manufacturers now produce fertilizers containing potassium, sodium, phosphorus and magnesium in the correct concentrations for sugar beet, based upon soil analysis of the fields (see Chapter 11).

Time of Application

Until quite recently, all potassium fertilizer (except that in kainite) was applied immediately before drilling sugar beet. Following the introduction in the early 1970s of blended fertilizers containing no nitrogen, first in the USA and then in Europe, potassium is increasingly spread during the autumn or winter before the sugar beet crop. This earlier application has several advantages: (i) less traffic on the soil prepared for

Table 4.5. Sources of potassium, sodium and magnesium.

	Concentration %		
	K_2O	Na	Mg
Muriate of potash	60	–	–
Agricultural salt	–	37	–
Kainite	11–13	15–22	3–3.5
Sylvinite	21	19	0.6
Nitrate of soda 16% N	–	27	–

To convert Na to NaCl, multiply by 2.5, and Mg to MgO, multiply by 1.7.

sowing (see Chapter 10); (ii) decreased spring workload; (iii) better incorporation of potassium in the soil and therefore better uptake; and (iv) no negative effect on seed germination.

Recommendations in the UK and the USA

Table 4.6 summarizes the most recent advice for the UK (MAFF, 2000). The amounts are based on a slightly above-average crop of 60 t roots ha^{-1}. It is assumed that the crop is supplied with sufficient sodium either from fertilizer or from that already present in soil (as is the case with some organic and silt soils).

Recommendations from various sugar beet-growing regions in the USA are also in Table 4.6. There is a very large range in amount of potassium recommended, particularly when the soil-test value is less than 60 mg K kg^{-1}. However, few soils where sugar beet is produced would have a value less than 60 mg. Further work would be needed to reconcile some of these disparities but generally the similarities of the median values are remarkable between the two countries.

Environmental Considerations

In common with other fertilizers, the crop does not take up all the potassium applied. Some remains in the topsoil and may be taken up by following crops and some may be leached to considerable depth. Where precipitation exceeds evapotranspiration for part of the year, as in some of the sugar beet-growing areas of western Europe, there is evidence of leaching of potassium fertilizer. It is leached in significant quantities on sandy soils with little clay and organic matter. Potassium seems to have little, if any, adverse environmental impact but losses out of the root range decrease the efficiency of use and there is financial loss.

In a recent study at Rothamsted (Johnston and Goulding, 1992), it was found that the amount leached on medium-textured soil was small, being about 1 kg ha^{-1} for each 100 mm drainage. On silty clay loam at Rothamsted and sandy loam at Woburn, a portion of the potassium lost from the plough layer was retained in the subsoil (Johnston, 1986). Johnston and Milford (2001) believe that this was useful to following crops, especially deep-rooted ones like sugar beet. This would reduce still further the amount of potassium fertilizer leaking into the environment via ground and surface waters.

In another study, in The Netherlands (Neeteson and Ehlert, 1989), on the major soil types where sugar beet is grown there, it was found that potassium leaching was not occurring. Potassium inputs were based on: (i) soil analysis; and (ii) offtake in crops. On lighter soils, the authors were concerned that potassium and chloride may leach

Table 4.6. Amounts of potassium fertilizer recommended for sugar beet in UK and USA.

UK	mg K l^{-1} soil				
	0–60	61–120	121–180	181–240	241+
Index	0	1	2–	2+	3 and above
Category	Deficient	Low	Moderate	High	
Amount of potassium advised (kg K_2O ha^{-1})	150	125	100	75	0
USA	mg K kg^{-1} soil				
	0–60	61–120	121–180	181–240	241+
Range of amount of potassium advised (kg K_2O ha^{-1})	166–285	50–151	0–65	0–31	0

during the winter period both from sugar beet tops and autumn-applied fertilizer. Fortunately, in contrast to nitrogen and phosphorus, potassium has no known deleterious effects on the quality of natural waters (Syers, 1998).

SODIUM

Sodium in Soil

Total

In common with potassium, sodium is present in many of the parent-material minerals from which soils are formed. During weathering and other soil-forming processes, sodium is released. The quantity present in any particular soil depends on the concentration initially present in these minerals.

Igneous rocks generally contain the most sodium and soils formed from these usually range in concentration of total sodium in the order of 1–2%. In contrast, in soils formed from sedimentary rocks (where minerals have often had much physical and chemical alteration), total sodium concentration is much less and may range from 0.02 to 0.1%. In many light sandy soils where sugar beet is grown, total sodium may fall below 0.02%.

Thus, for the average furrow slice, total sodium present may range from 0.6 to 60 t Na ha^{-1}. Soil forming processes release a small fraction each year. This, together with atmospheric deposition, organic additions and fertilizers, is the part usable by sugar beet.

Available

Sodium is taken up by sugar beet as the Na^+ ion from soil solution, just like the K^+ ion. The exchangeable form plus that already present in soil solution accounts for nearly all of that available to the crop in the course of the growing season for both ions. However, in marked contrast to potassium, sodium is never fixed within the lattice structure of clay in a form not readily accessible for exchange.

Thus, for sodium, a simple measure of the amount soluble plus exchangeable in soil accurately predicts what is available to sugar beet. Studies show that the uptake of sodium during a growing season parallels the depletion from the soil. Tinker started measurements of plant uptake and soil depletion of potassium and sodium at Broom's Barn in 1963, reported by Draycott *et al.* (1970a). These experiments tested factorial combinations of both elements, necessary to distinguish between their relative importance in beet nutrition and their mode of action in soil and plant. It was shown for the first time that both elements were taken up in large quantities and greatly increased growth and yield in a similar way to each other. Soil depletion of sodium mirrored uptake. This was not the case with potassium, which was much more complex, potassium being released from the slowly exchangeable pool during the growing season, particularly with no potassium fertilizer.

Almost any aqueous solution of a salt with a strong base will remove available sodium from soil. The common procedure now in the UK is to shake soil with NH_4NO_3 solution. Sodium is weakly held by soil colloids and readily exchanges with ammonium ions.

Great care must be taken during such analyses because common glass, filters, water, dust, etc. contain much sodium. It easily finds its way into the measuring solution. Additionally, many light sugar beet soils contain only a few parts per million of exchangeable sodium so the slightest contamination confuses the result.

Loss and accumulation

In sugar beet-growing regions where annual precipitation exceeds evaporation and transpiration, sodium in the ionic form is rapidly leached from soil and sodium released from minerals during weathering and from fertilizer and organic matter does not accumulate. The situation is completely different in climates where evaporation exceeds precipitation; sodium does accumulate in soil and the concentration may be large enough to damage crops.

That sodium does not accumulate in soils in the UK is well illustrated by analyses at Barnfield (Warren and Johnston, 1962a). Applying sodium annually for a century only increased the amount in soil by 10 p.p.m. Na exchangeable in NH_4OAc. Williams (1971) measured sodium concentration in drainage water at Saxmundham and Woburn; at Saxmundham it was 7–45 p.p.m. Na and 20 p.p.m. on average, and at Woburn it was 7–28 p.p.m. and 11 p.p.m. on average. Tinker (1967a) reviewed the factors affecting the movement of sodium in soils in the UK and, from theoretical considerations and experimental results, found that sodium applied in fertilizer was rapidly leached from soil even in drier areas of the country. Winter rainfall was sufficient to remove most sodium applied in fertilizers in the course of 2 years.

Walsh (1970), reviewing sodium in Irish soils in relation to deposition in rain and response by sugar beet, found that soils contained 70–350 p.p.m. Na because annual deposition was from 45 to 335 kg Na ha^{-1} (at Broom's Barn rainfall deposits less than 12 kg Na $year^{-1}$ (Draycott *et al.*, 1970a). Walsh considered that soil and rainfall analyses were insufficient for predicting sodium need. Gallagher (1967) found that, despite the sodium supplied in rainfall, the element was still needed as a fertilizer in some parts of Ireland because rainfall leached sodium rapidly. It is likely that proximity to the sea accounts for some of the differences.

Problems of Excess

Characteristics of soils

In complete contrast with north-west Europe, sodium and other salts accumulate in the drier sugar beet-growing regions of the world. These salt-affected soils may be divided into three groups: saline, sodic and saline-sodic. Criteria for these designations were first defined by Richards (1954) and, since then, definitions have been further refined (Anon., 2001). Table 4.7 gives the criteria based on analyses of a saturated paste of soil and pure water. Grouping is useful in developing management strategies for these soils. Details are beyond the scope of this book and the reader is referred to Richards (1954), van Schilfgaarde (1974),Hanson *et al.* (1999) and Skaggs and van Schilfgaarde (1999) for more details.

Saline soils have accumulated sufficient soluble salts (K, Na, Ca, Mg) to adversely affect crop growth. This condition can occur naturally or develop following irrigation with water containing a large concentration of soluble salts. Saline soils have a pH of less than 8.5. Such conditions may be rectified with the application of sufficient quantities of water to leach the salts from the soil.

Sodic soils have greater than 15% of the exchange capacity occupied by sodium. The pH of these soils exceeds 8.5 and may be as high as 10. Sodicity has dramatically damaging physical effects on soil structure. Clay and humus colloids are dispersed into individual hydrated particles instead of remaining flocculated. The disruption makes it impossible for seedlings to emerge and establish a satisfactory population.

Saline-sodic soils reflect both high sodium and large concentrations of soluble salts. The pH of these soils is less than 8.5. Initially, the problem on these soils is the same as on saline soils, and soluble salts affect germination and growth. If the soluble salts are removed, then the soil is sodic and physical problems are dominant.

Table 4.7. Classification of salt-affected soils (from Richards, 1954; Anon., 2001).

Soil class	Electrical conductivity ($dS\ m^{-1}$)	Exchangeable sodium (%)	pH
Saline	≥ 4.0	< 15	< 8.5
Sodic	< 4.0	≥ 15	> 8.5
Saline-sodic	≥ 4.0	≥ 15	≤ 8.5

Remediation of sodic or saline-sodic conditions includes treatment with gypsum and leaching.

Effects on plants and salinity tolerance

Sodicity affects the physical structure of soil, seriously limiting emergence, growth and final yield. Salinity reduces growth by increasing the osmotic potential of soil solution and thus inhibiting water uptake. An excess of fertilizer can have a similar, though transitory, effect, as shown in Fig. 2.8 earlier.

Plants vary widely in their tolerance of salinity; a rating system for tolerance to salinity was reported by Hanson *et al.* (1999). Sugar beet is among the most tolerant, no doubt because it is a halophyte. Hanson *et al.*'s advisory publication suggests a threshold conductivity – the conductivity above which yield is suppressed – of 7 dS m^{-1}. Above this value, yield is suppressed by 5.9% for each unit increase of conductivity. Application of 5–10% more water than the crop uses aids in preventing an accumulation of soluble salts in the rooting zone.

More recently, Kaffka *et al.* (2002) investigated the effects of salinity in two areas of California. The range of conductivity measured in the laboratory was 2–23 dS m^{-1}. Sugar yields were depressed almost linearly with conductivities above 10 dS m^{-1}. Field maps of conductivity were produced using electromagnetic induction and this technique was useful in identifying the causes of variable response to irrigation.

Sodium in Sugar Beet

At harvest

Table 4.8 shows the concentration of sodium in the tops and roots of sugar beet at harvest and the quantity in the crop. The concentration in the tops is up to 20 times greater than in the roots. If roots only are removed from the field, nine-tenths of the total plant sodium is returned to the soil. In a summary of much early work, when yields of roots were usually less than 50 t ha^{-1}, Draycott (1972) found that a mature crop contained about 60 kg Na ha^{-1}, 7 kg ha^{-1} being in roots and 53 kg ha^{-1} in tops.

Subbarao *et al.* (1999) recently attempted to replace potassium for two cultivars of red beet in nutrient culture with 0, 75, 95 and 98% Na. The work was done in relation to nutrition of plants for the possible establishment of long-term bases on the lunar or Martian surfaces. Sodium replaced nearly 95% of the total plant potassium at the 98% substitution solution concentration; however, biomass production was reduced slightly with one of the cultivars and considerably with the other cultivar. Substituting sodium for potassium did not affect the leaf chlorophyll, photosynthetic rate or osmotic potential of either cultivar. The authors concluded that, for some red beet, sodium could safely replace 95% of the normal tissue potassium without decreasing production.

Effect on growth

An early theory that sodium applied to sugar beet acted by mobilizing soil potassium reserves and increasing the potassium status

Table 4.8. Concentration and quantity of sodium in sugar beet at harvest.

Na concentration (% dry matter)		Quantity Na in crop (kg Na ha^{-1})			
Tops	Roots	Tops	Roots	Total	Reference
1.2	0.08	50	8	58	Average of early work – Draycott, 1972
3.0	0.05	50	10	60	Draycott and Farley, 1971
3.9	–	195	–	–	Moraghan and Ananth, 1985
2.0	0.10	90	10	100	Bravo *et al.*, 1989

of the plant was shown to be incorrect by Adams (1961c) reporting on analyses done earlier by Hale at Rothamsted. By measuring the uptake of sodium and potassium, it was clear that sodium applications increased sodium uptake, not potassium. Sodium itself acted as a nutrient and increased yield. Sodium and potassium were distributed differently in the plant and at harvest: only 6% of the total sodium was in the root, compared with 33% of the potassium. The conclusion was that sodium was a nutrient for sugar beet and not a potassium substitute.

Tinker (Draycott *et al.*, 1970a) made similar experiments at Broom's Barn, where the soil is much lighter than that at Rothamsted, testing a wider range of sodium and potassium. Samples of the crop taken in summer and at harvest confirmed Adams's finding that it contained most sodium in August, much of which was in the tops, and that the total amount had decreased considerably by harvest. Periodic soil samples from the experiments showed that the amount of sodium in the crop was balanced by a corresponding decrease in the exchangeable soil sodium and in August the crop contained about 155 kg Na ha^{-1}. Corresponding calculations for potassium in the crop not given fertilizer and the decrease in exchangeable potassium in the soil were not as closely related; plant uptake of 210 kg K_2O ha^{-1} decreased the exchangeable potassium in the topsoil by only 68 kg K_2O ha^{-1}. When the crop was given fertilizer, the plant uptake and soil depletion were in good agreement – 294 and 270 kg K_2O ha^{-1}, respectively. The stability of the exchangeable soil potassium in plots not given fertilizer may have reflected the transfer of potassium to and from a non-exchangeable pool or uptake from the subsoil.

Draycott and Farley (1971) analysed the growth and nutrient uptake of sugar beet, with and without sodium fertilizer, from the early seedling stage until late harvest when sugar accumulation had ceased. Sodium increased the dry-matter yield of tops and roots throughout the whole growing period. It also increased the sugar yield at each of three harvests in October–December but the size of the increase was about the same on each occasion. Sodium appeared to increase sugar yield by several independent effects. Early in the year it greatly increased the leaf-area index, which coincided with maximum solar radiation and day length (the number of leaves per plant was unaffected but the area of each leaf was increased). Another mechanism by which sodium increased sugar yield was by increasing the proportion of total dry matter partitioned to roots. In increasing the amount of dry matter in the roots, it increased the yield of sugar, for root dry-matter yield and sugar yield are very closely and positively correlated (Draycott *et al.*, 1972b). Sodium also improved the sugar percentage of fresh roots.

Use of soil analysis to predict response

In early work by Adams (1961c) where sodium was extracted with NH_4NO_3 or 0.1 M HNO_3, no relationship was found between soil sodium and response. With the materials and equipment available in the 1950s, soil sodium was probably difficult to measure accurately. However, Adams remarked on a close relationship between soil potassium and the response to sodium, which is understandable due to the large degree of interchangeability of the two elements in beet nutrition. Holmes *et al.* (1961) simultaneously proposed using soil potassium analyses to predict sodium requirements.

Tinker (1967a) extended Adams's work and showed that, for a group of UK sugar beet soils, exchangeable sodium was weakly held by colloids. It was also a good measure of what plants could take up in pot experiments with the soils. This held out hope for the first time that exchangeable soil sodium measurements might be able to predict response. Draycott (1969) continued the investigation in field experiments. All plots were given a standard dose of 125 kg K_2O ha^{-1}, so responses to sodium were diminished. However, he found that all significant responses to sodium were on fields where soil contained less than 25 mg Na kg^{-1}. In six detailed field experiments at Broom's Barn (soil containing 11–20 mg Na kg^{-1}), uptake and soil depletion were measured monthly from May to November (Draycott and Farley, 1971; Farley and Draycott, 1974) with and without sodium fertilizer. Exchangeable sodium depletion closely matched uptake and the crop responded greatly to sodium fertilizer. Peak uptake was usually in June, when the fertilized crop contained 150 kg Na ha^{-1}.

Draycott and Durrant (1976b) located 20 fields with small concentrations of exchangeable sodium (4–18 mg Na kg^{-1}) and measured the response to sodium fertilizer. These experiments were coupled with uptake studies in pots of the same soils in the glasshouse. Uptake in field and glasshouse was linearly related to soil sodium over the whole range of values, supporting the view that a simple soil measurement predicts the amount available to sugar beet. Draycott and Bugg (1982) worked on a wider range of soil sodium in an attempt to establish both a relationship between response and soil sodium and a threshold above which sodium fertilizer is not needed.

Figure 4.5 is based on this and much earlier work. Above 50 mg Na kg^{-1}, no crop responded significantly. All large responses were on soils below 25 mg Na kg^{-1}. Table 4.9 therefore contains our proposals for the future use of sodium fertilizer.

Further evidence that sugar beet does not respond to sodium application when the soil contains more than about 50 mg Na l^{-1} is provided by fieldwork by Christenson *et al.* (1990). Three years of trials on silty clay soils in Michigan tested 0–174 kg Na ha^{-1} where soil contained 60 mg Na l^{-1}. Neither yield nor quality of the crop was affected in any of 3 years. Moraghan (1979) found a response to applied sodium on a soil with 10 mg Na kg^{-1} when potassium was not given. However, when potassium was given, the effect of the applied sodium disappeared. In another report, Moraghan (1984) showed no response to applied sodium, which was not surprising because the soil contained 200 mg Na kg^{-1}. There have been no reports of response to applied sodium in other regions of the USA. Apparently these more arid soils have sufficient sodium for the growth of sugar beet.

Application in the EU and the UK

Beckers (1999) gave an overview of sodium usage on sugar beet in the 15 EU countries, showing how the amount recommended (0–200 kg Na ha^{-1}) was related to latitude and soil sodium concentration. He also drew

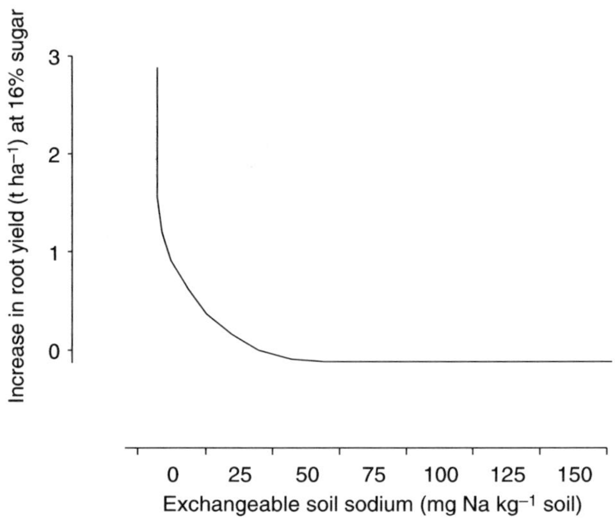

Fig. 4.5. Relationship between exchangeable soil sodium and root-yield response to sodium fertilizer.

Table 4.9. Summary recommendation for sodium fertilizer worldwide based on exchangeable soil sodium.

Soil sodium (mg Na kg^{-1})	Amount of sodium needed by sugar beet in fertilizer (kg Na ha^{-1})
0–25	150
25–50	75
> 50	0

attention to work in Austria, which showed that the more sodium used to replace the need for potassium, the stronger the additional growth stimulation. Strangely, sodium has not been recommended for sugar beet in Austria according to Beckers's report.

Durrant *et al.* (1974b) discovered that sugar beet seed was sensitive to sodium chloride solution. Soaking in a molar solution killed 6% of seeds. Water uptake during germination was slowed by increasing concentration, e.g. from 0 to 0.2 M NaCl. Durrant *et al.* (1978) later demonstrated in field and laboratory experiments that sodium fertilizer improved the water status of sugar beet especially in drought conditions. It also improved the leaf-area index early in the growing season and the efficiency of leaves under conditions of moderate water stress.

Despite much experimental evidence showing that sodium fertilizer increases sugar beet yield and decreases the need for potassium, there is resistance to its use on some soil types through fears of deterioration in soil structure. Field experiments with sugar beet were made, testing all combinations of autumn and spring applications of sodium and potassium. Fields were chosen with soils of loamy silt, silty loam, sandy clay loam and clay loam textures (Draycott *et al.*, 1976). Microplot and controlled-environment studies were also made with the same soils to examine the effects of sodium on seedling emergence and growth.

Assessments of soil physical state following sodium application revealed no effect in the year sugar beet was grown or in the following spring, when cereals were grown. Measurements of the physical properties of soils treated with sodium suggested that applications of several times the recommended amounts of sodium fertilizer would still not damage soil structure. For example, a soil with a cation exchange capacity of 10 cmol kg^{-1} would need more than 2 t NaCl ha^{-1} to damage soil. However, sodium fertilizer increased the osmotic suction of soil solution, which, under some circumstances, e.g. dry springs or giving the fertilizer close to the time of sowing, decreased germination and seedling growth. For this reason and not because it has a detrimental effect on the soil physical condition, sodium fertilizer is best given in the autumn or some weeks before sowing.

SUMMARY OF POTASSIUM AND SODIUM

The requirements of potassium and sodium by sugar beet to produce optimum yield are large and the need for both elements is well established. The two elements act in similar ways in sugar beet and they are partly interchangeable. Some soils are supplied with one or both of the elements and fertilizer is not required in some parts of the world. In Europe, where the majority of sugar beet is grown in intensive-cropping systems, potassium must be given. Few soils can make good the offtake in crops sold away from the farm. Optimum applications of potassium were shown in Table 4.6 in relation to soil concentration. It is also recommended that application be balanced against offtake. Factors for working out the surplus or deficit are given in Appendix A.

Work in the 1990s in UK (Jarvis and Bee, 1996; Milford *et al.*, 2000) suggested that responses to potassium are less than previously (Draycott, 1993). Johnston and Milford (2001) thought this was because current sugar beet utilizes more subsoil potassium. Armstrong *et al.* (1999, 2000) believed that uptake was not linearly related to root or sugar yield, which had been assumed before. They postulated that, with large yields of sugar beet, sugar is so concentrated in the tap root that potassium concentration is decreased. Detailed experiments are needed to investigate this further because the principle may apply to other nutrients.

Long-term experiments (see Chapter 9) show that it is important for sugar beet to have an adequate background supply of potassium in soil to perform well. This is because freshly applied fertilizer needs several years to become fully distributed through soil so that it is accessible to roots. Ideally this should be nearly 200 mg kg^{-1} soil.

In humid climates soils contain too little sodium for sugar beet to perform fully and fertilizer must be given. Soil analysis (provided it is done accurately – not easy in the experience of the authors) predicts the requirement, as was shown in Fig. 4.5 and Table 4.9. In contrast, in drier climates there is often an excess. To grow good sugar beet and other crops, techniques of combating too much salinity are necessary. Fortunately, sugar beet is one of the most tolerant of all common crops.

5
Calcium and Magnesium

Both calcium and magnesium are major (or macro-) nutrients because they are taken up in quantity by plants. Macronutrients are those contained in plant dry matter at a concentration greater than 500 mg kg^{-1} (Anon., 2001). In the case of sugar beet, amounts in a hectare of crop are similar to those of phosphorus and sulphur but less than those of nitrogen and potassium. Healthy sugar beets contain much more, as shown below, in leaves and roots.

Deficiency symptoms of calcium are rarely seen because sugar beet must be grown in soils of near-neutral pH. Such soils are kept near the neutral point either by calcium-containing minerals in the parent material or by regular additions of lime. Thus, in temperate climates, the soil complex is usually dominated by Ca^{2+} ions in quantities far in excess of crop requirement. Calcium-deficiency symptoms are occasionally seen where an imbalance of cations on the soil complex has been caused, e.g. by sea-water flooding or an excess of a cationic fertilizer (see Chapter 4).

Deficiency symptoms of magnesium appear quite commonly on sugar beet where soils are sandy or parent materials contain little. On all loamy soils, clays and organic soils symptoms are rarely seen because sufficient is released during weathering. Only where roots are damaged (e.g. by pests) or restricted (e.g. by compaction) or another cation is in excess (e.g. potassium or sodium) is magnesium deficiency seen. On sandy soils, minerals contain very little magnesium and, where intensive cropping has been practised on stockless farms, magnesium is continually removed and must be replaced by fertilizer.

CALCIUM

Calcium in Soil

The amount of calcium in soil varies enormously depending on the parent material, degree of weathering and amount of leaching. The concentration in most soils ranges from 0.1 to in excess of 3%. Old soils, soils derived from acidic parent material and those that have been highly weathered and leached have the smallest calcium content. Soils formed from alkaline or calcareous materials contain much more calcium and those that contain in excess of 3% Ca are defined as calcareous (Anon., 2001). These are easily identified by effervescence with addition of a few drops of molar mineral acid.

During soil formation, calcium is derived from calcium-containing feldspar (anorthite), pyroxene (augite) and amphibole (hornblende). Calcium is also present in calcitic and dolomitic limestone, chalk, apatite (calcium phosphate) and gypsum (calcium sulphate ($CaSO_4.2H_2O$). In arid and semi-arid soils, gypsum may accumulate in a layer within the root zone of sugar beet. Throughout the world where soil is formed from chalk or limestone, total calcium concentration ranges from 3 to 25% dry weight.

Calcium released from minerals into soil solution is quickly adsorbed on the exchange complex. Consequently, the amount of calcium is usually greater in soils containing clay. In a survey of Michigan soils, Robertson *et al.* (1976b) reported that fine-textured soils contained from 1700 to 3600 mg Ca kg^{-1} soil, while coarser soils ranged from 625 to 2400. Calcium in soil solution ranges from 30 to 300 mg l^{-1} for non-calcareous soils and up to 700 mg l^{-1} in calcareous soils.

Causes of Deficiency

Soils where sugar beet is produced generally contain sufficient calcium to provide for the nutrient need of the growing crop. Barber (1995) showed that calcium moves to the root surface by mass flow – the movement of solutes associated with a net movement of water. It follows that the quantity of water moving to the root surface will be influenced by the rate of transpiration. The transpiration rate generally ranges from 200 to 900 g H_2O g^{-1} dry weight of plant tissue.

A weighted average calcium concentration from tops and roots is calculated by Draycott (1972) to be 4000 mg Ca kg^{-1} dry plant tissue in the sugar beet crop. The ratio between soil-solution concentration (30–300 mg l^{-1}) and weighted average in plant tissue is 133 to 13. Any ratio of plant to solution concentration less than the transpiration ratio represents a case where ions would accumulate at the root surface. These considerations show that the primary cause of calcium deficiency in sugar beet is not usually related to supply from the soil, but rather to uptake, translocation, utilization in the plant or, most commonly, an excessive supply of another cation.

Chapter 7 below shows the damage caused to leaves by calcium deficiency. Whether this is related to precipitation of insoluble calcium compounds in the leaves is open to debate. In a review, Gallagher (1975) pointed out that plants that accumulate large quantities of oxalic acid tend to contain large quantities of calcium oxalate crystals. He stated that 'oxalic acid is a strong chelating agent and once reacted with calcium it becomes somewhat immobilized in plant cells'. However, some reports indicate that calcium can be reutilized from the calcium oxalate if plants are under extreme calcium stress. Van Egmond (1979) suggested that calcium deficiency (tip burn) during periods of rapid growth and high nitrogen fertilization is explained by an overproduction of oxalate in expanded young leaves. He suggested that disappearance of deficiency symptoms later in the season might be the result of redistribution from the root. Conversely, Mostafa and Ulrich (1976a) found no relation between the severity of deficiency and calcium oxalate in plant tissue. They inferred that deficiency was associated with an incomplete uptake of calcium from solution, a large demand by storage roots and inefficient translocation of calcium to the leaf tips. They explained that, if calcium immobility inhibits transport of calcium from older plant parts to younger plant parts, the formation of calcium phosphates could be a limiting factor under deficiency and not under sufficiency conditions. There appears to be no resolution of the issue of the role of phosphates and oxalates in calcium deficiency. At the practical level, this mechanism is not a factor in the nutrition of sugar beet. We feel that resolution of this issue would be of importance in the development of germ-plasm and varieties that are less susceptible to calcium deficiency.

Role of Calcium in Sugar Beet Plants

The main role of calcium in plants is in providing stability to cell walls by formation of calcium pectate in the middle lamella (Epstein, 1972). In some plants calcium polysaccharides are components of cell walls. Epstein also found that amylase was one of the enzymes where calcium is a necessary cofactor. Furthermore, calcium appears to 'detoxify' other ions and counteracts the effect of low pH on nutrient uptake. Calcium is necessary for the organizational integrity and function of membranes within cells. Cell division and elongation depend on an adequate supply of calcium. Since calcium is translocated in the xylem but not in the phloem, calcium is rather immobile and not readily redistributed within the plant.

More recently, Kauss (1987) reported that other enzymes require calcium for activation. Bush (1995) reported that calcium regulates ionic balance, mobility, gene expression, carbohydrate metabolism, mitosis and secretion. Calcium may not be the only regulator of these processes, but evidence is accumulating that points to the significant role of calcium as a regulator.

Amount in the Sugar Beet Crop

Wallace (1945) found that healthy sugar beet leaves contained 2.65% Ca in dry matter whereas deficient leaves contained only 0.66%. Ulrich and Hills (1969) reported that deficient leaf blades contained 0.1–0.4% Ca and non-deficient leaves greater than 0.4%. Draycott (1972) reported on a 5-year study at Broom's Barn. The data in Table 5.1 show that sugar beet contained 12–31 kg Ca ha^{-1} in roots and 18–67 in tops. The mean was 22, 41 and 63 kg ha^{-1} for roots, tops and total, respectively. Cooke (1967) reported that 80–100 kg Ca ha^{-1} was removed in a 33 t ha^{-1} crop. Viets and Robertson (1971) suggested that a 60 t ha^{-1} crop would contain 40–220 kg Ca ha^{-1}. Recalculated data from Robertson *et al.* (1976b) suggested a more modest estimate of 80 kg Ca ha^{-1}.

Table 5.1. Concentration and quantity of calcium in sugar beet at harvest at Broom's Barn.

	Mean	Range
Concentration (%)		
Tops	1.00	0.70–1.60
Roots	0.24	0.16–0.35
Quantity in crop (kg ha^{-1})		
Tops	41	18–67
Roots	22	12–31
Total	63	30–98

The variability in the amount of calcium contained in a sugar beet crop may be related to the variety grown (Finkner *et al.*, 1958) but there is a paucity of information on varietal differences. At Colorado State University, Schmehl and his students conducted an excellent study concerning the accumulation of nutrients in sugar beet plants. Figure 5.1 shows the accumulation of calcium in various plant parts over the course of the growing season (Bravo *et al.*, 1989). The pattern of uptake is similar to the growth of the crop. They reported the total uptake for the crop to be 120 kg for a 60 t crop. This value includes calcium lost to leaf senescence and unharvested roots. In an earlier report from the same study, Eslami M *et al.* (1988) noted that sugar beet leaf senescence accounted for approximately 70 kg Ca ha^{-1}. If tops are removed, total offtake may approach 80 kg ha^{-1}, more than half being in tops at harvest.

Dynamics of Calcium in Sugar Beet

There is much evidence that sugar beet needs a continuous supply of calcium for growth and development. For example, when Ulrich and Mostafa (1976) transferred sugar beet seedlings from a complete nutrient solution to one where calcium was withheld, rootlets and tops soon failed to develop. When the same transfer was done at the eight-leaf stage, the rootlets became stubby and swollen at the tips. The upper portion of nearly fully developed leaf blades developed cupping or hooding, an effect typical of calcium deficiency. As each new leaf developed, the blade became smaller until only a black tip remained at the apex of the petiole. This was called 'tip burn' and each new petiole had the symptom. Addition of calcium corrected deficiency on new growth, but did not eliminate symptoms on old growth.

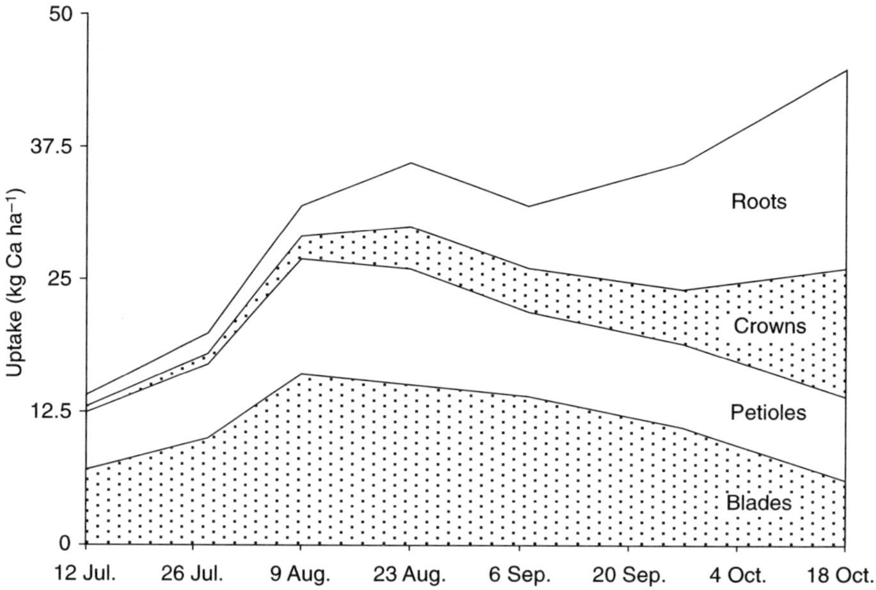

Fig. 5.1. Calcium uptake (after Bravo *et al.*, 1989).

Soils usually contain ten times as much calcium as potassium in solution. However, sugar beet plants at harvest contain approximately three times as much potassium as calcium (Draycott, 1972). There appear to be two interrelated factors associated with this phenomenon. First, calcium is absorbed only by young root tips, where the cell walls of the endodermis are unsuberized, whereas other cations are absorbed along the entire length of the root (Clarkson *et al.*, 1968, 1971). Secondly, and interrelated to the first, Berry and Ulrich (1968) showed that sugar beet plants depleted the concentration of potassium, sodium and magnesium in solution to 1 mg l^{-1} before deficiency symptoms appeared, whereas calcium deficiency symptoms occurred when the concentration in solution was 50 mg Ca l^{-1}.

Mostafa and Ulrich (1974) evaluated the effect of the concentration or activity of calcium in solution on the uptake of calcium. They showed that sugar beet plants developed calcium deficiency symptoms, even though the concentration and/or activity of calcium in solution was relatively great. They concluded that deficiency symptoms were probably caused by cation interactions rather than calcium availability. Earlier, Berry and Ulrich (1970) studied the effect of potassium on uptake of calcium by sugar beet seedlings grown in solution culture. Calcium deficiency symptoms developed progressively as potassium concentration was increased. However, it was shown that a sufficiency of both elements was necessary for translocation of calcium throughout the plant and complete absence of deficiency symptoms. If potassium was limiting, calcium was taken up by the roots, but not translocated to the leaves.

Calcium and magnesium also interact, each inhibiting uptake of the other. Mostafa and Ulrich (1976b) observed that sugar beet did not grow well when the calcium : magnesium ratio was 0.33 or less. This was true regardless of concentration of the respective ions. They further showed a mutual inhibitory effect of calcium and magnesium on ion uptake. Their conclusion was that the ratio of calcium : magnesium in the nutrient solutions might limit calcium uptake by sugar beet.

The effect of cation ratios on crop growth and yield has been the subject of numerous studies over the past 60+ years. Generally, significant effects occur only when the exchangeable calcium : magnesium ratio on a molar basis is significantly less than 1. This situation is known to occur only on soils derived from serpentine, but only small areas of the world have such soils, e.g. Cornwall (UK), Scotland, the eastern USA and Canada. Imbalance due to an excess of magnesium has been suggested but never proved on soils formed from dolomitic limestone.

Extrapolating from research on other crops, it seems unlikely that there is a calcium : magnesium ratio that gives best yield. McLean *et al.* (1983) conducted a long-term study to evaluate the effect of cation ratios on yield of maize, soybean, wheat and lucerne. Cation concentrations were adjusted annually by the addition of appropriate salts. They concluded that a specific cation ratio for ideal crop production does not exist. Rather, each cation should be supplied in adequate, but not excessive, amounts and soil and plant analyses should be utilized to provide guidelines, as described in Chapter 4 for potassium and sodium and below for magnesium.

MAGNESIUM

Magnesium in Soil

Total magnesium in soil is usually in the range 0.1–1.0%, being present in primary minerals such as biotite, serpentine, hornblende and olivine. In addition, clay minerals, such as chlorite, vermiculite, illite and smectite, have magnesium as part of their structure. All these minerals and clays release the element during weathering, usually in quantities sufficient for most crops.

Some soils may contain over 10% magnesium when formed on a parent material of dolomitic limestone (a mixed carbonate of magnesium and calcium) or magnesite (magnesium carbonate). Also, in some arid and semi-arid regions, epsomite ($MgSO_4.7H_2O$) and other salts may be present, resulting in large total soil magnesium concentration.

Sugar beet and similar plants take up magnesium from soil solution as Mg^{2+}. This exchangeable or 'available' concentration in soils usually ranges from 10 to 500 mg Mg l^{-1} soil. Magnesium is present on the exchange complex as well as in soil solution, and a simple extraction procedure with an aqueous salt solution (commonly ammonium nitrate or acetate) reflects the amount readily available to sugar beet and other crops, as described below.

Medium- and fine-textured soils contain more magnesium than sandy soils, so deficiencies are seen more often on sandy soils than on loams and clays. For example, Robertson *et al.* (1979a) noted that clay and clay loam soils ranged from 250 to 500, loam and sandy loam from 150 to 450 and loamy sand and sands from 30 to 175 mg Mg kg^{-1}.

Role in the Plant

A major function of magnesium in plants is its role as the central atom in chlorophyll. It also plays an indispensable role in protein synthesis as a bridging element for the aggregation of ribosome units. Another important role of magnesium in plants is enzyme activation and energy transfer. Groups of enzymes activated are phosphatases and carboxylases. Energy transfer involves ATP and a group of enzymes referred to as adenosine triphosphatases (ATPases), which use magnesium ATP as a substrate. Magnesium plays a wide role in reactions associated with respiration and photosynthesis.

Amount in Sugar Beet

Ulrich and Hills (1969) defined a critical concentration of 0.1% Mg in dried leaf-blade tissue but suggested that deficiency symptoms may appear below 0.2% Mg. A summary of results from a number of studies supports the idea that magnesium deficiency can be expected below about 0.2% (Table 5.2). Draycott (1993) in a further summary reported that leaf blades contained 0.6% Mg in the spring, declining to 0.2% late in the summer. Leaves with magnesium deficiency symptoms contained 0.1–0.2% Mg in dry matter.

Table 5.2. Concentration of magnesium in dry matter of sugar beet leaf blades with and without deficiency symptoms.

	Leaf magnesium (%)		
	With symptoms	Without symptoms	Reference
	0.096	0.546	Wallace, 1945
	0.150	0.390	Hale *et al.*, 1946
	0.170	0.490	Björling, 1954
	0.096	0.546	Jacob, 1958
	0.010–0.030	0.100–0.700	Ulrich, 1961
	0.149–0.030	0.217–0.559	Birch *et al.*, 1966
	0.110	0.480	Bolton and Penny, 1968
	0.100–0.200	0.200–0.650	Draycott and Durrant, 1970b
Range	0.010–0.219	0.100–0.700	
Mean	0.120	0.444	

Bolton and Penny (1968) found that the average amount of magnesium removed by sugar beet tops and roots was 21, 37 and 46 kg ha^{-1} with treatments of 0, 50 and 100 kg Mg ha^{-1}, respectively. Jacob (1958) in a review of magnesium as a plant nutrient gave the uptake of sugar beet as 35 kg ha^{-1}. Warren and Johnston (1962b) found that sugar beet removed only 12 kg without magnesium and 15 kg with magnesium fertilizer, but their yields were small.

Adams (1961c) measured the uptake of magnesium by sugar beet periodically from June to October. The concentration increased rapidly until September and was then fairly constant. Uptake was 27 kg ha^{-1} (assuming 74,100 plants ha^{-1}). Durrant and Draycott (1971) found that maximum uptake of magnesium was in August, but this decreased slightly by harvest in November due to leaf senescence. On these soils, which were prone to magnesium deficiency, total uptake was only 25 kg Mg ha^{-1}.

Bravo *et al.* (1989) also measured the uptake of magnesium over the course of a growing season. The results (Fig. 5.2) show a rapid increase in the amount until the middle of August and then a slower accumulation as the amount in crowns, petioles and leaf blades decreased. The amount stored in roots increased as the season progressed. These data show that the crop contained between 35 and 40 kg Mg ha^{-1}. Taking into account the magnesium in senescent leaves and fibrous roots contributes to the higher values reported (Eslami M *et al.*, 1988). In good commercial sugar beet production we expect the total uptake of magnesium to range between 40 and 50 kg Mg ha^{-1}, a little more than half being in the taproots.

Predicting Response to Magnesium by Soil Analysis

As pointed out earlier, magnesium deficiencies are more common on sandy soils than on finer-textured soils. The incidence of magnesium deficiency in the UK increased rapidly over the 23-year period 1946–1969 (Draycott, 1972). This was attributed to a large decrease of farmyard manure used, higher yields and intensive cash-cropping on sandy soils.

In general, magnesium deficiency may appear on sandy soils containing little in the exchangeable fraction. Table 5.3 from Draycott and Allison (1998) shows the relationship between exchangeable soil magnesium and the recommended magnesium application. Soil containing less than 50 mg kg^{-1} should have magnesium applied. They reported that soils 'with less than 15 mg kg^{-1} require special treatment because there are likely to be severe deficiency symptoms and significant yield depression'. In the range of 15–25 mg kg^{-1} application should improve and from 25 to 50 mg should maintain the magnesium status of the soil. Periodic application is needed to maintain magnesium supply for sugar beet.

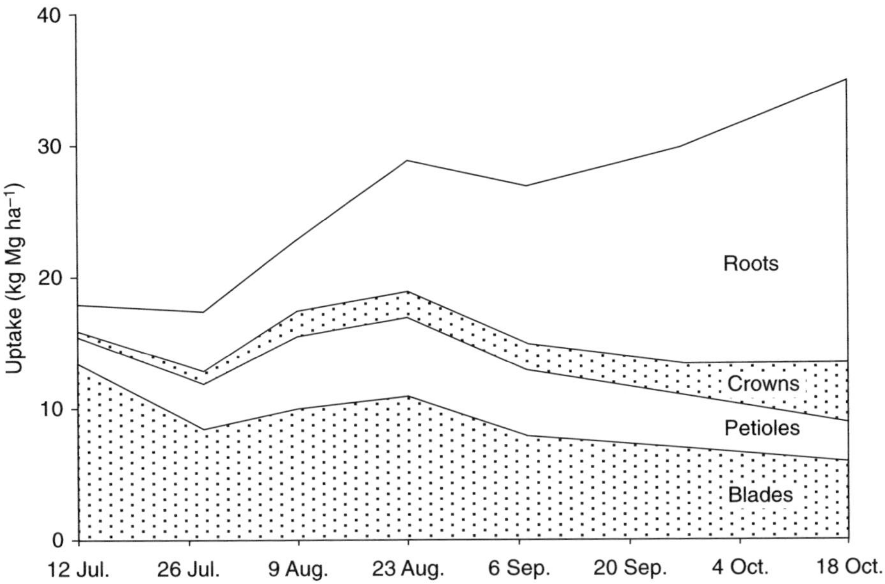

Fig. 5.2. Magnesium uptake (after Bravo *et al.*, 1989).

Table 5.3. Magnesium-fertilizer requirement as determined by the amount of exchangeable soil magnesium extracted with molar ammonium nitrate solution.

Concentration in soil (mg Mg kg^{-1} soil)	Fertilizer (kg Mg ha^{-1})
0–15	100
15–25	75
25–50	50
50+	0

Sugar beet production in the USA is predominantly on soils adequately supplied with magnesium. In those cases where magnesium is required, recommendations based on research for other crops are used, based on extraction with molar ammonium acetate solution adjusted to pH 7.0. In Michigan, magnesium is recommended when any one of three conditions is met (Christenson *et al.*, 1992): (i) the extractable concentration is less than 35 mg kg^{-1} of soil; (ii) extractable potassium exceeds magnesium by a factor of 3 or (iii) on a chemical-equivalent basis, magnesium is less than 3% of the extractable potassium plus calcium plus magnesium. On non-acidic soils, 10–20 kg Mg ha^{-1} is recommended for a band placement near the seed and 50–100 kg Mg ha^{-1} for broadcast material. Dolomitic limestone is preferred on magnesium-deficient soils needing lime in the USA, as in the UK.

Forms of Magnesium Fertilizer

Once it had been established that UK sugar beet on light soils responded economically to magnesium, research into various fertilizers followed. Table 5.4 shows the principal sources tested.

Kainite was and still is widely used in the UK and Europe for sugar beet. The sodium and potassium it contains are very useful for sugar beet. Table 5.4 shows that it contains a little magnesium (4.5%). Unfortunately, even a large application of kainite does not provide sufficient magnesium for deficient crops as discovered in field experiments begun by P.B.H. Tinker testing 900 kg ha^{-1} (Draycott and Durrant, 1969a, b, 1970a, b).

Field and glasshouse work established that kieserite (17% Mg) would provide readily available magnesium in the quantity

Table 5.4. Principal sources of magnesium fertilizers.

Source	Approximate composition	% Mg
Kieserite	$MgSO_4.H_2O$	17
Epsom salt (*Bittersalz*)	$MgSO_4.7H_2O$	10
Kainit	NaCl, KCl, $MgSO_4$	4.5
Calcined magnesite	MgO	48
Dolomitic limestone	$MgCO_3$, $CaCO_3$	11
Sugar-factory waste lime	$CaCO_3$, $Ca(OH)_2$, H_2O	0.2

required. Both kainit and kieserite have to be imported into the UK, so, in a search for locally obtainable sources of magnesium, dolomitic limestone (11% Mg) was tested in long-term field experiments. The availability of the element was found to be very pH-dependent. In acid soil the magnesium was available, inexpensive and long-lasting. Unfortunately, in soils at pH 7 and above, very little was available and sugar beet showed all the symptoms of magnesium deficiency (Draycott and Durrant, 1972a, b).

Later research investigated the use of calcined magnesite (48% Mg). Providing the calcining (conversion of carbonate to oxide by heating) was at a low temperature and for as short a period as possible, magnesium was readily available (Draycott and Allison, 1998). It is now widely used in blended fertilizers for sugar beet.

Interactions with Other Nutrients

Tinker (1967b) and Draycott and Durrant (1969a) made field experiments testing combinations of magnesium, nitrogen, potassium and sodium and found no significant yield interactions on average between magnesium and the other nutrients. However, all had large effects on the composition of the crop and on the percentage of plants with symptoms. Nitrogen decreased the number of plants with magnesium deficiency symptoms, but magnesium fertilizer increased yield by the same amount as when nitrogen was not applied. Similarly, sodium increased symptoms, but both nutrients were needed for maximum yield. In a long-term experiment at Woburn, Bolton and Penny (1968) found that potassium decreased the magnesium concentration in sugar beet and other crops but giving magnesium did not increase yield. This may have been because the concentration of exchangeable soil potassium was small and the response to potassium fertilizer was much larger (and more variable) than the small extra yield expected from magnesium.

SOIL ACIDITY AND LIMING

General Effects of Acidity on Plants

Although reports of calcium deficiency are rare, successful crop production relies on an adequate supply of the element in the form of lime ($CaCO_3$ or $Ca(OH)_2$) to overcome soil acidity. In acid conditions many plant nutrients are unavailable in concentrations needed for plants to thrive. Other elements, particularly aluminium and manganese, become available in toxic concentrations. For example, the concentration of manganese in soil solution increases 100-fold for each unit of pH decrease. Also the concentration of aluminium increases rapidly below pH 5.5, so that the soils exchange capacity is saturated with aluminium at pH 4. Few crop plants can tolerate metal concentrations such as these (Foy, 1984).

It is often mistakenly assumed that the action of liming is to provide calcium (or magnesium) to aid the growth of the crop. This is not so in most circumstances. Liming provides the hydroxyl ions needed to neutralize acidity:

$$CaCO_3 + H_2O = Ca^{2+} + HCO_3^- + OH^-$$

The calcium (or magnesium) ions are useful by-products of the main purpose of liming – to increase pH.

Effects of Soil pH on Sugar Beet

Sugar beet is very sensitive indeed to low pH and will only yield fully in soil near the neutral point. Acidity is still one of the major causes of poor performance of the crop when grown on naturally acidic soils.

Generally, germination and emergence are satisfactory on moderately acid soil but seedlings soon grow slowly and cotyledon leaves have red margins and are unusually erect. Many roots die and those left will often be 'stubby' and spatulated. Stands will be spotty with 'good' beets interspersed with stunted and missing beets, in soils with a pH less than 5.5.

Over 50 years ago Hale *et al.* (1946) reported typical acid injury, with blades and petioles uniformly pale yellowish green and plants with the margins of leaves rolling inwards. They found toxicity symptoms occurring in the range of 1200–3000 mg Mn kg^{-1} dry weight. Brown *et al.* (1968) found that apparently healthy sugar beet plants contained up to 5600 mg Mn kg^{-1}. In their greenhouse study, the objective was to investigate the cause of bronzing on two acid soils, thinking it might be manganese toxicity. However, the cause of bronzing was potassium deficiency; the symptoms were eliminated by the addition of potassium fertilizer.

Care needs to be exercised in evaluating such data because soil (dust) contamination of leaf tissue may increase manganese concentration. In neither of the above two cases was there any indication of the leaves being washed. Coincidentally with elevated manganese concentration due to soil contamination, iron and aluminium concentration also increases. There were no values given for either of these two metals in the above papers. Results from Brown *et al.* (1968) showed in excess of 600 mg Mn kg^{-1} in sugar beet leaf grown on limed soils in the field. However, the leaves were not washed prior to analysis.

Many of the biochemical effects of aluminium on plants are probably associated with alteration of the root-membrane structure due to the binding of aluminium to the membrane (Foy, 1984). Aluminium toxicity results in reduced root growth and the remaining roots are stubby and brittle. There is little branching and these remaining roots do not absorb nutrients or water efficiently. Aluminium toxicity may also induce phosphorus, calcium, magnesium and molybdenum deficiencies (Foy, 1984).

Terry *et al.* (1975) found that manganese toxicity reduced the numbers of cells per leaf and the average leaf-cell volume. Plants also had smaller leaf and root weight in acid conditions.

Optimum pH

Based on our many observations in the UK and USA, the crop grows best on soils of pH between 6.5 and 8.0. Viets and Robertson (1971) reported that sugar beet grew well on soil with pH 6.5, grew poorly at pH 5.5 and had virtually no growth at pH 4.5. Ulrich and Ohki (1956a) reported an experiment in solution culture showing that sugar beet grew best at pH 7.0 (Fig. 5.3). At pH 4.0 leaves were small, dark green and smaller in number than at a higher pH. At pH 9.0 yields of roots and tops were reduced. Leaf colour and nitrogen status of the plants were not affected by pH. Increasing pH decreased phosphorus concentration, but even at 9.0 there was adequate phosphorus in the tops.

In recent work (P. Wilting, The Netherlands, 2002, and O. Hellgren, Sweden, 2002, personal communications) in both field and controlled-environment experiments, sugar beet performs best when the soil or root medium is near the neutral point (pH 7.0).

Winter leaching of calcium and use of ammonium fertilizers increase the acidity of soil over time, so careful monitoring of pH is essential for sugar beet culture. Soils should be kept from becoming strongly acidic. Not only does sugar beet not grow well on such soils, but it is also more difficult to make large adjustment in soil pH quickly.

Response to Lime

Draycott (1972) stated, 'Surprisingly few thorough investigations have been made to determine optimum pH for sugar beet.' The comment still applies today. Morley Davies

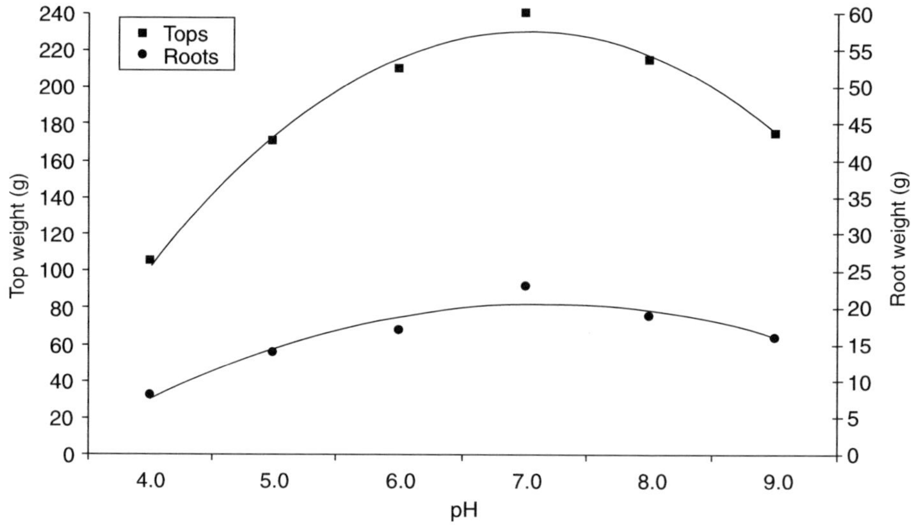

Fig. 5.3. Effect of pH of solution culture on weight of sugar beet tops and roots (after Ulrich and Ohki, 1956a).

(1939) reported on the effects of lime on a sandy soil with a pH of 4.9 (Table 5.5). McEnroe and Coulter (1964) surveyed over 3000 farms in Eire to determine the effect of pH on sugar percentage and yield of sugar beet (Table 5.6). Increasing soil pH was associated with an increase in both sugar concentration and yield. While the authors pointed out that there might be confounding factors in these results, the positive effect of increasing soil pH on sugar yield is important.

Lime from sugar beet-processing plants has been stockpiled on site in Michigan for over 100 years. Environmental regulations and limited storage space compel processors to remove the spent limestone and cease to stockpile on site. Using this material for liming soils is an excellent means of 'disposing' of this by-product. However, soils in the sugar beet-growing region are generally alkaline, requiring no lime, and transportation costs to areas with acid soil limit removal. Application to alkaline soils may be an option, but overliming of soils may be a concern on coarser-textured soils. Christenson *et al.* (2000) addressed this issue in a 6-year trial at seven sites representing four different soil series. The results showed clearly that there were no negative effects on the yield or nutritional status of either sugar beet or other crops commonly grown in the area. However, since these studies were conducted on medium- and fine-textured soils, the question of overliming sandy soils remains.

Table 5.5. Effect of calcium carbonate on yield of sugar beet and soil pH 24 months after lime application on a very acid soil.

Amount of $CaCO_3$ (t ha^{-1})	Yield of roots (t ha^{-1})	Sugar (%)	Soil pH
0	11.3	16.7	4.9
2.8	22.9	17.1	5.3
5.6	25.9	17.1	5.5
11.2	25.4	17.2	5.9

'**Table 5.6.** Results of a survey in Eire of soil pH, sugar percentage and yield (from McEnroe and Coulter, 1964).

Soil pH	Sugar concentration (%)	Sugar yield (t ha^{-1})
4.5–4.9	15.2	4.3
6.0–6.5	15.4	4.4
6.6–7.0	15.6	4.8
$>$ 7.0	15.8	4.9

Liming Materials

The chemical definition of lime is calcium oxide (CaO). In agriculture and this book, lime is any material used to increase the pH of acid soils. Common liming materials include natural deposits of calcitic ($CaCO_3$) and dolomitic ($Ca.MgCO_3$) limestone, chalk ($CaCO_3$) and refuse liming materials from various chemical processes; factory lime from sugar beet processing fits into the latter category.

The primary quality of lime is related to the neutralizing value of the material. In the USA, calcium carbonate equivalent is a measure of the capacity of the material to neutralize acidity and is calculated as if all of the carbonate is in the form of $CaCO_3$ (Anon., 2001). In the UK, the neutralizing value is calculated as calcium oxide equivalent. The effectiveness of lime is also controlled by particle size. A rigid specification for the particle size distribution of liming materials is given in the footnote of Table 5.7. Christenson *et al.* (1988) suggest, as a minimum, that 85% of the material should pass an 8-mesh sieve and 25% a 100-mesh sieve. Factors that influence the rate of dissolution include crystalline make-up, density, surface area, porosity and surface coatings on the particles.

Factory lime is a good quality material since it is both fine and has a good neutralizing value. Factory lime averages about 50% $CaCO_3$ and 50% water when fresh. Dewatering at some factories now produces a lime that is much easier to spread, with 70% $CaCO_3$ and 30% water.

Methods of Determining Lime Requirement

There are three general methods of determining lime requirement. At the beginning of the 20th century a system was developed basing lime requirement on soil pH and texture. The basic premise of this procedure is that total acidity is related to the amount of clay and organic matter in the soil. This method gave unsatisfactory results on groups of dissimilar soils. This led to research that developed buffering systems to measure total acidity. These methods have been adopted over much of the USA. On highly leached mineral soils, Kamprath (1970) suggested basing lime requirement on neutralizing exchangeable and soluble aluminium. This approach has been adopted in many areas of the world where such soils exist. Draycott (1993) outlined a procedure for determining the lime require-

'**Table 5.7.** Amount of ground limestone needed to bring mineral soils to a desired pH as determined by the SMP buffer method (after Watson and Brown, 1998).

Buffer pH	Desired soil pH 7.0 (t ha^{-1})	Desired soil pH 6.5 (t ha^{-1})	Buffer pH	Desired soil pH 7.0 (t ha^{-1})	Desired soil pH 6.5 (t ha^{-1})
6.8	3	3	6.1	19	16
6.7	5	5	6.0	21	18
6.6	8	6	5.9	24	20
6.5	10	9	5.8	26	22
6.4	12	10	5.7	28	24
6.3	15	12	5.6	31	26
6.2	17	14	5.5	33	28

Limestone of 90% calcium carbonate equivalent ground so that 40% passes a 0.15 mm, 50% passes a 0.25 mm, 70% passes a 0.85 mm and 95% passes a 2.36 mm sieve. Plough depth of 20 cm.
SMP, Shoemaker, McLean and Pratt.

ment of sugar beet fields by pH and soil texture. This method holds for groups of similar soils, such as in the sugar beet-producing regions of the UK. The relationships in Table 5.8 for the UK and the USA show remarkable similarity.

Shoemaker *et al.* (1961) developed a system utilizing a buffer solution composed of paranitrophenol, potassium chromate, calcium chloride, calcium acetate and triethanolamine adjusted to pH 7.5. The solution is added to a suspension of soil and water and allowed to equilibrate and the resulting pH is measured. Generally accepted rates of limestone in relation to soil buffer pH are given in Table 5.7. The procedure may be inaccurate on soils with low lime requirement (< 4 t ha^{-1}), soils with over 10% organic matter and soils with a predominance of kaolinitic clays. This method has been adopted across large sections of the USA where soils meet the above requirements. It is referred to as the Shoemaker, McLean and Pratt (SMP) test.

Adams and Evans (1962) developed a procedure for determining lime requirement on soils containing predominantly kaolinitic clays. The procedure relates pH to base saturation. Lime recommendations are formulated to give a predetermined base saturation.

As mentioned above, a general procedure of neutralizing soluble and/or exchangeable aluminium is often used on highly leached mineral soils. Kamprath (1984) discusses the response to applied lime on these highly weathered soils.

Liming in Practice

In commercial sugar beet production, good liming practice is fundamental to success. Not only does optimum pH ensure the availability of major and minor nutrients and decrease toxicity due to excess as already described, but it also favours microbial activity and promotes better soil structure. For soils that are inherently acidic a strategy is necessary to reach and maintain the pH at 6.5–7.0 (Draycott and Messem, 1979).

To be effective the strategy must take into account that lime needs to be mixed thoroughly with soil. Surface application without tillage is ineffective just before sugar beet. We therefore suggest checking the soil pH 18 months before sowing the crop and liming areas with the amounts shown in Table 5.8. Tillage (preferably ploughing) will then incorporate the lime. To be absolutely sure that the sugar beet show no acidity prob-

Table 5.8. Amount of calcium carbonate needed to increase soil pH from measured value to pH 7.0 (from Christenson *et al.*, 1988; Draycott, 1993).

Soil pH	$CaCO_3$ required for 20 cm depth of soil (t ha^{-1})			
	Soil texture			
After Draycott	Light	Medium	Organic	Peats
4.0	9	15	23	39
4.5	8	12	16	26
5.0	7	9	10	14
5.5	5	7	5	8
6.0	3	4	0	0
After Christenson *et al.*	Loamy sand	Sandy loam	Clay loam and loam	Organic
4.5–4.9	9	12	15	11
5.0–5.4	8	9	12	7
5.5–5.9	7	8	9	2
6.0–6.4	4	5	7	0

lems, soil needs to be checked and limed again if necessary six months before sowing and ploughed again.

The amount of lime applied effectively in one application is also limited by the amount that can be mixed with the soil plough layer. Research shows that a range of 6–12 t ha^{-1} may be applied and mixed by harrowing and ploughing. When greater quantities of material are required, half should be applied before ploughing with the remainder applied after ploughing.

SUMMARY OF CALCIUM AND MAGNESIUM

Calcium and magnesium are important cations for the production of sugar beet. Both are divalent, being held on the exchange complex, and are generally well supplied on most soils where sugar beet is produced. Magnesium may be deficient on sandy soils, especially where the subsoil is also sandy. Calcium deficiency is rare and relates to uptake and translocation rather than supply to the root. Mass flow brings an over abundance of calcium to the root surface, even on soils with pH as low as 5.5.

The role of calcium in the plant is both structural and functional. It is essential in the formation of cell walls and appears to activate some enzymes. Magnesium is the central atom in the chlorophyll molecule and is involved in many enzyme reactions. A supply of both elements is needed over the entire growing season. Sugar beet contains approximately 80 kg calcium and 50 kg magnesium ha^{-1}, divided about equally between roots and tops.

Dolomitic limestone is preferred for dealing with magnesium deficiency on acid soils. Kieserite, calcined magnesite and potassium-magnesium sulphate can be used on all soil types. Magnesium deficiency may also be treated by foliar application with magnesium sulphate (Epsom salts). However, foliar application should be regarded as a 'last resort' approach to managing this nutrient.

Maintaining a near-neutral pH of soil is essential for sugar beet production. Application of lime should be made early in the rotation prior to growing sugar beet, so that there is time for it to be fully mixed with the soil. The amount of lime needed may be determined by the pH and soil texture or by using a buffer method. Both utilize tabular data based on sampling or testing in the field.

6
Micronutrients or Trace Elements

The macro- or major nutrients needed by sugar beet have been dealt with in previous chapters. This chapter deals with those additional elements needed in relatively tiny but important amounts known as micronutrients or trace elements. Quantities of macronutrients are usually taken up in the range of 50–500 kg ha^{-1} whereas amounts of micronutrients taken up are measured in g ha^{-1}. The concentration of micronutrients in plants is usually < 100 mg kg^{-1} of dry plant material (Anon., 2001).

Those needed by sugar beet are boron (B), chlorine (Cl), copper (Cu), iron (Fe), manganese (Mn), molybdenum (Mo) and zinc (Zn). Elements needed by some other plants are cobalt (Co), nickel (Ni), selenium (Se) and silicon (Si) but these have not been investigated much for sugar beet. By far the most widespread shortages for sugar beet are of boron and manganese. Consequently much research has been on these two and relatively little on the rest.

Table 6.1 shows the concentration of micronutrients in healthy sugar beet and approximate amounts taken up by tops and roots at harvest. Where these are not readily available from the soil, additions must be given or crops will show deficiency symptoms and/or do not yield fully. Chapter 7 shows nutrient-deficiency symptoms and concentration of nutrients for deficient plants.

Sugar beet is not equally susceptible to a shortage of all micronutrients. Under conditions of low supply, the crop is most responsive to boron, iron and manganese, moderately responsive to copper, molybdenum and zinc and least responsive to chlorine (Table 6.2). Occasionally copper or zinc deficiency may be found but molybdenum deficiency is rare. To the best of our knowledge, there has never been a case of chlorine deficiency reported from the field.

Micronutrient deficiencies are generally corrected with fertilizer materials applied either to soil or to plant foliage. A list of common materials is given in Table 6.3. Soil type and reaction, along with environmental conditions, exacerbate the incidence of micronutrient deficiencies (Table 6.4).

BORON

Amount in Soil

Total

Traces of boron are found in various silicate minerals, where it substitutes for silicon. Most important is the mineral tourmaline, which is widespread in volcanic rocks; this contains 3–4% boron. Weathering releases boric acid, H_3BO_3, very slowly. Boron is also

Table 6.1. Average concentration of micronutrients in tops and roots of healthy sugar beet and quantity in the crop at harvest.

	Concentration in dry matter (mg kg^{-1})		Quantity in crop, tops plus roots (g ha^{-1})
	Tops	Roots	
Boron	40	15	335
Chlorine	2,000	1,000	19,000
Copper	7	1	44
Iron	200	100	1,900
Manganese	50	30	520
Molybdenum	7	5	80
Zinc	20	10	190

Table 6.2. Response of sugar beet to micronutrients under soil or environmental conditions favourable to a deficiency (Lucas and Knezek, 1972).

Probability of response	Micronutrient
Low	Cl
Medium	Cu, Mo, Zn
High	B, Fe, Mn

Table 6.3. Sources of micronutrient fertilizers used on sugar beet.

Nutrient	Chemical formula	Concentration (%)	Method of application
Boron			
Sodium borate	Various	11–17	Soil
Boric acid	H_3BO_3	17	Soil or foliar
Solubor	$Na_2B_{10}O_{16}.10H_2O$	20	Foliar
Copper			
Copper sulphates	Various	13–53	Soil or foliar
Copper chelates	Various	9–13	Soil or foliar
Copper oxychloride	$CuOCl_2$	42	Soil or foliar
Copper oxide	CuO	80	Soil
Iron			
Ferrous sulphate	$FeSO_4.7H_2O$	19	Soil or foliar
Iron chelates	Various	5–14	Soil or foliar
Manganese			
Manganese sulphate	$MnSO_4.3H_2O$	26–28	Soil or foliar
Manganese nitrate	$Mn(NO_3)_2$	22	Foliar
Manganous oxide	MnO	41–68	Soil
Manganese chelate	MnEDTA	12	Foliar
Molybdenum			
Sodium molybdate	$Na_2MoO_4.2H_2O$	39	Soil or foliar
Ammonium molybdate	$(NH_4)_6Mo_7O_{24}.4H_2O$	54	Soil or foliar
Frits	Various	Various	Soil
Zinc			
Zinc sulphate	$ZnSO_4.H_2O$	35	Soil or foliar
Zinc oxide	ZnO	78	Soil
Zinc chelate	Various	9–14	Soil or foliar

EDTA, ethylenediaminetetraacetic acid.

Table 6.4. General factors contributing to micronutrient deficiencies in sugar beet (modified from Lucas and Knezek, 1972).

Boron	Manganese
Soils low in B (see Table 6.6)	Naturally poorly drained soils
Alluvials	Organics
Podzols	Groundwater podzols
Organics	Loess
Light sands	Low humic gleys
Moderate to heavy precipitation	Marly or shelly soils
Neutral or alkaline reaction	Slightly acid to alkaline reaction
Dry weather	High soil Fe, Cu or Zn and dry weather
High light intensity	Low light intensity and soil temperature
	Herbicides
Iron	Copper
Soils low in available Fe	Soils low in copper (see Table 6.9)
Free $CaCO_3$, high HCO_3^-	High soil P, N or Zn
Moisture and temperature extremes	
Large amounts heavy metal and P in soils	
Heavy manuring on alkaline soils	
Poor aeration	
Molybdenum	Zinc
Soils low in Mo	Soils low in Zn
Acid sands	Humic gleys
Acid podzols	Alluvials
Acid organic soils	Loess
High free Fe	Organics
Bog or fen	Low soil organic matter and low temperatures
	High soil P and amount of applied N
	Calcareous and compacted soils

present in soil adsorbed to clay minerals and iron and aluminium oxides and in soil organic matter. All provide plant-available boron, which is then subject to leaching, both as H_3BO_3 and as borate ion ($H(BO)_4^-$).

In common with many nutrient elements, total soil boron concentration depends first on the nature of the parent material and secondly on precipitation. In wet climates total soil boron is usually in the range of 2–100 mg kg^{-1} (Gupta, 1979). Sandy soils where sugar beet is produced frequently contain 2–6 mg kg^{-1}, whereas clay soils may contain up to 60 mg kg^{-1} (Adriano, 2001). Archer and Hodgson (1987) measured total boron in a large number of UK soils and found that it ranged from 7 to 119 mg kg^{-1}, with a mean of 33 mg kg^{-1}.

Inputs

Sugar beet fields may receive small amounts as impurities in major-element fertilizers, which often contain a few mg B kg^{-1}. Those fields subject to sewage-sludge disposal may receive larger amounts nowadays because of widespread domestic and industrial use of boron in detergents and bleaches. Atmospheric deposition of boron is very small, both wet and dry, estimates ranging from 50 to 300 mg B ha^{-1} year^{-1} in the UK (Wadsworth and Webber, 1980). In the sugar beet context, most boron inputs are fertilizer applied on to fields specifically for the crop, often of the order of 2–3 kg B ha^{-1}.

Available

Plant-available boron is associated with decaying organic matter and inorganic forms originating from minerals and clay. Some is fairly tightly bound, being adsorbed by hydrous oxides of aluminium and iron and by clay minerals. Similarly boron is also rather tightly bound to soil organic matter.

The portion that is fully available for uptake by sugar beet is present in soil solution and can be easily measured, as described below. In acid and neutral soil, it is present as boric acid (H_3BO_3) and in alkaline conditions as the borate ion ($H(BO)_4^-$).

Both easily leach in climates where precipitation exceeds evapotranspiration. Not surprisingly, deficiencies are seen most regularly on sandy alkaline soils because of small total boron, leaching (especially in acid conditions), increased adsorption in alkaline conditions, and a tendency for boron to precipitate as relatively insoluble calcium borate. Deficiency is uncommon on finer-textured soils, particularly those at the neutral point or below.

Dry summers in humid regions exacerbate boron deficiency in sugar beet either directly due to decreased flow of soil solution (mass flow) to the root surface or indirectly due to slowed microbial activity. In some crops, boron-deficiency symptoms may disappear on new growth after receiving rainfall, presumably with increased mass flow accounting for some of the recovery.

A very good and widely used soil test for available boron reliably predicts where sugar beet is likely to show deficiency (see below for values). This entails boiling the soil with water under reflux, extracting the plant-available fraction of boron. Addition of a little calcium chloride (0.01 or 0.02 M $CaCl_2$) produces a colourless extract that can be used in a colorimetric determination. The method was introduced by Berger and Truog (1939) and improved slightly by Offiah and Axley (1988).

Deficiency and Toxicity

In common with most, but not all, trace elements needed for satisfactory growth of sugar beet, shortage in plant tissue is accompanied by visible symptoms. These symptoms can be identified without supporting chemical analysis. This is because unique symptoms appear not only on the leaves (as with most element deficiencies) but also on the petioles, crowns and roots of sugar beet. Brandenburg (1931) first showed that boron deficiency was the cause of 'heart rot' and 'dry rot'. Heart rot is the term applied when the growing point becomes blackened and dies. Dry rot describes the symptoms on the tap root, which usually appear subsequently. Boron deficiency not only decreases yield but damages tap roots, decreasing their value and keeping quality. Photographs of plants with typical problems are shown in the plate section at the front of this book.

Significant quantities of boron may be added in irrigation water, which may be sufficient to cause toxicity to growing crops. Boron toxicity is most common in arid regions and in crops that have been irrigated extensively (Nable and Paull, 1991). Sugar beet is generally tolerant of boron, perhaps because it can sequester the element in cell walls (Rozema *et al.*, 1992). Hanson *et al.* (1999) suggest a threshold concentration of 4.9 mg B l^{-1} in irrigation water. This would amount to 4.9 g B m^{-3} of irrigation water. Careful management of such situations is critical so that boron does not accumulate to toxic proportions in the soil.

Field bean (*Vicia faba* and *Phaseolus vulgaris*), soybean (*Glycine max*) and cereals, such as wheat and barley, are damaged when boronated fertilizer is inadvertently applied. The critical toxicity concentration in wheat is about 100–250 mg B kg^{-1} dry matter (Paull *et al.*, 1988).

Amount in Sugar Beet

Of all crops, sugar beet has one of the largest requirements of boron. It must be able to take up sufficient element to maintain a minimum concentration in tissue or growth is disrupted in a spectacular manner, as described in Chapter 7.

Table 6.5 shows the concentrations of boron expected in healthy and deficient sugar beet. They are remarkably consistent from country to country. Interestingly, Brandenburg's (1931, 1939) original analyses of over 70 years ago still apply today.

Leaf tissue is the most useful for such diagnoses, boron concentration varying closely with supply and symptoms. Root boron is not a good guide. Ideally, leaf concentration should exceed 30 mg B kg^{-1} and deficiency symptoms can be expected below 20 mg B kg^{-1}.

Table 6.5. Concentration of boron in sugar beet dry matter.

Stage of growth	Plant part	Boron concentration ($mg\ B\ kg^{-1}$) Showing deficiency	Without symptoms	Country	References
–	Leaf	4–28	25–52	Germany	Brandenburg, 1931,
	Root	13–14	13–15		1939
August/September	Leaf blade	19–35	20–426	California, USA	Eaton, 1944
	Root	6–30	2–28		
	Whole plant	13–33	10–139		
June	Old leaf	13–29	22–73	UK	Hamence and
August		10–37	26–75		Oram, 1964
November		16–36	26–52		
June	Young leaf	25–37	29–90		
August		8–22	25–40		
November		19–24	15–40		
Range	Leaf	4–37	10–90	UK	Draycott, 1972
	Root	6–14	2–28		
Means	Leaf	20	40		
	Root	13	15		
Mid-season	Leaf, fully expanded	< 20	26–80	Michigan, USA	Robertson and Lucas, 1981
June–July	Leaf blades	–	55–75	Germany	Kluge, 1990
Whole season	Petioles	–	30–35		
Harvest	Roots	–	15–20		
12 weeks after sowing	Leaf blades	–	34–72	Michigan, USA	Christenson *et al.*, 1991a
	Leaf	20–25	–	Pakistan	Tariq *et al.*, 1993
Six leaves	Leaf	20–30	40	UK	Hill, 1999

Prediction of Boron Requirement by Soil Analysis

Smilde (1970) made a comprehensive study of the value of soil analysis for predicting where sugar beet needs boron. Results of plant analyses from field and pot experiments showed a close relationship between hot-water-soluble soil boron and leaf boron over a wide range of soil pH and organic matter.

Soil and leaf boron concentrations were also closely related to the percentage of affected plants. No heart rot appeared when the hot-water-soluble boron concentration was greater than 0.35–0.40 mg B kg^{-1} or when the leaf boron concentration was greater than 35–40 mg B kg^{-1}. There was no loss in yield of tops, roots or sugar from boron deficiency when the soil boron exceeded 0.50 mg B kg^{-1}.

Leaf boron decreased and heart rot increased as soil pH and organic matter increased. For soils with 0.30–0.35, 0.20–0.29 and less than 0.20 mg B kg^{-1}, the best dressings were 5, 10 and 15 kg sodium borate ha^{-1}, respectively. Sodium borate markedly increased the sugar percentage of sugar beet growing in deficient soil in pots but the effect was much less pronounced in the field.

Gembarzewski and Korzeniowska (1990) in Poland tested 11 methods of extracting boron, copper, iron, manganese, molybdenum and zinc simultaneously from soil and looked for correlations with uptake. Relationships were poor, the only extractant of some usefulness being molar HCl. It would appear that, for boron-deficiency forecasting, the standard hot-water method should be used.

Robertson *et al.* (1975) made a thorough survey of concentrations of available boron in Michigan soils, using the hot-water-soluble boron test. Neither soil features nor soil pH appeared to be related to the concentration of soil boron. Topsoil boron ranged from 0.16 to 0.95 mg B kg^{-1} soil. The large values were found where soil had been treated with boron in the recent past. Subsoils varied from 0 to 0.54 mg B kg^{-1}. In their survey of paired samples from cropped and uncropped areas, there was two-thirds as much boron in the cropped as in the uncropped samples. Soil-test concentrations were not increasing as a result of long-term application of boron.

Several surveys in various parts of the world have tried to determine the extent of boron shortage in sugar beet-growing areas. Over half a million soils were analysed for available boron in a survey in southern Germany by Fürstenfeld and Bürcky (2000). Between 50 and 90% needed boron application. Their recommendation was for 2 kg B ha^{-1} for fields with less than 0.65 mg B kg^{-1} soil, 1 kg B ha^{-1} for fields with 0.65 to 0.95 mg B kg^{-1} and nil for those with greater than 0.95 mg B kg^{-1}. They added the proviso that, in a long dry period in early summer, the last category should receive a foliar spray of boron. Similarly, Rodriguez and Tomic (1984) analysed soils in Chile and also classified over half as boron-deficient.

Table 6.6 summarizes much information concerning hot-water-soluble boron and deficiency, which needs correcting, as described below. In constructing this table, we have taken into account the patchy within-field nature of boron deficiency, where mean soil concentrations are 0.50–1.00 mg B kg^{-1}.

Supplying Boron to Sugar Beet

Early work

Once Brandenburg (1931) had shown that heart rot was caused by boron deficiency, workers in several countries made experiments to determine how best to apply boron (usually sodium borate) to prevent the problem (reviewed by Draycott, 1972). The standard application became 20 or 25 kg sodium borate ha^{-1}. Sodium borate in its fully hydrated form is $Na_2B_4O_7.10H_2O$ (11% B). Thus the amount of the element applied averaged about 2.5 kg B ha^{-1}.

Different times of application to soil and crop were tested, with solid and solutions of sodium borate. The general conclusion was that, provided the application was before June, liquid and solid were equally effective, whether applied to soil or foliage.

Current usage

Later experiments have only confirmed the above. Work in the USA showed that boron could safely be applied before ploughing and major nutrient fertilizers (containing all the crop needs except nitrogen) became 'borated'. These can be applied in autumn, in winter or just ahead of ploughing in spring on sandy soils with equal efficacy. Similarly, some NPK fertilizers for seedbed application are borated. Experiments in many countries show that these soil applications should provide about 2 kg B ha^{-1}.

Voth *et al.* (1979) in Michigan measured boron concentration in sugar beet tissue and

Table 6.6. Relationship between hot-water-soluble boron, expression of deficiency by sugar beet and amount required.

Soil B (mg kg^{-1})	Symptoms expected	Application required (kg B ha^{-1})
0–0.25	Very severe in most conditions	3
0.26–0.50	Severe at high pH	2
0.51–1.00	Some plants may be deficient	1
1.10–2.00	None deficient	0
2.10–4.00	High concentration	0
> 4	May be toxic	0

response to boron. Without boron leaf tissue contained 26 mg B kg^{-1} dry matter and up to 95 mg B kg^{-1} with boron applied. Despite this result, no deficiency symptoms were seen at any site and the authors concluded that this was because the tissue concentrations were always above the 20 mg B kg^{-1} threshold below which symptoms appear. Studies in India (Narayan *et al.*, 1989) on boron-deficient soils found large increases in yield from 1–3.5 kg B ha^{-1} applied to soil. Response to one foliar spray was much smaller but with two sprays resulted in yields similar to soil application.

For foliar sprays, a very soluble borate has been produced (trade name Solubor). This contains 21% B and is applied at 7–10 kg ha^{-1} (1.5–2 kg B ha^{-1}). It is of equal efficacy to soil applications provided it is sprayed in the early stages of growth before plant demand is great (early June at the latest in Europe). In the USA the findings are similar; much smaller amounts of boron (0.15–0.2 kg B ha^{-1}) are used in foliar applications in the Great Lakes.

Surveys of boron applications

V. Shorrocks (UK, 1970, personal communication) surveyed the use of boron worldwide and we recently surveyed its use in EU countries. Sugar beet in most countries has a need on certain soil types. Amounts applied are in the range 1.5–3.0 kg B ha^{-1}, solid or spray applications both being common. Generally, the hotter the climate and the more alkaline the soil reaction the larger the boron need and application. In some countries the whole area of sugar beet receives boron routinely.

Amount of Boron Removed from Soil by Sugar Beet and Residual Effects of Boron Applications

Some crops are sensitive to an excessive supply of boron (whether from newly applied boron fertilizer or from residues in the soil); barley and field bean often show boron-toxicity symptoms where boron fertilizer has been used by mistake. Consequently there have been several investigations to determine the residue from boron applications given to sugar beet. According to Brandenburg (1939) a normal crop of sugar beet extracts 0.3–0.4 kg B ha^{-1}. Thus, of the 2 kg boron ha^{-1} usually applied for sugar beet, 1.5 kg ha^{-1} remains in the soil. Leaching experiments made by Krügel *et al.* (1938) in Germany showed that 75% of this was leached out of the soil during the first winter, and by the third year after application little remained. In the climate of western Europe, they concluded that there was no evidence that boron accumulated in the soil. Robertson *et al.* (1975) showed that cropped soils had two-thirds as much extractable boron as uncropped soils. More recently, Bravo *et al.* (1992) confirmed the offtake of boron at 0.3 kg B ha^{-1} for a yield of 53 t roots ha^{-1}.

Very few long-term experiments have measured the effects of continued dressings of micronutrients on crops and soils, but one field experiment in Norway (Ødelein, 1963) continued for 23 years. Annual dressings of 0.2 kg B ha^{-1} $year^{-1}$ prevented boron deficiency, but 0.1 kg B ha^{-1} $year^{-1}$ was partly successful. Giving 1 kg B ha^{-1} $year^{-1}$ injured some of the crops grown after a few years, especially on unlimed soil. Hamence and Oram (1964), van Luit and Smilde (1969) and Chabannes (1959) in England, Holland and France, respectively, have investigated the fate of boron applied for sugar beet. These reports indicated that little boron remains 2 years after application, presumably because it is subject to rapid leaching during winter.

Forms of Boron for Soil and Foliar Applications

Table 6.7 lists the common forms of boron applied for sugar beet. For comparison, amounts supplied in 25 t of farmyard manure ha^{-1} and in the mixed salt sylvinite are given. Different forms of sodium borate are used throughout the world for correction of boron deficiency. A more soluble form (Solubor) is used for foliar application. Methods of application are discussed in Chapter 11.

Table 6.7. Forms of boron applied to sugar beet.

	B (%)	Application (kg ha^{-1})	Amount of B applied (kg ha^{-1})
Sodium borate (borax) $Na_2B_4O_7.10H_2O$	11	20	2.0
Boric acid H_3BO_3	17	10	1.7
Granubor Mixture of $Na_2B_4O_7.5H_2O$ and $Na_2B_8O_{13}.4H_2O$	15	7–20	1.0–3.0
Fertibor $Na_2B_4O_7.5H_2O$	15.2	7–20	1.0–3.0
Solubor $Na_2B_{10}O_{16}.10H_2O$	21	7	1.5
Sylvinite	0.37 (±0.22)	600	2.2 (±1.3)
Farmyard manure	40 mg kg^{-1}	25 t ha^{-1}	1.0

Response by Sugar Beet to Boron

Many experiments have compared sugar beet grown with and without boron where soil supply is small and where boron deficiency appears on untreated sugar beet. Beneficial and economic effects on yield usually accompany the marked visual effects already discussed. Draycott (1972) reviewed early work and Allison (1996) more recent information. Commonly, yields of roots, tops and sugar are all increased greatly. The sugar percentage is often improved. Surprisingly, other aspects of root quality, such as juice purity, do not appear to be affected.

Experiments comparing different amounts of boron and those comparing solid with spray have rarely shown any significant differences. It would appear that, provided some boron is applied before sowing or during the early stage of growth, deficiency is completely eliminated.

The incidence of boron deficiency appearing on sugar beet in the Great Lakes and Red River Valley is quite low. Some studies have shown no response to applied boron over a range of soils where sugar beet is grown (Cattanach, 1991; Christenson *et al.*, 1991a; Giles *et al.*, 1991). In particular, Voth *et al.* (1979) showed a significant yield response to applied boron; however, a decade later no response was shown in trials conducted in the same area (Christenson *et al.*, 1991a). Earlier Tandon (1979) had shown a differential response to boron in solution culture among four varieties. The role that genetics plays in sugar beet needs for boron is in need of further research.

MANGANESE

Forms in Soil

Total

The amount of total manganese present in soil is largely dependent on the amount present in minerals of the soil's parent material. This is because organic matter, in all its forms, contains only traces of manganese; also little is lost by leaching from temperate arable soils. Weathering of minerals present in most rock types yields a little manganese but its plant availability is immediately governed by pH, redox conditions and other soil characteristics, discussed below.

The total concentration of manganese present in various soils, recently reviewed by Whitehead (2000), ranged from 40 mg Mn kg^{-1} to 40,000 mg Mn kg^{-1}. Adriano (2001) compiled data from a number of sources and reported similar values for a large number of soils. Temperate-zone soils on which sugar beet is grown can be expected to contain 50–500 mg Mn kg^{-1}. Few sugar beet soils are highly leached, so, when manganese-containing minerals weather, manganese remains in the root zone, some in clays (smectite), some as oxides and hydroxides and some in plant-available forms.

Several common sugar beet soils in the UK contain relatively large quantities of total manganese as the insoluble oxide MnO_2. This can be seen by the naked eye as black granules 0.5–2 mm in diameter in subsoils within root range. It is thought that these nodules were formed by precipitation reactions during periglacial conditions following melting of ice sheets that covered the region. Unfortunately, such manganese is not in a plant-available form.

Other small additions of manganese to sugar beet soils are through atmospheric deposition (50–300 g Mn ha^{-1} $year^{-1}$), as impurities in NPK fertilizers (from a few grams to 1 kg Mn t^{-1}) and where sewage sludges are used (see Chapter 8). Farmyard manure contains variable quantities, commonly supplying 1–2 kg Mn ha^{-1} in an average application of 25 t ha^{-1}.

Plant-available

In common with iron and copper, manganese is present in soil in more than one oxidation state. The most common and stable in soil are divalent Mn^{2+} and tetravalent Mn^{4+}. Only Mn^{2+} is stable in solution and this is the form taken up by plants such as sugar beet.

The tetravalent form precipitates as the oxide MnO_2 (as in the nodules mentioned above), rendering it completely unavailable. Thus transitions that occur in soil between the valency states markedly influence availability. In the lower valency state (Mn^{2+}) manganese is considered reduced, in the higher valency state (Mn^{4+}) oxidized (Brady and Weil, 1999).

Thus, when oxygen supply is decreased in soil (reducing conditions), as in wet soils containing decomposable organic matter, manganese becomes more available. Reduction is also brought about by metabolic reducing agents produced by plants and microorganisms in soil.

In some situations, sugar beet during the early stages of growth often shows mild symptoms of deficiency during dry periods in spring. A prolonged period of rain causing temporarily waterlogged soil increases manganese availability and symptoms disappear due to the balance between divalent and tetravalent moving in favour of the plant-available divalent state.

Other observations have also been made where manganese deficiency occurs in the early spring when the soil is cold and wet, and Farley and Draycott (1978) explored the use of manganese pelleted with seed. Warm weather decreases the incidence of the problem, increasing the amount of ions brought to the root by mass flow as the crop grows more rapidly.

It is a prerequisite of successful sugar beet production that soils are kept near the neutral point (pH 7.0). Thus liming necessary before the crop often results in manganese deficiency because the divalent form is taken out of solution, particularly where there is also a large amount of organic matter present (Farley and Draycott, 1973). A major factor seems to be an interaction with sesquioxides (iron–aluminium oxides/hydroxides) as pH increases above 5.8, which reduces the solubility of manganese in the soil (Mehlich, 1957). This does not occur when liming less acid soils since the activity of the sesquioxides is very low. Many fenland, peat or 'muck' soils are naturally alkaline due to presence of chalk or fossil-shell deposits. These soils contain little divalent manganese and sugar beet shows the most severe forms of deficiency every year without suitable applications of manganese, described below.

The effect of soil pH on manganese solubility is marked, increasing concentration in solution 100-fold for each pH unit decrease (Lindsay, 1972). Generally sugar beet does not show deficiency below soil pH 6.5. Toxicity is seen when sugar beet is grown by mistake on certain acid soils. Soil pH in the range 3.5–5.5 severely damages sugar beet growth. Much manganese is present in solution in the divalent form and is taken up by seedlings in toxic amounts with spectacular results (Draycott, 1972). Leaves become bright yellow and stunted. Analysis of such plants shows manganese concentrations of more than 10–20 times the normal concentration; 1000–2000 mg Mn kg^{-1} in dry leaf compared with 70 mg Mn kg^{-1} in healthy plants (Table 6.8).

Incidence of Deficiency

Field experiments testing response to applications of manganese (whether to soil or plant) generally conclude that only where symptoms appear on leaves does the crop respond. Thus, in the absence of deficiency symptoms, it can be assumed that the crop has a satisfactory supply. This is of great practical value, perhaps more so than soil or plant tissue analysis, because the symptoms are unique and easily recognized. See Chapter 7 for further discussion and photographs.

Concentration in Tissue and Movement within Plants

Concentration

Table 6.8 shows how much manganese is present in sugar beet with and without deficiency symptoms. Deficient leaves usually contain 10–30 mg Mn kg^{-1} of dry matter and healthy leaves 40–100 mg Mn kg^{-1}. When plants are analysed at intervals during the growing season, manganese concentration tends to decrease (Last and Bean, 1990). It is suggested that this is because manganese concentration is reduced by the dilution effect of growth without additional translocation of manganese to the tissue.

Translocation in sugar beet plants

Henkens and Jongman (1965) used radioactive Mn (^{54}Mn) to investigate translocation of the element in sugar beet plants. In fact, their experiments demonstrate very well that it does not move from leaf to leaf. The concentration of manganese in sugar beet leaves from plants where one leaf was treated with manganese solution compared with leaves from untreated plants showed that after 20 days only the treated leaf contained the applied manganese. The manganese did, however, move down into the roots.

Field experiments were made which confirmed these findings; newly formed leaves did not benefit from spray given earlier, with the result that deficiency symptoms reappeared. One spray scarcely increased yield, whereas two sprays gave substantial increases of over 10%. A general recommendation for treatment is to make one spraying and in 7–10 days another if the new growth is still showing symptoms.

We have observed what may be called a 'shadow effect' when foliar application of manganese is made. A leaf exhibiting deficiency would not be affected if shielded from the spray by an overlapping leaf. When

Table 6.8. Manganese concentration in plants with and without deficiency symptoms.

Plant part		With symptoms	Without symptoms	References
		mg Mn kg^{-1} dry matter		
Leaf	Range	12–17	46–110	Summary of early work, Draycott, 1972
	Mean	14	70	
Leaf		20–30	Above 30	MAFF, 1976
Leaf		16–22	60–76	Farley, 1980
Leaf	May	30	162	Last and Bean, 1990
	June	20	58	
	July	23	76	
	October	32	35	
Leaf		20	74	Last and Bean, 1991
Root		8	11	

portions of leaves were shielded, those portions of the leaves did not recover from deficiency symptoms while the unshielded portions did.

Methods of Soil Analysis and Response to Manganese

Several methods of soil analysis for 'available' soil manganese were tested by Draycott and Farley (1973) and related to the amount of manganese in dry matter of sugar beet. Molar NH_4OAc at pH 7 probably extracts only the immediately available manganese. Adding a mild reducing agent to this extractant includes some potentially available manganese. The manganese held as organo-metallic complexes on soil organic matter is also potentially available. Ammonium dihydrogen phosphate solution probably extracts some of this manganese, plus the easily reducible and available fractions. It was found that the buffered NH_4OAc plus the reducing agent hydroquinone gave the best relationship with plant manganese, accounting for 64% of the variance.

Various methods of soil analysis were put to the test in 30 field experiments to discover if any could forecast where sugar beet would be deficient, the extent of deficiency and yield response to added manganese (Farley and Draycott, 1976). Several methods of analysis were moderately successful in predicting where symptoms would appear but none the yield response. Plant analysis and deficiency symptoms were much more reliable guides to the severity of a shortage of manganese. Of all the methods tried, the amounts of manganese extracted (10–300 mg Mn kg^{-1}) by ammonium acetate/hydroquinone were moderately correlated with concentrations of Mn (10–50 mg Mn kg^{-1}) in sugar beet leaves (Farley and Draycott, 1973).

Reisenauer (1988) compared a wider range of extractants for various crops (not specifically sugar beet) grown in a large number of conditions. He concluded that, from both pot and field experiments which compared soil analysis with crop uptake, soil analysis had limited ability to predict plant-available manganese. Germida *et al.* (1985) had more success using a simple microbial bioassay to determine plant-available manganese in soil.

In the USA Robertson and Lucas (1981) extracted soil-'available' manganese with 0.1 M HCl. They found that, where less than 5 mg Mn kg^{-1} was extracted, deficiency was likely on mineral soils and on certain organic soils. Deficiency was directly related to pH, increasing in severity with increasing pH. They recommended foliar spray and/or banded manganese at sowing. Depending on the soil analysis, 5–20 kg Mn ha^{-1} was recommended as a soil application. For a foliar application, 1–3 kg Mn ha^{-1} was suggested.

In a companion study, Salcedo *et al.* (1979) reported that the correlation between extractable manganese and plant uptake decreased in the following order: 0.3 M H_3PO_4 > steam/NH_4OAc > 1.5 M $NH_4H_2PO_4$ > 0.1 M HCl > NH_4OAc > 0.005 M chelate solution. The linear coefficient of determination (r^2) ranged from 0.92 to 0.69. The critical value for 0.3 M H_3PO_4 was shown to be 12 mg kg^{-1}, and it was 14 mg kg^{-1} for steam/NH_4OAc and 1.5 M $NH_4H_2PO_4$.

Our experience suggests that, in addition to soil analysis, the history of manganese deficiency and soil type should be used to predict future problems. Routine preventive treatments are part of many growers' fertilizer programmes. The cost–benefit ratio is favourable for this routine practice.

COPPER

In Soil

Total

In common with other metal trace elements, the total amount of copper present in soil depends primarily on the nature of the parent material. It is present in variable amounts in sedimentary and igneous rocks. The latter contain copper as the sulphide, which releases divalent cupric (Cu^{2+}) or monovalent cuprous (Cu^{+}) ions. Both cations form complexes with organic matter and the Cu^{2+} form is also adsorbed by iron and manganese oxides.

Quantities of total copper found in temperate soils are generally in the range 2–100 mg Cu kg^{-1} dry soil. In some arable soils where sugar beet is grown copper may accumulate from sewage sludge and poultry manures and where pigs are reared outdoors. Smaller inputs are from farmyard manure (0.3 kg Cu ha^{-1}), fertilizers (1–20 g Cu t^{-1}) and atmospheric deposition (500 g ha^{-1} $year^{-1}$ in parts of Europe).

Significant amounts are also applied where copper-based fungicides are applied to vegetable crops and, of course, where it is applied as a nutrient for sensitive crops, such as cereals, grown in rotation with sugar beet. Bearing in mind that copper accumulates in topsoil, where cropping is intensive, a running total of copper inputs should be kept to avoid unnecessary applications to sugar beet, which could lead to toxicities.

Available

In most sugar beet soils, provided they are not waterlogged, copper available or potentially available to the crop is in the divalent form, Cu^{2+}. Much of this, however, is complexed by organic matter more tightly than any other divalent micronutrient cation. Some of the copper may be in an adsorbed form and only minute amounts in an entirely inorganic form. However, most (76–99%) of the copper in solution is organically bound (Hodgson *et al.*, 1965, 1966).

Thus a simple, buffered salt solution (such as molar ammonium acetate or nitrate) does not extract plant-available copper. Instead, it has become practice to use solutions that extract not only water-soluble but also complexed copper. Most common are solutions containing ethylenediaminetetraacetic acid (EDTA). The usefulness of such results may be improved by also taking into account pH and/or soil organic-matter concentration. Experiments described below have tested the value of such soil measurements to predict the copper requirement of sugar beet. Table 6.9 summarizes current views on soil analysis for copper and the effect of other soil factors.

Concentration in Sugar Beet Plants

Summarizing copper concentrations in early reports, Draycott (1972) suggested that deficiency might be expected when dried leaves contained below 6–7 mg Cu kg^{-1}. Plants well supplied with copper contained up to 20 mg Cu kg^{-1} in leaves. Pizer *et al.* (1966) reported smaller values in crops treated with copper, with a total uptake of 40 g Cu ha^{-1} in tops plus roots.

More recent field experiments in the UK where deficiency was thought likely, found 6 mg Cu kg^{-1} in dried leaves at harvest and 3 mg Cu kg^{-1} in dried roots (Allison *et al.*, 1996b). Applying copper had little effect on these concentrations.

Effects of Copper on Sugar Beet Yield

Copper sprays applied to foliage can control fungal diseases that damage leaves. Thus it is important to separate the nutritional and fungicidal effects. In particular, in climates where *Cercospora beticola* dramatically affects sugar beet yields, the fungicidal properties of a copper spray will greatly outweigh its nutritional value. Confusion is avoided where applications are to soil. In addition to

Table 6.9. Response to copper in relation to soil copper and organic matter.

Soils with less than 10% organic matter	Peaty soils with greater than 10% organic matter	Response to copper application
mg Cu kg^{-1} in EDTA extract		
< 1.0	< 2.0	Likely
1.0–1.6	2.0–3.0	Possible
> 1.6	> 3.0	Unlikely

suppression of fungal diseases, foliar application of copper has been reported to suppress nematode activity (Graham and Webb, 1991).

Several groups of field experiments have been made on fields where response was thought likely. Some crops are more sensitive than sugar beet and show deficiency symptoms unique to copper, such as cereals with 'white tip'. Fields where sugar beet was likely to respond have also been located by soil analysis by the EDTA test.

In early experiments, responses were usually small but significant, though crops on some fields did not respond (van Schreven, 1936; Lachowski, 1961; Pizer *et al.*, 1966; MAFF, 1976; Tills and Alloway, 1981; Alloway and Tills, 1984).

In recent experiments supported by soil and plant analysis (Allison *et al.*, 1996b), fields on loamy sand and peat or in sand were chosen, where response was thought likely. Copper was extracted from soil samples with 0.05 M EDTA solution at pH 7.0. The range of values was from 0.7 to 3.7 mg Cu kg^{-1}. It was thought that 15% of the UK crop was on potentially responsive fields.

Copper was applied as a foliar spray of oxychloride or as a proprietary formulation, Cutonic copper. The oxychloride was at 2.25 kg ha^{-1} (1.13 kg Cu ha^{-1}) and the Cutonic supplied 62 g Cu ha^{-1}. The crops responded to copper on two of the loamy sands but not on any of the peaty soils. Copper uptake at harvest averaged 58 g Cu ha^{-1} and giving copper had little effect on uptake. These experiments confirmed the general conclusion that sugar beet is able to take most of the copper that it needs from soil reserves. Only tiny quantities are involved and these are often already present in soils where sugar beet is grown. McGrath and Loveland (1992) thought that less than 10% of UK soils were likely to be deficient.

Commercial Use of Copper

Initially $CuSO_4$ (Table 6.3) was used as a soil application at 20–50 kg ha^{-1} (5–12 kg Cu ha^{-1}). This is still a useful treatment with long lasting value. A foliar spray of the sulphate at a lower amount is also effective. Copper oxychloride is also used as a solid for soil treatment at 5–10 kg ha^{-1} or as a foliar spray (1 kg ha^{-1}). Cupric oxide (CuO, 80% Cu) can be applied effectively to the soil at 5 kg ha^{-1}.

In current farming practice sugar beet may receive a prophylactic spray of copper. This is often tank mixed with other chemicals so the copper must dissolve easily or be in liquid form. Commonly, specially formulated products (e.g. 'Cuprokylt', a solid form of copper oxychloride) may be sprayed once or twice at 0.25 kg ha^{-1} (0.1 kg Cu ha^{-1}) during spring or early summer. Use of chelated copper products is also widespread due to their convenience for tank mixing. In all cases some form of wetting agent has been desirable in field experiments described above.

ZINC

In Soils

Zinc neighbours copper in the periodic table of elements and their presence in soils and functions in plants are somewhat similar. One major difference between them is that zinc is present in soil simply as the divalent ion Zn^{2+}, in contrast to copper and some other micronutrient metals, which are in several valency states. Zinc must be taken up by sugar beet in very small but important quantities for the plant to function properly and for the crop to produce full yield.

Total amounts of zinc present in soils are of the order 10–300 mg kg^{-1} dry soil (Berrow and Burridge, 1980); this zinc originates mainly from parent materials. Other sources are atmospheric deposition, fertilizers, organic manures, sewage sludges, industrial wastes and zinc-based fungicides. Zinc is an important metal in many industrial processes, the effluents of which, if included in sewage, may carry toxic amounts to arable land (see Chapter 8).

Sugar beet takes up zinc as Zn^{2+}. Part of this is in water soluble and exchangeable forms but some is adsorbed on hydrous oxides, clays and soil organic matter. Thus, to obtain some measure of the amount of zinc available to sugar beet, several workers have used buffered chelating agents, as described below.

In Sugar Beet

Concentration in healthy leaves is about 20 mg Zn kg^{-1} and in roots 10 mg Zn kg^{-1}. Total uptake by leaves and roots is of the order of 0.2 g Zn ha^{-1} (Draycott, 1972). This places zinc between copper and manganese in concentration and uptake by sugar beet crops.

Zinc deficiency has occasionally been identified on some crops other than sugar beet on sandy soils with both high pH and phosphorus status in Ireland and Scotland, but has not so far been encountered in England. There are no reports of zinc deficiency in UK sugar beet.

Not only do soils contain natural supplies of the element, but also atmospheric deposition exceeds crop removal, even with a high-yielding crop. One application of 25 t farmyard manure ha^{-1} also contains many times the zinc taken up by sugar beet. There is little information, however, on the plant availability of zinc from these sources.

In other parts of the world, there are many reports of zinc deficiency on a wide range of crops. In some countries, sugar beet is known to be deficient on certain soil types. The limited amount of trial work with zinc is summarized below.

Forms of Zinc Used to Treat Crops

The most common zinc fertilizer is zinc sulphate (Table 6.3), used as a soil (20 kg ha^{-1} of the salt) or as a foliar (5 kg ha^{-1}) application. Zinc oxide is used as a soil application but is most effective when finely ground. More often, zinc is now applied as a chelate, some examples being zinc lignin polycarboxylate (7% Zn), zinc lignosulphonate (6% Zn) and zinc EDTA (14% Zn).

Field Experiments with Zinc

Boawn and Viets (1956) and Boawn *et al.* (1960a, 1961) reported one of the first comprehensive studies of zinc nutrition of sugar beet in the state of Washington, USA. The fine sandy loam soil was known to contain little available zinc, so applications of $ZnSO_4$ were tested. Without zinc tops contained 12 mg Zn kg^{-1} in dry matter, and 22 mg kg^{-1} with 18 kg $ZnSO_4$ ha^{-1}. Deficiency appeared when leaf blades contained less than 10 mg Zn kg^{-1} in June–August. Boawn and colleagues considered that a satisfactory concentration was 20 mg Zn kg^{-1} in mid-season. The concentration of zinc in roots was 7 mg kg^{-1} without zinc and 12 mg kg^{-1} with 18 kg Zn ha^{-1}.

Studies in both Michigan and the Red River Valley found very little likelihood of response to applied zinc. Judy *et al.* (1964) reported no effect on yield or quality of sugar beet roots on a site where yield response to applied zinc was evident on field beans. Lamb and Cattanach (1990) found no response to applied zinc in the Red River Valley on soils testing low in zinc and would have zinc recommended for some other crops. Referring back to Table 6.2, it should be noted that sugar beet has a medium rating in respect of response to zinc. Generally sugar beet does not respond to applied zinc on sites where maize and field beans will show deficiencies.

Christenson *et al.* (1992) in Michigan, USA, used 0.1 M HCl to extract zinc from mineral and organic soils. They produced a table of the amount of zinc to be applied depending on the soil test combined with soil pH. For the lowest category of soil zinc, the application advised was 1 at pH 6.6, 3.5 at pH 7.0 and 7 kg Zn ha^{-1} at pH 7.6, respectively. Finely ground ZnO was effective as a soil application, especially if banded near the seed. Alternatively, Christenson *et al.* (1992) suggested foliar sprays of the sulphate.

Bravo *et al.* (1992) in Colorado, USA, measured concentration of zinc in blades, petioles, crowns and roots of sugar beet throughout the growing season. Blades contained most, 25 mg kg^{-1} in May, falling to 9 in October. Roots contained 15 in May and 4 mg kg^{-1} in October. Total plant uptake at harvest was about 0.1 kg Zn ha^{-1}. This compares with 0.2–0.3 kg Zn ha^{-1} recorded by Boawn *et al.* (1960a) and 0.6 kg Zn ha^{-1} by Robertson and Lucas (1981).

In a recent report from Iran by Bakhsh Kelarestaghi *et al.* (2002), zinc sulphate on its own (at 40 kg ha^{-1}) and in combination with other nutrients (boron, magnesium, copper and iron) was tested as a soil application before sowing. Zinc alone gave the largest response, up to 10% increase in sugar yield. Iron also appeared to be increasing yield when applied as Fe-EDTA.

IRON

In Soil

Iron is by far the most abundant of the metals present in soil and is released from many minerals during weathering processes. Both ferric (Fe^{3+}) and ferrous (Fe^{2+}) forms are present in minerals and soils. Under reducing conditions (waterlogging) ferric iron tends to be converted to the ferrous form.

Soil colour is a good indicator of the oxidation–reduction that has taken place in the soil. Brighter browns and reds are suggestive of oxidized iron (Fe^{3+}), while dull greys and greens are the reduced form (Fe^{2+}). Colour is a useful indicator of the suitability of the soil environment for sugar beet root growth. Dark-coloured soils may indicate problems of permanent waterlogging of subsoils, accompanied by limitations with physical structure of both top and subsoils (Chapter 10).

These types of soils need artificial drainage, such as tile. Sugar beet is successfully produced on these types of soil in the Great Lakes region of the USA, where this type of artificial drainage has been installed. When production is attempted without the installed drainage, increased crop failure or low yields result.

Total iron present in soils is generally 100–1000 times that of the other metal micronutrients, copper, manganese and zinc. Values of 12,000–100,000 mg Fe kg^{-1} have been reported (Whitehead, 2000). Relatively large amounts are also deposited from the atmosphere annually (1 kg Fe ha^{-1} $year^{-1}$) and applied as impurities in fertilizers (particularly triple superphosphate) and organic manures. Thus, in comparison with the small amount needed in plant nutrition, there is unlikely to be a shortage of total iron in soil.

The concentration of iron in solution on well-drained soils is not sufficient to furnish iron needs via mass flow and/or diffusion (Lindsay, 1974). It would appear that iron deficiency should be more widespread than is observed. Organic matter contains materials that serve as chelating agents for iron and other metal ions. These indigenous substances are released from decaying organic matter or root exudates or synthesized by microbial activity. Hodgson *et al.* (1966), Geering *et al.* (1969) and Lindsay (1974) have demonstrated an important role of these chelates in the solubility of several micronutrients including iron. The concentration of iron maintained by these chelates appears to be sufficient when both mass flow and diffusion participate in the transport process.

Thus, it is only on a very few soils in some countries where iron-deficiency symptoms have been reported in sugar beet leaves and there is a paucity of clear experimental evidence that the crop needs treatment.

Response to Iron

In the greenhouse

Nagarajah and Ulrich (1966) reported one of the few investigations of the iron nutrition of sugar beet. The plants were grown in culture solution with 11 different amounts of iron (from 0.01 to 10 mg Fe l^{-1}). Plants were harvested when those in the smallest five of the 11 iron treatments showed iron-deficiency symptoms.

Leaf yield was decreased when the concentration of iron in the dry matter of the laminae was less than 55 mg kg^{-1}. The stage of maturity of the leaf sampled had little effect. Laminae ranging in symptoms from severe yellowing and necrosis to a light green contained from 20 to 50 mg Fe kg^{-1} in dry matter. Normal laminae contained from about 60 to 150 mg kg^{-1}.

Welkie and Miller (1989) also grew sugar beet plants in culture solution varying in strength of iron concentration from 0 to 2.0 mg Fe l^{-1}. Symptoms were pronounced with 0 and 0.02 mg Fe l^{-1}, and chlorophyll concentration in the apical leaves was markedly decreased, as was the leaf size and dry weight of leaves and roots. A solution concentration of 0.2 mg Fe l^{-1} removed most of the deficiency but 2.0 mg Fe l^{-1} gave the best results.

Winder and Nishio (1995) examined the effect of iron deficiency on stress response in sugar beet leaves grown hydroponically. Plants were grown with and without iron in a strictly controlled environment. A shortage of iron caused marked reduction in leaf chlorophyll and this decreased carbon dioxide fixation linearly. RNA synthesis was halved in iron-deficient leaves.

In the field

The influence of iron sulphate ($FeSO_4.7H_2O$) on the yield of sugar beet in Poland was investigated in 82 field experiments by Lachowski and Wesolowski (1964). At 40 kg $FeSO_4$ ha^{-1}, the yield of roots was increased by 10% and tops by 5%. Yield was increased on leached sandy soils containing less than 4.5 g Fe kg^{-1} in the plough layer.

Iron deficiency of soybean is widespread on alkaline soils in the north central region of the USA. Generally, soil application of inorganic salts produces unsatisfactory results and, while soil-applied iron chelate may be effective, cost limits its use. To improve the availability of $FeSO_4$, Goos *et al.* (2001) mixed the sulphate with various waste plant materials prior to mixing into the soil. Spent sugar beet molasses improved the chelate-extractable iron in soil compared with sulphate alone, and slightly decreased iron-deficiency symptoms.

In field-grown crops, opinions differ on the value of soil and/or plant analysis for iron in predicting or diagnosing deficiency. Archer (1980) did not think analysis of either soil or plants offered any help. In contrast, some soil-analysis laboratories extract available iron with chelate solution. Deficiency is expected with 0–5 mg Fe kg^{-1} and shortage is possible with 6–10 mg Fe kg^{-1}. For sugar beet, 1–3 kg Fe ha^{-1} is the suggested treatment with 0–5 mg Fe kg^{-1} soil and 0.2–1 kg Fe ha^{-1} with 6–10 mg Fe kg^{-1} (Natural Resource Management Laboratories, 1994).

Fertilizer sources of iron are listed in Table 6.3. Of the chelated sources available on the market, FeEDDHA appears to be the most effective. More discussion of this aspect is given in Chapter 11.

CHLORINE

Warburg and Lüttgens (1946) first showed that chlorine was necessary for the evolution of O_2 from isolated chloroplast fragments during photosynthesis. It was noted that the rate of photosynthesis was depressed by the absence of chlorine, but could be restored by the addition of the element. Arnon and Whatley (1949) considered it unlikely that chlorine was a cofactor for photosynthesis since chlorine had not been proved to be essential. However, Broyer *et al.* (1954) published a study confirming that chlorine met the criteria for essentiality. Ulrich and Ohki (1956b) demonstrated the essentiality of chlorine for sugar beet. Terry (1977) demonstrated that the principal effect of chlorine deficiency was to lower the cell multiplication rates in leaves. Using sugar beet as the test crop, Findenegg *et al.* (1989) noted huge growth reductions in chlorine-deficient plants.

Fortunately, the tiny amount needed by field-grown sugar beet is usually present in soil already. Part is derived from parent materials, but much more from atmospheric and agricultural inputs. In maritime climates where sugar beet is grown, atmospheric inputs are often large. Measurements in the UK show depositions near the coast of 100 kg Cl ha^{-1} $year^{-1}$ but at 200 km inland they may be 20 kg ha^{-1}. In the Midwest of the USA, only about 1 kg Cl ha^{-1} $year^{-1}$ may be deposited (Fixen, 1993). Fixen *et al.* (1987) showed a good probability of a yield response by wheat to applied chlorine when the extractable amount was less than 30 kg ha^{-1} in the surface 60 cm of soil.

Another source of huge amounts of chlorine is in the KCl fertilizer commonly used for sugar beet and other arable crops. An application of 100 kg K_2O ha^{-1} as KCl supplies 80 kg Cl ha^{-1}. Similarly, where NaCl is used for sugar beet, relatively large quantities of chlorine are supplied, far in excess of the plant's requirement.

Several workers have produced deficiency symptoms of chlorine in water culture (Ulrich and Hills, 1969; Terry, 1977) but never in the field. Deficiency symptoms appear as an interveinal chlorosis on the middle-aged leaves, somewhat similar to those of manganese deficiency (Ulrich and Hills, 1969). Deficiency leaves contained 0.01–0.04% and healthy leaves 0.8–8.5% Cl. Sauchelli (1969) reports that the leaves take on a mottled chlorosis, which becomes visible only in transmitted light; these affected areas later appear light green and are depressed.

MOLYBDENUM

Molybdenum is an essential element in plants and soils. It is a constituent of several plant enzymes involved in oxidation and reduction. Sugar beet must take up a few mg Mo kg^{-1} dry matter; otherwise deficiency symptoms, similar to those of nitrogen, appear on leaves.

The element is present in soils in minerals, as the molybdate ion (MoO_4^{2-}), adsorbed on to iron and aluminium oxides and in organic matter (Whitehead, 2000). Plants take up the element as the molybdate anion from soil solution. It is not easy to measure the quantity available because the amounts present are very small in most soils. The solubility of molybdenum in soils increases with increasing pH, probably explaining the infrequency of molybdenum deficiencies in the field.

There have been few investigations of the molybdenum nutrition of sugar beet. Ulrich and Hills (1969) described the symptoms of deficiency and found that healthy leaves contained 0.2–20.0 Mo kg^{-1} of dry matter, whereas deficient leaves contained 0.01–0.15 Mo kg^{-1}. Henkens and Smilde (1966) found increased yields in pot experiments from molybdenum given to sugar beet plants grown in molybdenum-deficient soil (previous sugar beet plants grown in the same soil showed distinct symptoms of molybdenum deficiency). Sodium molybdate ($Na_2MoO_4.2H_2O$ – 39.6% Mo) was tested and compared with several glassy frits (2–3% Mo). The amounts tested were between 0.06 and 0.8 kg Mo ha^{-1}.

Dry-matter production was increased considerably by molybdenum and there was a negative relationship between dry-matter yield and severity of molybdenum deficiency. Sodium molybdate gave the largest yield and plants receiving the equivalent of 0.6–0.8 kg Mo ha^{-1} had no symptoms of the deficiency. The frits varied in their ability to correct deficiency but had a greater residual effect than the molybdate, as shown by further cropping. A compilation of sources of molybdenum fertilizer is given in Table 6.3.

Nowicki (1969) investigated the effect of 2, 4 and 8 kg Mo ha^{-1} on yield, health and processing quality of sugar beet grown on acid, neutral and alkaline soils in Poland. It had no effect on yield but showed some tendency to decrease the incidence of disease and to increase the quality of the roots. More recently, Bravo *et al.* (1992) measured molybdenum concentration in leaf blades, petioles, crowns and roots at four harvest dates. It was considered necessary for leaf blades to contain 0.2 mg Mo kg^{-1} dry matter to avoid deficiency throughout the growing season.

SUMMARY OF MICRONUTRIENTS

Micronutrients are required in small amounts by plants, usually less than 1 kg ha^{-1}. However, this small quantity does not make these nutrients less important. In regard to the importance of an element, 'that the nutrient most limiting is the most important element for crop growth' would certainly apply in the case of micronutrients. Fertilization with micronutrients in deficiency situations usually increases yield. Conversely, none of the micronutrients appears to have an appreciable effect on per cent sugar or juice purity.

Of the seven micronutrients needed by sugar beet, boron and manganese deficiencies are more common than the other five. Consequently, there has been more research concerning these two nutrients. The solubility of both elements decreases with increasing pH. Since sugar beet grows best on near-neutral soils, it is not surprising that the incidence of deficiency is high for these nutrients.

Boron is a component of silicate minerals and, when released, may leach from the soil. In humid regions, boron does not accumulate appreciably in soil. Manganese availability in soil is controlled by soil pH and redox potential (measurement of degree of aeration). The reduced form of manganese (Mn^{2+}) is most available, while the oxidized form (Mn^{4+}) is essentially unavailable. Consequently, weather conditions can influence the availability of manganese.

Correction of boron deficiency may be with soil or foliar application. Boron may be broadcast prior to sowing or placed with the starter fertilizer near but not in contact with seed. Manganese is not effective when broadcast prior to sowing. Placement with the starter fertilizer is a means of effectively using this fertilizer. Foliar application is the only recourse if not applied with starter fertilizer and is the usual method of treatment in most countries.

Sugar beet is susceptible to iron deficiency. Similar to manganese, iron exists in soil in two valence states. The oxidized form (Fe^{3+}) is sparingly soluble, while the reduced form (Fe^{2+}) is considerably more soluble in soil. Soil or foliar application may be employed to correct this deficiency.

Copper, molybdenum and zinc deficiencies may occur on sugar beet, but the incidence is very low. Correction of deficiency may be by application of fertilizer to soil or by foliar sprays. In the case of molybdenum, seed application has been used for other crops but there are few studies of seed application for sugar beet.

There are no reports of chlorine deficiency in the field. In many areas, sufficient chlorine is deposited in precipitation to meet crop needs. The use of chlorine-containing fertilizers – KCl, NaCl and NH_4Cl – often adds sufficient nutrient to meet crop needs and in arid regions there is sufficient chlorine added from irrigation water.

7
Nutrient Deficiencies

Deficiencies of nutrients result in a range of symptoms, which are somewhat unique for each nutrient. Chlorosis (yellowing), mottling, stunting and deterioration of the growing point and leaf tissue are among the signs observed. Care needs to be exercised in the evaluation of symptoms, since weather conditions, chemical and insect damage and disease may mask, confuse or exacerbate symptoms. Identification of deficiency becomes increasingly difficult when there is more than one cause of abnormal growth or appearance.

The 16 elements generally considered essential to field crops are carbon, hydrogen, oxygen, nitrogen, phosphorus, potassium, calcium, magnesium, sulphur, boron, chlorine, copper, iron, manganese, molybdenum and zinc. However, an adequate supply of sodium is also necessary for sugar beet production, making it a 17th essential element. Carbon, hydrogen and oxygen comprise the largest concentration in the plant, but are not considered as mineral nutrients. Carbon and oxygen are absorbed from the air, while hydrogen is removed from the water molecule through metabolic processes in the plant. Specific deficiency symptoms are not recognized for these elements and there are no diagnostic tools to identify a deficiency.

The macronutrients for all plants are nitrogen, phosphorus, potassium, calcium, magnesium and sulphur. Sometimes calcium, magnesium and sulphur are regarded as secondary fertilizer nutrients. All are present in plants at concentrations greater than 500 mg kg^{-1} of dry plant tissue: henceforth the descriptor 'macro' (Anon., 2001). Micronutrients include boron, chlorine, copper, iron, manganese, molybdenum and zinc, and are usually present in concentrations less than 500 mg kg^{-1} dry plant tissue (Anon., 2001). The unique role of sodium in sugar beet nutrition is discussed in Chapter 4 but deficiency symptoms have not been seen. This element is considered to be a macronutrient for sugar beet.

All nutrient deficiencies decrease the growth of sugar beet, but not all deficiencies affect sugar production. When all other nutrients are present in sufficient quantities, nitrogen deficiency increases sugar concentration in the root. If deficiency appears early in the season, loss of root yield is not offset by increase in sugar concentration. However, when nitrogen deficiency appears between 4 and 6 weeks prior to harvest, the increase in sugar concentration usually compensates for any loss in root weight. The suggestion has been made that sulphur and sometimes phosphorus deficiencies may increase sugar concentration, but the research evidence supporting this contention is not strong. It does not seem advisable to allow these deficiencies to develop as a means of enhancing sugar concentration. The data indicate that deficiencies of all other nutrients tend to cause a decline in sugar yield. Any factor limiting top growth early in the season nearly always limits root weight as well as sugar production.

Plant Tissue Sample Preparation

Plant tissue analysis may assist in the identification process for any nutrient deficiency symptom. Samples should be taken from unaffected as well as affected areas of the field for comparison. It is important to select specific tissue when collecting samples for analysis. A guide for sampling, along with the critical nutrient range for nutrients, is given in Table 7.1.

Sugar beet leaves and petioles are usually contaminated with soil and dust particles, whether grown in the field or glasshouse. In the field, plant surfaces may also be coated with elements from foliar sprays. Pesticides, particularly fungicides, and foliar-applied nutrients leave residues on the plant surfaces. Removal of such contamination is essential for accurate assessment of the nutrition status of the plant. Sonneveld and van Dijk (1982) made a detailed study of methods of preparation of plant tissue for analysis. They recommend washing in a mild 2% (w/w) solution of a phosphorus-free detergent solution, followed by thorough rinsing with pure water.

The procedure must be done quickly on turgid tissue with minimum exposure to the washing solution to prevent loss of soluble elements from the plant tissue (particularly boron and potassium). Mills and Jones (1996) suggest that dust contamination can be detected by observing the aluminium, iron and silicon concentrations found in the tissue. When all three elements are > 100 mg kg^{-1}, dust contamination should be suspected. In the authors' experience,

Table 7.1. Plant part tested, constituent analysed, critical concentration and ranges where deficiencies are and are not expected to occur (adapted from Ulrich and Hills, 1969).

Nutrient	Plant part	Constituent analysed	Unit	Critical concentration	Deficiency symptoms: Yes	Deficiency symptoms: No
Nitrogen	Petiole	Nitrate	mg kg^{-1}	1000	70–100	350–35,000
	Blade	N	%	2.1	1.9–2.3	2.2–4.0
Phosphorus	Petiole	HPO_4	mg kg^{-1}	750	150–400	750–4,000
	Blade	HPO_4	mg kg^{-1}	–	250–700	1,000–8,000
	Blade	P	%	0.3	–	–
Potassium						
$>$ 1.5% Na	Petiole	K	%	1.0	0.2–0.6	1.0–11.0
	Blade	K	%	1.0	0.3–0.6	1.0–6.0
$<$ 1.5% Na	Petiole	K	%	[a]	0.5–2.0	2.5–9.0
	Blade	K	%	1.0	0.4–0.5	1.0–6.0
Calcium	Petiole	Ca	%	0.1	0.04–0.10	0.2–2.5
	Blade	Ca	%	0.5	0.1–0.4	0.4–1.5
Magnesium	Petiole	Mg	%	–	0.010–0.030	0.10–0.70
	Blade	Mg	%	–	0.025–0.050	0.10–2.50
Sulphur	Blade	SO_4-S	mg kg^{-1}	250	2–13	500–14,000
Boron	Blade	B	mg kg^{-1}	27	12–40	35–200
Chlorine	Petiole	Cl	mg kg^{-1}	0.4	0.01–0.04	0.8–8.5
Copper	Blade	Cu	mg kg^{-1}	–	–	–
Iron	Blade	Fe	mg kg^{-1}	55	20–55	60–140
Manganese	Blade	Mn	mg kg^{-1}	10	4–20	25–360
Molybdenum	Blade	Mo	mg kg^{-1}	–	0.01–0.15	0.20–20.0
Zinc	Blade	Zn	mg kg^{-1}	9	2–13	10–80

[a]Critical value not useful when the petioles contain less than 1.5% Na.

manganese values will also be inordinately high with dust contamination. Cherney and Robinson (1982) suggest that titanium (Ti) concentration may be used as an indicator of soil or dust contamination. They found that nitric perchloric acid and sulphuric selenous acid adequately extracted titanium. Neither nitric acid nor dry-ash procedures extracted titanium satisfactorily. However, analysis for titanium is readily included, using plasma-emission spectrometry. Any added cost would be worthwhile, given the nature of the decisions that will be made from the data.

Nutrient Deficiencies

Separation of the various symptoms may be aided by the following categories:

Uniform yellowing	Nitrogen, sulphur, molybdenum
Stunted greening	Phosphorus
Leaf scorch	Potassium, magnesium
Growing-point damage	Boron, calcium
Yellowing with green veining	Manganese, iron, chlorine, copper, zinc

Nitrogen

Since nitrogen is a mobile element in the plant, deficiency symptoms appear on the older leaves. The initial symptom is a general light-green colour (yellowing) and often the leaves are smaller in size than normal (Plate 1). A series of increasingly nitrogen-deficient leaves is shown in Plate 2.

Leaves of nitrogen-deficient plants grow horizontally from the crown (Plate 3). In addition to horizontal growth, the leaves take on a long, narrow shape (Plate 4). In many cases, the new leaves in the centre of the plant are smaller in size and dark green in colour. Chlorosis continues to develop as the plant grows.

The continued development of symptoms depends on nitrogen supply. In cases of little additional supply, the leaves continue to turn yellow, wither and die. An additional supply regenerates leaves in the crown. New leaves are small and dark green and they grow horizontally. Increased supply, but insufficient to completely correct deficiency, results in new leaves growing larger, covering the older leaves and filling the space between the rows. Detection of deficiency in these cases is more difficult. Frequent scouting of fields alerts the grower to any problems in sufficient time to take corrective action.

Leaf symptoms of nitrogen deficiency may be confused with virus yellows. This disease exhibits a patchy yellowing on the individual leaves, as compared with the general yellowing of nitrogen deficiency. Nevins and Loomis (1970) have shown that nitrogen deficiency decreases chlorophyll concentration and the rate of photosynthesis in the old but not in the new leaves. General conditions for the development of nitrogen deficiency are described in Chapter 2.

Sugar beet in soils with little organic matter and in coarse sandy soils is susceptible to inadequate nitrogen supply. Excessive rainfall after nitrogen application and inadequate supply from fertilizers are also conditions contributing to nitrogen deficiency. Often deficiency is in patches in the field, due to differences in mineral nitrogen supply brought on by variation in the soil (Draycott, 1972). The effects of inadequate water supply, soil compaction and pest damage are described in Chapter 12.

Phosphorus

Phosphorus deficiency is probably the most difficult disorder to diagnose in the field. The main symptoms of phosphorus deficiency are a reduction in leaf size, gradual development of a deep green colour followed by red or purple coloration of the leaves and stunting due to slow growth. The leaves may develop a metallic lustre from greyish green to blue-green.

The development of deficiency in seedlings often shows yellow cotyledons with the first pair of leaves green (Plate 5). Stunting of the plant is also a symptom, as shown with the eight-leaf plant in Plate 6. Comparison of the eight- to ten-leaf plants in Plate 7 – one is phosphorus-deficient and one has an adequate supply – emphasizes this point. A leaf-colour series (Plate 8) shows a range of symptoms, including purple coloration, brown veining and smaller leaf size.

In long-term experiments where phosphorus fertilizer has been withheld for many years, leaves have a purple coloration. However, development of purple colouring may not be a good indicator of phosphorus deficiency in sugar beet (Ulrich and Hills, 1969). The formation of a group of coloured compounds known as anthocyanins cause this coloration in the leaves of plants. Any environmental factor that favours an increase in sugar content in the plant tissue will promote the formation of anthocyanins. Among the factors that will increase sugars in plant tissue are low temperature, drought or low nitrogen supply. Residues of herbicide may also cause purple coloration.

Potassium

One of the earliest reports described initial potassium deficiency as a 'scorch' on the margins of the leaf blade (Hale *et al.*, 1946). The scorch may be described as a leathery tan colour. Initially, symptoms show leaf-margin scorching accompanying a smooth leaf surface (Plate 9). More severe conditions cause the symptom to extend all the way to the midrib, but not involving the veins of the leaf (Plate 10). The youngest leaves may be hooded, similar to those with calcium deficiency (Plate 11). The plant shown here was grown without sodium. When sodium is adequately supplied, the severity of the potassium deficiency is markedly reduced (Plate 12).

Draycott (1972) observed:

> Scorch symptoms develop after chlorosis or may develop independently of it. Necrosis generally follows, both forming an unbroken border around the leaf and lobes between the veins. The necrotic tissue is dull or reddish brown; it is tough but soft to handle and it does not crumble or disintegrate like the necrosis associated with magnesium deficiency.

Cook and Millar (1953) reported that the scorch moves toward the centre of the leaf. It primarily affects the recently mature leaves on the plant. The centre leaves remain green and may have a hooded appearance. Dark, longitudinal lesions may form on petioles of the older leaves (Ulrich and Hills, 1969).

Potassium deficiency is most likely to develop on sandy, high organic matter and organic soils. Where the coarse fraction of sandy soils is formed from quartzite, supply will be less than if the fraction is formed from the potassium feldspar, arkose. Dry conditions, cool temperatures and poorly aerated or compacted soils limit uptake.

Magnesium

Magnesium deficiency symptoms on sugar beet are unique and not easily mistaken for any other deficiency. Deficiency first develops as small, pale yellow areas near the outside margins of recently mature leaves (Plate 13). The youngest mature leaves are affected. These lemon-yellow areas spread between the veins towards the midrib. Necrosis starts near the edge of the leaf and gradually expands to include most of the interveinal tissue except for the triangular area near the base of the leaf (Plate 14). Leaves in the final stages of magnesium deficiency show black necrotic tissue between the veins, followed by brown necrosis (Plate 15). Late-season deficiency often shows a 'washing out' of necrotic tissue resulting in holes in the leaves (Plate 16). Magnesium deficiency (as well as potassium) should not be mistaken for virus yellows. The main difference between magnesium deficiency and virus yellows is that the latter shows vein clearing or vein yellowing in younger leaves of the plant. Older leaves on plants with virus yellows have a characteristic thickened and brittle feel.

Symptoms of magnesium deficiency usually appear in July and August, but may not be seen until September (Draycott, 1972). Coarse-textured soils, particularly those with subsoils as coarse as or coarser than the surface soil, are most prone to magnesium deficiency. These coarse-textured soils, when heavily fertilized with potassium, may show magnesium deficiencies since potassium suppresses magnesium uptake. Deficiencies are likely when soil-test potassium plus fertilizer potassium (kg ha^{-1}) is more than three times greater than soil-test magnesium (Christenson *et al.*, 1992).

Calcium

Calcium deficiency affects the growing point of sugar beet. As a result, most of the symptoms are in the crown and on new leaves. Ulrich and Hills (1969) grew sugar beet in solution culture to develop deficiency symptoms. Key symptoms are shown in Plates 17–20. Centre leaves have severe tip burn and the blades of the older leaves are crinkled and deformed and fail to expand (Plates 17 and 18). Young leaf blades are also hooded (downward cupping) and have a crinkled appearance (Plate 19). As deficiency progresses, the leaf blades may be reduced in size to a mere stub of blackened tissue at the end of the petiole (Plate 17).

A range of symptoms on leaf blades from a deformed leaf to a black-tipped stub, are shown in Plate 20. 'Tip burn' symptoms are seen on plants in fields where plant growth is vigorous. They usually disappear with a decrease in the supply of nitrogen from the soil or a change of climate or growth stage.

When calcium deficiency remains chronic, the growing point is often permanently damaged and lateral shoots develop abnormally. Such a condition can produce damage to the conducting system and cambium tissue of the storage root. This results in concentric rings of darkened tissue when the storage root is cut in cross-section. This may be followed by a permanent wilting of the top and then death of the plant. A low supply causes poor root development and, in severe cases, the roots fail to grow.

Deficiency as described above is seldom seen, as most mineral soils are well supplied with available calcium (Mengel and Kirkby, 1978). The supply of calcium is usually adequate in soils that are limed and where sugar beet thrives best. Mass flow carries more calcium to the root surface than is needed by the growing sugar beet in most soils with pH values above 5.5. An exception is where sea-water flooding has left a large residue of sodium, which antagonizes the uptake of calcium.

Sulphur

Ulrich and Hills (1969) and more recently Connors (2000) have described sulphur deficiency symptoms for sugar beet. The initial appearance of sulphur deficiency is very similar to that of nitrogen deficiency. There is a general yellowing across the entire leaf with both nutrients. There is no veining with either nitrogen or sulphur deficiency. When sulphur is in short supply, the leaves of the entire plant change gradually from green to light green and then to light yellow with a faint tinge of green remaining (Plate 21). Since sulphur is not as mobile as nitrogen, the new centre leaves in sulphur-deficient plants become light green to yellow, rather than dark green as in nitrogen deficiency. Leaves remain erect, the petioles and blades are brittle, breaking readily when compressed by the hand. A progression of increasing severity of sulphur deficiency symptoms is shown in Plate 22. With severe sulphur deficiency the petioles develop brown, longitudinal lesions. In the field there is a general yellowing in patches, similar to nitrogen deficiency (Plate 23).

Connors (2000) suggests that a slow crop growth response to applied nitrogen should arouse suspicions of sulphur deficiency. Plant tissue analysis should be used to further confirm the problem. Sulphur deficiency is more pronounced with large nitrogen inputs and is accompanied by a wide nitrogen : sulphur ratio ($> 17:1$) in young tissue. Organic matter is the primary source of sulphur from soils. Hence, sulphur deficiency is more likely on sandy soils with low organic matter (see Chapter 3).

Boron

Boron deficiency was thought to be a disease and was referred to as heart rot and dry rot. Brandenburg (1931) first showed that boron deficiency was the cause of these two 'rot' conditions. Heart rot is the term applied when the growing point blackens and dies while dry rot describes the symptoms on the tap root that usually appears after heart rot is observed.

One of the best descriptions of the anatomical effects of boron deficiency on the root crop is that of Rowe (1936). She found that the apical meristem of the shoot, the youngest leaves and the newly developed cambia were the most sensitive to boron deficiency and these were the first to degenerate. Cells of the vascular rings in the process of differentiating and sporadic groups of parenchyma cells adjacent to conducting elements were also sensitive to deficiency. Later stages of the deficiency were characterized by decay of cambial and adjacent parenchyma cells, together with complete disintegration of the phloem. The root tip did not degenerate but merely ceased to grow. A concentration of 0.17 mg kg^{-1} in the culture solution was enough for normal growth and development. Recovery in boron-starved plants involved the activation of axillary buds at the top of the beet, each of which developed its own system of secondary vascular rings.

Boron deficiency can be identified without supporting soil and plant tissue analysis because of unique symptoms on the leaves, petioles, crowns and roots. Symptoms may appear on either or both the roots and aboveground growth. Boron is relatively immobile in plants and symptoms appear on growing points and in meristematic tissues. Boron deficiency not only decreases yield but also damages the roots, decreasing their value and keeping qualities.

One of the first symptoms of boron deficiency is a blackening/dying of the growing-point/centre of the crown (Plate 24). Mature leaves may have netting and cracking symptoms (Plate 25). The most prominent symptoms in a field-grown plant include blackening of the centre of the crown and proliferation of small leaves, along with a prostrate habit, and cracked and corky areas on the petioles (Plate 26). Conductive tissue in the roots deteriorates, showing heart and crown rot (Plate 27). Plants may appear wilted even when grown in moist soil or in aerated culture solution. Young leaves wilt the most, instead of the least, as they do in true water shortage. In time, they collapse and fail to develop.

Ulrich and Hills (1969) report that boron deficiency appears to have a twofold effect: (i) the conducting tissue is damaged, causing wilting and occasionally the exudation of syrupy materials from the leaf blades; and (ii) the meristematic tissues of the growing-point collapse and die. A further effect is on root growth, as fibrous roots fail to develop in solutions deficient in boron. This may be the primary cause of wilting in the tops. Cambium tissues of storage roots darken, cells collapse, growth stops and the storage root become subject to decay.

Deficiencies are associated with alkaline soils. The symptoms are often greater on sandy soils than on finer-textured soils. Dry weather will promote the appearance of deficiency (see Chapter 6). As affected plants age, many leaves die and fall off. If growing conditions improve, plants usually begin to grow again from secondary growing points. Tissue round the shoulder of the taproot also begins to decay in advanced stages of deficiency. Such roots are invaded by fungi and begin to rot. Hull (1960) described this secondary attack and some of the fungi involved.

Stoker and Tolman (1941) working in Oregon, USA, first showed the need for boron to ensure normal development of the seed stocks. The main flowering stem is stunted and the growing point dies. Similarly, if other stems shoot, although they are taller than the first one, their growing points also die. The growing points of laterals along these stems also give rise to stunted growths, which appear as small rosettes of discoloured bracts, and eventually these growing points also die.

Manganese

Field experiments testing the response to applications of manganese (whether to soil or plant) generally conclude that only where symptoms would otherwise appear on leaves without treatment does the crop respond. Thus, in the absence of deficiency symptoms, it can be assumed that the crop has a satisfactory supply. This is of great practical value, perhaps more so than soil or plant tissue analysis, because the symptoms are unique and easily recognized.

Manganese deficiency, known as 'speckled yellows', is characteristically interveinal, not affecting the veins (Plate 28). Further development of the speckling results in a translucent nature to the spots (Plate 29). The yellow spots become necrotic and holes develop as a result of loss of tissue (Plate 30). In the field, leaves have a characteristic upright posture due to the petioles growing nearly vertically and the laminae rolling inwards (Plate 31).

Symptoms appear most commonly on plants from May onwards and disappear in August, although they may appear and disappear at any time. The severity of symptoms fluctuates during this period but, on average, declines from June to September.

Deficiencies are usually found on neutral and alkaline soils, which are usually dark at the surface and have a grey subsoil colour. Organic soils and dark-coloured sandy loam and loam soils are also very prone to manganese deficiency. Deficiencies usually occur on these soils at pH values as low as 5.8, while on other mineral soils they occur at pH values above 6.5 (Vitosh *et al.*, 1994). In glaciated regions, the deficiency is seldom found on soils formed from glacial till and moraine materials. Manganese deficiency may be exacerbated by soils with elevated iron, copper and zinc concentrations. Dry weather, low light intensity and low soil temperature will also promote the deficiency (Lucas and Knezek, 1972).

Iron

Greenhouse

Hewitt (1953) found that sugar beet plants in sand culture given iron as citrate or magnetite grew normally. Without iron, plants developed interveinal mottling in young leaves, followed by acute chlorosis and interveinal necrosis. Nagarajah and Ulrich (1966) found that plants in culture solution without iron developed chlorosis first on the young leaves in the heart of the plant. Severely affected leaves became completely bleached and developed necrotic spots. On recovery, such leaves formed a network of prominent green veins, which often characterizes iron deficiency in the field.

When plants are grown in water culture, iron deficiency symptoms may be caused by excessive concentrations of manganese salts. Hewitt (1948) confirmed this effect with sugar beet plants but found that metals other than manganese also caused iron deficiency. He showed that the toxic effects of excess manganese could be readily distinguished from true iron deficiency by the nature of the symptoms.

Field

In the UK, iron deficiency symptoms occur sporadically, usually in May or June, on sandy calcareous soils (Draycott, 1972). Iron chlorosis has not been observed under field conditions in the USA. The deficiency would be expected to occur on both calcareous and non-calcareous soils. Nagarajah and Ulrich (1966) showed that iron-deficient beets increased the acidity of culture solutions and released riboflavin from the roots. They suggested this as a mechanism that allows it to remove iron where other crops may not. Cool, wet weather may accentuate chlorosis and the deficiency may be aggravated by irrigation with water containing a high concentration of HCO_3. Iron deficiency is not widespread in the USA and is rarely seen in the UK.

Ulrich and Hills (1969) reported that symptoms appear very quickly when young seedlings are transferred to iron-free solutions, or when iron is withheld from older plants (Plate 32). The veins remain green, standing out against the yellow interveinal areas (Plate 33). An iron-deficient plant stands out in the field compared with plants without iron chlorosis (Plate 34). Eventually, the bleached blades become necrotic, which causes them to cup upward. If iron is reabsorbed before the blade tissues become permanently damaged, the fine veins become green and prominently netted (Plate 35). This symptom is often assumed to be associated with iron deficiency but is actually associated with recovery from iron deficiency.

Zinc

In its early stages, deficiency appears as a light-green coloration of the larger leaves near the centre of the plant (Plate 36). Small pits develop between the veins on the upper surface of the blades as chlorosis becomes more intense (Plate 37). The small pits enlarge in an irregular pattern as more tissue collapses. The entire area between the veins gradually becomes dry, leaving the primary veins prominently outlined, turgid and green (Plate 38). Plate 39 shows a whole plant exhibiting zinc deficiency. The light-green colour of the larger leaves is in the centre of the plant, along with the development of chlorosis and necrosis. The entire leaf blade becomes necrotic except the main veins. The petioles exhibit an upright growth habit in advanced stages of deficiency.

Sugar beet seedlings require zinc from an external medium almost immediately after germination (Ulrich and Hills, 1969). Seedlings transferred to zinc-free solution will develop deficiency symptoms even before the cotyledons have been fully formed or the first true leaves appear. Apparently, the zinc supply from the seed is just sufficient for germination and starting growth, but no more. Thereafter, the seedling must absorb zinc from the soil if zinc deficiency is to be avoided.

Zinc deficiencies appear on alkaline soils, with both calcareous and non-calcareous parent materials. Often deficiencies appear when the subsoil is exposed as in where tile drainage has been installed. High concentrations of phosphorus in plants have been shown to restrict zinc movement within the plant, resulting in accumulation in the roots and deficiency in the top. However, sugar beet appears to be a very good forager for zinc, and deficiency symptoms are not very common. Field bean (*Phaseolus vulgaris*) and maize (*Zea mays*) may show zinc deficiency while sugar beet will not when grown on the same soil (see also Chapter 6).

Copper

Van Schreven (1936) first described copper deficiency symptoms on sugar beet after growing plants in purified nutrient solutions. Symptoms developed after the plants had grown without copper for 3 weeks. Ulrich and Hills (1969) repeated this work and were able to produce symptoms only after purification of salts and water, confirming that sugar beet needed only small amounts of copper. They showed that deficiency develops as a mild chlorosis of the young, centre leaves similar to that of iron, chlorine and manganese deficiency (Plate 40). The symptoms progress from this mild chlorosis to a fine, green, netted veining, contrasting with light-yellow areas (Plate 41). Another view of the green veining is shown in Plate 42. Further development of copper deficiency is a bleached appearance of the leaf blade (Plate 43). This bleaching differs from the spotted necrosis that results from iron deficiency, the black spotting of manganese deficiency or the raised veining of chlorine deficiency. In contrast, nitrogen, sulphur and molybdenum deficiency tends to cause an overall yellowing of the plant.

Copper deficiencies on sugar beet are generally not observed in the field. Such deficiencies would first be expected on organic soils and sandy soils derived from quartzite. The solubility of copper decreases with increasing pH, reducing its availability (Chapter 6). In his early work, van Schreven (1936) reported that some older leaves were a blue-green colour. Root development was also retarded by copper deficiency and yields were much decreased, as was the sugar percentage. Another characteristic was that the necrotic areas on old leaves were greyish brown or greyish white and the dead leaves looked bleached.

Hull (1960) and Pizer *et al.* (1966) found no evidence of copper deficiency symptoms in sugar beet crops in England. Even in the trials on copper deficient soils, deficiency symptoms on leaves have not been reported. Ulrich and Hills (1969) also reported that copper deficiency on sugar beet has not been seen in the USA. In the intervening years, no other reports were found pertaining to incidences of copper deficiency on the sugar beet crop.

Chlorine

Several workers have produced deficiency symptoms of chlorine in water culture (Ulrich and Hills, 1969; Terry, 1977). However, there are no reports of the deficiency being observed in the field. Ulrich and Hills described the symptoms first as a chlorosis on the blades of the younger leaves near the centre of the plant (Plate 44). The interveinal areas of the leaf blades become light green to yellow, with the main veins remaining green and raised (Plate 45). When viewed against bright light, the leaf blade shows a netted mosaic pattern, branching out from the main veins. Early phases of this mosaic pattern are reminiscent of manganese deficiency. Sauchelli (1969) reports that the leaves take on a mottled chlorosis, which becomes visible only in transmitted light; these affected areas later appear light green and are depressed. As the symptoms develop, the interveinal areas appear as flat, yellow-green depressions, which become dry and are in sharp contrast to the adjacent area of 'raised' green veins. These advanced symptoms of chlorine deficiency are unique and are clearly distinguishable from other nutrient deficiencies (Plate 46).

As mentioned above, deficiencies of chlorine on sugar beet are unknown in the field. Precipitation, irrigation water, air pollutants, fertilizers and animal wastes provide sufficient quantities. Unirrigated, sandy soils located inland, away from the influences of chlorine cycling, and fertilized only with nitrogen are sites most likely to exhibit this deficiency. The main effect of chlorine on cereals in western USA is to reduce root and leaf disease infestation (Christensen *et al.*, 1981).

Molybdenum

Symptoms of deficiency are similar to those of both sulphur and nitrogen. As deficiency progresses, uniform yellowing with slight veining begins to develop into necrotic spotting (Plates 47 and 48). Pronounced necrotic spotting then develops along the veins (Plate 49). This pitting differs from the black sheen and spotting of manganese deficiency and the coalescing of the spots of zinc and sulphur deficiency. Petioles of deficient plants usually contain a large concentration of nitrate. While molybdenum deficiency has been reported in Europe, the problem is not widespread. This is related in part to greater availability of molybdenum in the slightly acid, neutral and alkaline soils in which most sugar beet is produced in the USA.

8
Organic Manures, Green Manuring and Organic Production

Organic materials applied to soil include animal manures, sewage sludge, green-manure crops and other organic residues. All provide varying quantities of plant nutrients for sugar beet. Poultry manure and animal slurries are particularly rich in nutrients and must be taken into account in sugar beet nutrition. In addition to being mixed with straw and other bedding materials, manure is stored and handled in different ways, which may affect the nutrient content, particularly if composted.

In the USA and EU, sludge from sewage-treatment plants is often referred to as biosolids. The term is reserved for such materials and does not apply to other organic materials. Special regulations govern the application of biosolids due to the potential for pathogens and heavy-metal pollutants entering the food-chain. Much is used in the EU and USA before sowing sugar beet and is a rich source of nutrients and organic matter.

Green-manure crops may be either non-leguminous or leguminous species grown and ploughed down prior to sowing sugar beet. These crops serve to improve soil tilth, cover the soil over winter months (reducing erosion) and decrease leaching losses of nitrate nitrogen. They are considered briefly here, and in relation to set-aside requirements in the EU in the next chapter.

Other organic residues include cereal-grain straw, by-products of sugar beet processing (e.g. vinasse), sawdust and waste from food-processing plants. Waste material from chemical plants may also be applied to the soil. Care needs to be exercised when applying these if they contain heavy metals or other potentially toxic materials.

On a worldwide basis, accurate estimates of the percentage of sugar beet area treated with organic additions (animal manure, biosolids and green manure) are very difficult to obtain, but the amount of land treated with organics (animal manure in particular) has probably declined over the past 50 years. Crop production has become more specialized in certain areas and animal production in others.

In a survey in 2001 of 12 member countries of the Institut International de Recherches Betteravières (IIRB) we found that about a third of the sugar beet area received some form of organic manure. Strawy farmyard manure from cattle and pigs was most commonly applied for sugar beet, at about 20–40 t ha^{-1}. Liquid manures (slurry) from pigs and cattle were regularly used in some countries in Europe, often in large quantities. Highly nutritious solid poultry manures (usually at 8–10 t ha^{-1}) and sewage sludge in its various forms were locally important and provided much of the nutrient requirement of sugar beet.

When land is used for animal-manure application without accounting for the nutrients, there are also risks to the environment. Excess nitrogen associated with the practice may leach into groundwater supplies and EU directives are in place to counteract potential problems. Available phosphorus, as measured by chemical extraction, may increase to very high concentrations. Research in the USA and EU suggests downward movement when such concentrations are reached. This has led to 'best management practice' guidelines, formulated in Michigan, giving protection from litigation to growers utilizing manure. When the soil-test concentration increases above a certain value the guidelines suggest that no additional phosphorus be applied either from organic additions or from commercial fertilizer.

Draycott (1972) suggested that the farming community seemed to consider organic manures mainly as soil conditioners, but little account was taken of their nutrient content. Boyd (1959) found that the amount of mineral fertilizer used on sugar beet fields was the same with and without farmyard manure (Table 8.1). The situation has changed only slightly in the 40 years since the survey by Boyd. Chambers and Smith (1999) reported that the average fertilizer allowance for manure was 18 kg N, 18 kg P_2O_5 and 12 kg K_2O ha^{-1}. Considering that amounts up to 250 kg ha^{-1} total N from farmyard manure are often used, there is a large potential for greater credit being given for nutrients contained therein. New regulations being introduced in the UK (DEFRA, 2001a) will ensure that this is taken into account from 2003 onwards.

Table 8.1. Inorganic-fertilizer allowances for nutrients in organic manure made in the UK from surveys in the 1950s (Boyd, 1959) and 1990s (Chambers and Smith, 1999).

	N	P_2O_5	K_2O
		kg ha^{-1}	
1950s	0	2	6
1990s	18	18	12

Short-term experiments have investigated the mineral-fertilizer equivalent of animal manure, along with the optimal rates of mineral fertilizers for sugar beet when farmyard manure is applied. Studies where improvements of the physical properties of soil are recorded require long-term experiments. Most stress the importance of maintaining or improving soil organic matter if sugar beet is to perform well, as shown in Chapter 9.

Animal Manures

General studies on the use of animal manure

The application of animal manure for crop production was recorded as early as 900 BC and substantial yield increases were attributed to the practice. Much of the early writing described farming practices but it was not until the late 18th century that progress was made with research into the chemical and physical properties of manures applied to the soil.

According to Boyd (1959), numerous experiments with farmyard manure were conducted between 1890 and 1910. The numbers declined until the 1930s when a renewed interest developed. For the next 30 years there was an increase in the number of such experiments and published reports, particularly with sugar beet.

Nuckols (1942) reported work done in Nebraska, where manure was applied and sugar beet grown for 4 consecutive years. No differences were noted between cattle, sheep and horse manure applied at equivalent rates of dry matter. Maximum production was obtained with 27 t ha^{-1} (13.5 t ha^{-1} dry matter) of manure. No additional yield was obtained with 40 or 54 t ha^{-1} of wet manure. In another study, Hill (1946), working in Alberta, Canada, found that addition of farmyard manure increased yield and did not suppress sugar percentage in sugar beet. Even though phosphate fertilizer was added in some treatments, there was no systematic approach to determining the value of manure for supplying nutrients to the sugar beet crop.

Studies with poultry manure have been fewer in number than with farmyard manure. Manure from turkeys appears to mineralize on into the second and subsequent seasons after application. Malzer and Graff (1995) reported that the amount of nitrate leached in the second year after application was greater than in the first after application. Following up on this work, Lamb *et al.* (2001) conducted an experiment on pig and turkey manure for sugar beet production, comparing both a residual and a current effect. Preliminary results indicate that turkey manure caused excess carry-over of nitrogen to the second year after application. They also suggested that turkey manure was not as detrimental to sugar yield as had been initially thought. Giardini *et al.* (1992) and Pimpini *et al.* (1992) studied the effect of poultry manure and fertilizer treatments on yield and recoverable sugar. Root yields were increased by all rates of poultry manure, fertilizer and combinations. Recoverable sugar was negatively affected by all treatments except a small amount of poultry manure applied at 2 t ha^{-1}.

When farmyard manure is applied on a regular basis over a period of time, nutrient reserves may increase, reducing the amount of mineral fertilizer needed. Halvorson and Hartman (1975) evaluated the effect of repeated application of farmyard manure and mineral fertilizers on sugar beet grown on an irrigated site in eastern Montana, USA. Treatments were applied nine times over an 18-year period, with sugar beet grown as an indicator crop in the 19th year. Manure applied at 22.4 t ha^{-1} produced the highest yield of sugar (7.9 t ha^{-1}), while 112 kg N ha^{-1} from ammonium nitrate produced a similar yield (7.6 t ha^{-1}). They suggested that farmyard manure could be used successfully as a source of nitrogen for sugar beet production, greatly decreasing the need for fertilizer.

Eck *et al.* (1990) evaluated the residual effects of beef-feedlot waste (manure and soil) on sugar beet production. The study was initiated in 1969 and feedlot waste was applied through the 1984 season. Grain sorghum (10 years), maize (3 years) and wheat (3 years) were grown in the 16 years. The experimental area was fallowed in 1985 and sugar beet grown in 1986. Supplemental nitrogen on the sugar beet increased yield only on the control treatment (Table 8.2). Manure treatments all yielded more than the fertilizer treatment. It is interesting to note that a large application of feedlot waste supplied sufficient nutrients for sugar beet even though those treatments were applied in the first 3 years of the study.

Fertilizer equivalents

Most of the studies where the fertilizer equivalent of farmyard manure was assessed were conducted in the UK and reported between 1941 and 1969. Boyd (1959) suggested quite large values for the availability of nitrogen, phosphorus and potassium from animal manure, at 3.7 kg N, 5.0 kg P_2O_5 and 9.4 kg K_2O t^{-1} (Table 8.3). Values suggested by Crowther and Yates (1941), Grimes (1959), Adams (1962) and Draycott (1969) range from 1.3 to 2.0 kg N, 1.7 to 2.0 kg P_2O_5 and 3 kg K_2O t^{-1} of manure. Halvorson and Hartman (1975) reported a somewhat larger value for nitrogen.

Table 8.2. Long-term effects of fertilizer and feedlot waste on the nitrogen requirement of sugar beet (after Eck *et al.*, 1990).

	kg N ha^{-1}	
	0	134
	t sugar ha^{-1}	
Control, no fertilizer, no feedlot waste	7.4	8.6
NPK fertilizer	9.2	9.8
Feedlot waste	10.5	10.8
Feedlot waste plus fertilizer	10.6	10.7

Table 8.3. Nutrient equivalents of farmyard manure for sugar beet.

N	P_2O_5	K_2O	
	kg t^{-1}		References
–	1.7	3.0	Crowther and Yates, 1941
3.7	5.0	9.4	Boyd, 1959
2.0	2.0	–	Grimes, 1959
1.7	1.7	6.7	Adams, 1962
1.3	–	> 4.0	Draycott, 1969
5.0	–	–	Halvorson and Hartman, 1975
3.1	1.4	5.3	Widdowson and Penny, 1979
2.8	2.4	5.7	Average values

Optimal fertilizer with farmyard manure

A number of studies are summarized in Table 8.4. While the amount of nitrogen is fairly consistent, the amount of P_2O_5 and K_2O varies considerably. Among the factors that may explain this difference is the concentration of phosphorus and potassium in the soil. Soil texture also plays a role, the coarser-textured soils generally having a lesser cation supply than finer-textured soils.

In an early report, Boyd (1961) suggested that the effects of farmyard manure were independent of any mineral fertilizer applied. He showed a similar requirement for nitrogen with and without manure. However, Draycott (1969) showed that, where phosphorus and potassium were applied, farmyard manure decreased the nitrogen-fertilizer requirement of sugar beet by about 50 kg N ha^{-1}. In the work by Boyd (1961) no additional phosphorus and potassium were given. Response to nitrogen in manure is different when phosphorus and potassium are limiting than when supplied in adequate amounts.

Other methods for assessing nutrient availability from manure

Evaluation of the fertilizer equivalent of animal manure is difficult because conditions are confounded by the addition of several nutrients contained in the manure, favourable effects from the added organic material and other beneficial effects not completely characterized. In the conduct of research with manure, soils need to be characterized in respect of available nutrients so that correct

Table 8.4. Amount of fertilizer for optimum sugar yield, with and without farmyard manure (FYM).

		Optimum fertilizer			
	Amount	N	P_2O_5	K_2O	
FYM	(t ha^{-1})		kg ha^{-1}		Reference
With	22	0	0	0	Patterson and Watson, 1960
Without	–	112	94	188	
With	12	75	0	100	Adams, 1962
Without	–	125	63	200	
With	30	75	38	63	Draycott, 1969
Without	–	150	–	–	
With	22[a]	0	–	–	Halvorson and Hartman, 1975
Without	–	112	–	–	

[a]22 t ha^{-1} every other year for a total of nine applications.

amounts of fertilizer may be applied for comparison. Manure should be analysed for available nutrients prior to application in order to have a basis for comparison.

The general approach has been (and still is in some cases) to base the nutrient needs on a soil test and then subtract the amount of nutrients in manure from the fertilizer recommendation. Christenson *et al.* (1972) made fertilizer recommendations based on soil tests for phosphorus and potassium and on previous management for nitrogen. All the recommendations were developed from field trials on numerous soil types and over a number of years. Nitrogen recommendations based on previous management included credits for legumes ploughed down and manure applied. Fertilizer values for various manure sources are given in Table 8.5. Recommendations given in a soil-test report are then decreased by the amounts derived from the table.

This approach is used by a number of countries and agencies making fertilizer recommendations, and Table 8.5 shows comparisons between USA and UK data used in practice. However, there are limitations to the approach. In particular, no consideration is given to nutrient availability in the years following manure application. Patterson and Watson (1960) suggested that nutrients would be available to sugar beet in subsequent years. Another limitation to the approach is that little consideration is given to nutrient variation in manure from various storage handling systems. It is also assumed that manure is incorporated immediately after spreading, which leaves ammonia volatilization unaccounted for if incorporation is delayed.

Jacobs (1995a,b,c) developed a more elaborate system for evaluating nutrient contributions from animal manure for crop production while maintaining water quality. His object was to recycle nutrients from animals back to the soil in a responsible manner. Depending on the species of animal, 70–80% of the nitrogen, 60–80% of the phosphorus and 80–90% of the potassium fed to animals is excreted in urine and faeces (Jacobs, 1995a).

Jacobs (1995b) published comprehensive analyses of the manure applied, including dry matter, total nitrogen, NH_4^+-N and total P_2O_5 and K_2O. It is preferable to have each source analysed prior to application. Average values (Table 8.6) may be used; however, it needs to be recognized that there may be 100% variation in the values listed. Systematic record keeping is advised to include: (i) soil-fertility-test reports; (ii) dates of manure application; (iii) amount of manure applied; (iv) analysis of manure used; and (v) yield of past crops grown.

Table 8.5. Nutrient credits given for the first year after application of various manures in the USA (Christenson *et al.*, 1972) and UK (SBREC, 1995).

Manure	N	P_2O_5	K_2O
		kg t^{-1}	
USA			
Cattle, pig and horse	2	1	4
Sheep	7	1	9
Poultry	7	6	4
UK			
Cattle	1.5	2	4
Pig	1.5	4	3
Poultry			
Deep litter	10.5	11	10
Broiler litter	14.5	13	11

Nitrogen credits are based on NH_4^+-N plus mineralized organic nitrogen. Ammonia may be lost through volatilization if the manure is left on the surface of the soil. A guide for expected losses is given in Table 8.7. These values are similar to those given by Chambers and Smith (1999). Mineralized nitrogen is found by multiplying the mineralization factor (Table 8.6) by organic nitrogen (total N minus NH_4^+-N). Residual mineralized nitrogen from previously applied manure may be credited at 50%, 25% and 12.5% of the manure applied 1, 2 and 3 years before, respectively. Wadman and Ehlert (1989) reviewed ammonia loss to the atmosphere in relation to the environmental effects of organic manures applied before sugar beet in the EU.

In Jacobs's (1995a,b,c) system, 100% of the P_2O_5 and K_2O is considered to be available in the first season after application. Since sugar beet is not highly responsive to

Table 8.6. Average analyses of manure from various sources (Jacobs, 1995b).

Animal species	Manure type	Dry matter (%)	Total nitrogen	NH_4^+ nitrogen	Total P_2O_5	Total K_2O	Mineralization factor
			kg t^{-1} or kg 10,000 l^{-1}				
Dairy	Solid w/o bedding	18	4.5	2.0	2.0	5.0	0.35
	Solid w/ bedding	21	4.5	2.5	2.0	5.0	0.25
	Anaerobic liquid	8	29	14	22	35	0.30
	Flushed liquid	1	4.8	3.0	4.8	6.0	0.30
Beef	Solid w/o bedding	15	5.5	2.0	3.5	5.0	0.35
	Solid w/ bedding	50	11	4.0	9.0	13	0.25
	Anaerobic liquid	11	48	29	32	41	0.30
	Flushed liquid	1	4.8	2.4	11	6.0	0.30
Pig	Fresh w/o bedding	18	5.0	3.0	4.5	4.0	0.50
	Anaerobic liquid	4	43	31	32	26	0.35
	Flushed liquid	1	4.8	3.6	2.4	4.8	0.35
Poultry	Deep pit (solid)	76	37	22	32	23	0.45
	Solid w/o litter	45	17	13	24	17	0.35
	Solid w/ litter	75	28	18	23	17	0.30

w/o, without; w/, with.

Table 8.7. Estimated losses of nitrogen by volatilization of ammonia gas from surface-applied manure, followed by incorporation (Jacobs, 1995b).

Days to incorporation	Nitrogen lost (%)
0–1	30
2–3	60
4–7	80
>7	90

applied phosphorus, this approach seems reasonable. A soil-testing programme monitors changes in residual phosphorus in the soil and applied fertilizer is adjusted. Crediting all of the K_2O for sugar beet production is supported by work done by Groves *et al.* (1999). On a low-potassium status sandy soil, approximately 90% of the total potash in the manure was utilized by the sugar beet crop where manure was applied and incorporated in February prior to sowing sugar beet.

Chambers *et al.* (1999) developed a decision-support system to predict plant availability of nitrogen following organic manure applications. It draws together the latest UK research information on factors affecting manure-nitrogen availability and losses. The manure nitrogen evaluation routine (MANNER) requires analysis of manure to include total N, NH_4^+-N and uric acid-N. The routine then accounts for volatilization losses of NH_4^+-N and uric acid-N, along with leaching losses of nitrate nitrogen. The system can provide a reliable estimate of fertilizer-nitrogen value of farm manures spread under a range of conditions. The predicted fertilizer-nitrogen value from animal manure related fairly well with the amount of nitrogen needed by sugar beet.

Use of organic manures in nitrate-vulnerable areas/zones

In some parts of the world where sugar beet is grown, there are restrictions on the amounts and timing of application of organic manures due to their nitrogen content. For example, in the EU there is a directive in force regarding nitrate pollution in both surface and groundwaters which aims at improving water quality. In the UK tables have been published (DEFRA, 2001a) with guidelines for making the best use of manure while minimizing pollution risk.

There is an upper limit for the application of organic manure equivalent to 250 kg ha^{-1} total nitrogen each year averaged over the area of grass and 210 kg ha^{-1} over the area not in grass. The 210 kg N ha^{-1} limit is to be reduced to 170 kg N ha^{-1} in 2006. In the case of organic manures with high available nitrogen, such as poultry manure and slurries, application during August, September and October is banned, to reduce the risk of nitrate leaching on sandy and shallow soils.

Other effects of animal manures

Addition of animal manures has many beneficial effects on soil fertility. In particular it enhances microbial activity in soil due to the added substrate and nutrients. Waste products from the microbes bind soil particles together forming aggregates, and polysaccharides play a major role in this process. Generally, changes in soil structure are not seen in short-term studies and 5 or more years are often needed to obtain significant changes in soil structure, as described in Chapter 9.

In the Eck *et al.* (1990) study already described (Table 8.2), feedlot waste increased organic matter from 2.3 to 3.5%. Numerous experiments have shown the effects of manure on various components of soil structure. In one representative study, bulk density decreased and the mean diameter of water-stable aggregates increased (Tiarks *et al.*, 1974). In another study, manure increased water-holding capacity and decreased evaporation rate (Unger and Stewart, 1974). In a separate report from the study by Eck *et al.* (1990), Mathers and Stewart (1984) found that manure increased saturated hydraulic conductivity and reduced bulk density. These effects indicate improved soil tilth and are generally regarded as beneficial to plant growth. However, the studies did not show the direct effects of these changes on sugar beet yield.

Other workers have suggested that the addition of manure and other organic materials increases the moisture-supplying characteristics of coarser-textured soils. For example, Hoyt (1968) found that farmyard manure greatly increased the yields of roots, but analysis of the plants indicated that the increases in yield were not attributable to improved nutrition. Hoyt therefore concluded that the increases were due to an improvement in the physical condition of the soil, but the nature of the effects was not investigated. Salter *et al.* (1965) found that available water capacity was about 39% greater in farmyard manure plots.

Other effects of organic manures on sugar beet and related crops have been described. Mangold yields were increased due to improvements of plant establishment and survival as a result of manure application, as summarized by Draycott (1972). Farmyard manure improved the germination of sugar beet and increased seedling weights and final yield when compared with broadcast fertilizer (Warren and Johnston, 1961). Mann and Patterson (1963) also reported similar improvements in the establishment of globe beet with roots of marketable size developing earliest on the organic-manured plots.

There are numerous reports of experiments where organic manures have given larger increases in yield than an equivalent dressing of fertilizer. Frequent, large applications of organic manures cause profound changes in soil physical conditions. Some of these have been measured in long-term experiments, considered in the next chapter. However, when beneficial 'extra' effects occur in annual experiments, they are not so easily explained. Some workers have seen improvements in plant growth due to the production of indolyl-3-acetic acid (IAA), gibberellic acid, vitamins and phenolic compounds such as are found in lignins, the reduction of the effect of eelworms (nematodes) and the growth of saprophytic microorganisms, which reduce the populations of certain pathogenic organisms (Whitehead, 1963), but specific examples for sugar beet were not found in the literature.

Draycott (1972) mentioned that not all of the effects of animal manures are positive. Increased fanging (sprangling) of the roots is attributed to organic manures. Nelson and Ruppel (1970) conducted a detailed study on the causes of sprangling in sugar beet after manure application. Results from experiments with sterilized soil and manure produced as much sprangling as non-sterilized soil. They concluded that microorganisms in the manure or the soil were not directly involved in the process. The results suggest that the causal factor in manure is the presence of free ammonia, which kills the primary root, and two or more secondary roots replace it.

Another effect of manures is to supply excessive nitrogen. Eck *et al.* (1990) noted that excessive nitrates in the soil after large manure applications (134 and 268 t ha^{-1} $year^{-1}$ for 5 years) caused a decline in recoverable sucrose. Halvorson and Hartman (1975) also reported excessive nitrogen in the profile with manure application. Where excessive rates of manure were applied, nitrates leached below the rooting zone (Mathers and Stewart, 1984; Eck *et al.*, 1990). Mathers and Stewart reported that salt or ammonia damaged seedlings of crops planted soon after manure incorporation. Jarvis (1997) said that poultry manure, rich in nitrogen, should be applied at no more than 6 t ha^{-1} and followed in spring by no more than 30 kg N ha^{-1} as mineral fertilizer.

Sewage Sludge or Biosolids

Rules for application in the EU and USA

In 1980, the EU announced a directive for water quality for human consumption (80/778/EC) which set an upper limit of 50 p.p.m. (50 mg l^{-1}) for nitrate. This has had a marked influence on the use of all nitrogen fertilizers and organic manures, particularly materials such as sewage sludge (see Chapter 2). In the USA, applications come under provisions of the Clean Water Act of 1987. The US Environmental Protection Agency (USEPA) established rules for the disposal of biosolids, promulgated in USEPA (1994) and referred to as 'Part 503 Rules'. Similarly, in the UK, the minimum standard for sustainable sludge use on agricultural land is set out in Department of Environment's Code of Practice (1989). Discussion of these rules is beyond the scope of this book and only those aspects that directly affect sugar beet production are included.

Biosolids are grouped in respect of the potential for passing disease to humans through the food produced on the land. Class A biosolids have had the potential risk reduced and have no restrictions related to timing of crop harvest after application. Class B biosolids possess a greater potential risk and certain time restrictions between application and harvest are imposed. These are summarized in Table 8.8. With careful planning, application of the class B solids should be workable with sugar beet production. Vector control (flies, mosquitoes and other fauna) is necessary in the use of class B biosolids. Methods include reduction in volume by evaporation, and injection or incorporation of the material into the soil.

In many cases, biosolids contain metal pollutants that may be toxic to sugar beet seedlings and/or pose problems in the food chain. Those listed are arsenic, cadmium, chromium, copper, lead, mercury, nickel, selenium and zinc. Limits for concentration in sludges and land-loading rates are summarized in Table 8.9. Any biosolid with pollutant concentrations greater than the ceiling concentration (column 2) is restricted from land application and must be disposed of by other means. Biosolids with concentrations less than the pollutant limit (column 3) may be applied to land without restriction and no record keeping is required. Biosolids with one or more pollutants exceeding the pollutant limit, but below the ceiling concentration, involve limits on land application and require record keeping of the amount applied. The cumulative loading rate restricts the total pollutant that may be applied.

Nutrient availability from sewage sludge

Sewage sludge contains large amounts of plant nutrients and adds organic matter to soil. Like additions of manure, sludges supply nutrients in the short term. In the longer term, particularly in large amounts, additions of sludge may increase the

Table 8.8. Site restrictions from time of application to harvest of various crops as affected by management of class B biosolids (USEPA, 1994).

Management of biosolid	Do not touch soil	Touch soil	Grown in soil	Grazing animals
Incorporated or injected	30 days	14 months	38 months	30 days
Surface-applied Left on surface more than 4 months	30 days	14 months	38 months	30 days
Surface-applied Left on surface less than 4 months	30 days	14 months	20 months	30 days
Example crops	Wheat	Cabbage	Beet	Forage legumes
	Oat	Celery	Leeks	and grasses
	Barley	Cucumbers	Onion	
	Soybean	Aubergine	Potato	
	Maize	Lettuce	Radish	
			Melons	Rutabaga (swede)
			Squash	Sugar beet

Table 8.9. Concentration limits of pollutants present in biosolids governing application of biosolids to land (USEPA, 1994).

Element	Ceiling concentration[a] ($mg\ kg^{-1}$)	Pollutants limit[b] ($mg\ kg^{-1}$)	Cumulative loading[c] ($mg\ kg^{-1}$)
Arsenic	75	41	41
Cadmium	85	39	39
Chromium	3000	1200	3000
Copper	4300	1500	1500
Lead	840	300	300
Mercury	57	17	17
Molybdenum	75	–	–
Nickel	420	420	420
Selenium	100	36	100
Zinc	7500	2800	2800

[a]Concentration for all pollutants in biosolids. No biosolids with concentrations greater than these values may be applied to land.
[b]Biosolids with concentrations of one or more pollutants exceeding these values trigger requirements for record keeping and limits on total metal loading.
[c]Cumulative loading for sites where one or more of the pollutants exceed the limits in column 3.

organic-matter content of soil. Chemical analyses of sludge materials from various sources are given in Table 8.10. Compared with animal manures (Table 8.6), sludges contain considerably smaller quantities of potassium.

Experiments testing the effects of sludges for sugar beet have included comparisons with farmyard manure. Results are summarized in Table 8.11. Bunting (1963) found that sewage sludge increased yield less than manure. Probably this was due to sludges containing less potassium than the farmyard manure included in the study. In a study by Garner (1966), manure and sludges were applied annually for 4 years followed by sugar beet. These results represent the cumulative effects of manure and

Table 8.10. Concentration of nitrogen, phosphorus and potassium in various types of sewage sludges.

Type of sludge	N Range	N Mean	P_2O_5 Range	P_2O_5 Mean	K_2O Range	K_2O Mean	Reference
	% dry weight						
Activated	–	6.4	–	4.3	–	0.47	Bear and Prince, 1947; Barrow, 1955; Gilmour *et al.*, 2000
Aerobic	–	4.9	2.5–12.6	6.6	0.10–1.3	0.55	Sommers, 1977
Anaerobic	0.50–17.6	5.1	–	–	–	–	Sommers, 1977; Gilmour *et al.*, 1996
Digested	–	2.4	–	1.9	–	0.36	Bear and Prince, 1947; Bunting, 1963
Other	0.10–10.0	2.1	$<$ 0.20–7.6	2.9	0.02–1.0	0.24	Bear and Prince, 1947; Bunting, 1963; Sommers, 1977; Gilmour *et al.*, 2000
Overall average	–	4.1	–	5.5	–	0.46	

Table 8.11. Comparison of yields of sugar beet treated with sewage sludge and farmyard manure (FYM).

Treatment	Amount ($t\ ha^{-1}$)	Root yield ($t\ ha^{-1}$)	Reference
None	0	37.6	Bunting, 1963
Sludge	10	40.2	
FYM	16	45.2	
None	0	29.2	Garner, 1966
Single FYM	8	32.0	
Double FYM	16	34.0	
Single sludge	5	32.2	
Double sludge	10	34.7	

sludges. Both manure and sludge increased the yield of roots by similar amounts. However, the yield of tops from sewage sludge was considerably greater than with manure, suggesting that the sludge supplied more nitrogen.

Assessment of nutrient availability from sewage sludges is similar in development to that for manure. It is essential to predict the amount of nutrients, particularly nitrogen, which may be available from sludges. Gilmour *et al.* (2000) reported results from a multi-site study on predicting biosolid application rates for agriculture. This procedure is similar to the one discussed earlier by Jacobs (1995a,b,c) for animal manures. Required information includes the nitrogen needs of the crop, assessment of available nitrogen in the soil (other fertilizer nitrogen, green-manure crops, previous applied sludges and contribution from irrigation water) and analysis of the sludge. The analysis of sludge includes total nitrogen, ammonium-N, nitrate-N and per cent solids. The system gives credits for mineralization of organic nitrogen through the use of constants. The constants are given for 4 years for both irrigated and non-irrigated systems. However, the system accounts for nitrogen but not for phosphorus and potassium.

Composted Materials

Composting is the controlled biological decomposition and conversion of solid organic material into a stable humus-like substance. A sufficient supply of oxygen is essential because the process is aerobic. The main purpose of composting is to decrease the volume of waste materials and create a stable, usable product. Additional effects include greatly decreasing odours, killing of weed seeds present in the organic material, rendering impotent pathogens present and modifying harmful chemicals, making them less harmful.

Composting relies mainly on bacteria, actinomycetes and fungi, but yeasts, algae and protozoa may be involved. In the initial stages, bacteria play the main role and heat is generated from sugars, starches, fats and proteins. Actinomycetes, which give compost an earthy smell, along with other organisms, decompose cellulose and lignin. A stable compost is produced with little odour and low oxygen demand. This can be stored in piles, where it cures.

Even though there is much promise from the use of composts in sugar beet production, there are few reports on the subject. Draycott (1972) reviewed the work done until that time, primarily in Russia. He concluded:

> Farmyard manure and other organic materials were improved by composting with fertilizer. Additions of trace elements, *Azotobacter* and phosphobacteria during the process also improved the fertilizer value of the compost. The best period of composting was 6 to 7 months, after which the available nitrogen in the compost diminished.

Advantages from compost should be similar to that from manure. Improved soil tilth, increased nutrient availability and increased biological activity result from the added material.

Reports of disease and nematode suppression in vegetables and crops grown in containers are numerous. Information is now available that facilitates the formulation of container media capable of suppressing several soil-borne diseases (Hoitink *et al.*, 1991). Included are *Fusarium* spp., *Phytophthora* spp., *Pythium* spp., *Rhizoctonia solani* Kühn and other pathogens. Some studies have shown success in the field (Keener *et al.*, 2000). Hoitink *et al.* (1977) showed that increased percolation was brought on by compost application, with an associated reduction in *Phytophthora* root rots. Compost mulches may prevent crusting at the soil surface, giving positive effects of O_2 levels in the root environment. Highly saline composts (dairy manure) enhance *Phytophthora* and *Pythium* disease unless applied several months ahead of planting. Composts that release considerable amounts of NH_4^+-N boost the incidence of *Fusarium* diseases (Quarles and Grossman, 1995).

Correct management of composts is essential for maximum benefit from composted materials. More research on sugar beet disease suppression is needed to evaluate the effect of composting. Two approaches can be used to improve the opportunity to reap benefits from composts. First, curing composts renders the suppressive effects more consistent (Kuter *et al.*, 1988). Secondly, composts should be incorporated into field soils for several months before sowing (Lumsden *et al.*, 1983).

Straw

Straw from cereals may be burnt in the field (where the practice is allowed), removed for use elsewhere or incorporated into the soil after harvest. When straw is ploughed in, additional nitrogen is often needed for aerobic decomposition; generally about 20 N kg ha^{-1} is sufficient.

In long-term studies on the effect for sugar beet, small increases in yield were seen as a result of straw incorporation (Rayns and Culpin, 1948; Harvey, 1959; Patterson, 1960; Rayns, 1961; Short, 1973). There was essentially no evidence that responses to straw additions increased the longer the experiments continued. This suggests that soil tilth was not improved and any advantage from straw was from nutrients contained therein.

Since the above studies were conducted, production practices for both crops has changed. The yield of cereals has increased due to a number of factors. In the UK, part of the change is due to a switch from spring-planted to autumn-planted grains. New varieties of cereals respond to more intensive management, including higher rates of nitrogen, giving more straw to be handled. Management of sugar beet has also changed. Nitrogen application has declined (from 155 to 115 kg N ha^{-1}). Soils have greater residual phosphorus, potassium and other nutrients due to fertilization practices over the past 40 years.

Allison *et al.* (1992) reported on the effect of 1 year's straw incorporation on a subsequent sugar beet crop. This work showed that, at 120 kg N ha^{-1}, incorporation of straw had no effect on sugar yield. However, with 40 kg N, sugar yield was reduced by straw incorporation. In a subsequent study, Allison and Hetschkun (1995) employed straw incorporation over 5 years to measure the effect on the growth, yield and nitrogen nutrition of sugar beet. Plant population densities were not significantly affected by straw or nitrogen treatments. Stands averaged between 80,000 and 100,000 plants ha^{-1} over the 3 years of the study. Averaged across all amounts of nitrogen (0 to 200 kg ha^{-1}), there was no difference in yield of sugar between straw- and no-straw-incorporated treatments. At 120 kg N ha^{-1} – the recommended amount – the yield of sugar was 12.8 and 12.6 t ha^{-1} for incorporated and no-straw treatments, respectively. With less than 120 kg N ha^{-1}, sugar yield was lower when straw was incorporated. Indications from this study suggest that straw incorporation had no effect on soil mineral nitrogen after cereal harvest and that the soil mineral-nitrogen contents at sowing time for sugar beet were not significantly affected by straw treatment.

Incorporation of straw may reduce leaching of nitrogen in the short term, such as over winter after cereal harvest (Allison, 1989). This is supported by the work done by Powlson *et al.* (1985). Ocio *et al.* (1991) followed changes in biomass nitrogen as a result of incubating ^{15}N-labelled straw with soil. They found that about two-thirds of the increase in biomass nitrogen resulting from straw incorporation was derived from the incorporated straw. When additional nitrogen was added to incubations, the amount of nitrogen derived from the straw fell to one-third. Allison *et al.* (1992) suggested that this could account for the decreased leaching of nitrogen when straw was incorporated in soil with high mineral nitrogen. The reduction in leaching would be directly related to the amount of mineral nitrogen present in the soil at the time of incorporation.

Other Wastes

Some view agricultural land as a potential disposal receptacle for industrial wastes. Caution should be exercised when using cropland for industrial waste. These materials may contain heavy-metal pollutants or chemicals toxic to crops. Studies have not been made on the effects of many industrial wastes, but we have seen many examples of the damage caused to sugar beet.

Garner (1966) tested several domestic-waste materials, comparing them with farmyard manure. The nutrient content of the waste materials was 0.83–0.88% N, 0.57–0.82% P_2O_5 and 0.34–0.37% K_2O. These materials increased the yield of sugar beet by about 22% of the increase seen with farmyard manure. This was probably related to the lower nutrient content of the domestic waste. Christenson *et al.* (2000) evaluated the effect of lime from sugar beet processing on the growth and nutrition of several crops. The results, discussed in Chapter 5, suggested that there was no effect on the yield or nutrition of sugar beet, soybean, maize, wheat or field bean on medium- and fine-textured soils not needing lime to correct the pH.

Several by-products of beet sugar factories find their way back to the land, varying from factory to factory and country to country. These are left after most of the sugar has been extracted from roots (écumes, raffinate) or after distillation of alcohol (vinasses). They all contain small quantities of organic matter but are a useful supply mainly of major nutrients (nitrogen, phosphorus, potassium, sodium, calcium and magnesium).

Green Manuring

Ploughed-in green crops provide not only nutrients but several benefits for following sugar beet, including reduced soil erosion, improved soil tilth, reduced pressure from pests, reduced nitrate leaching and improved yields. The crop is ploughed down in winter or early spring prior to sowing sugar beet. Care must be exercised to prevent the crop from attaining excessive growth, thus depleting the moisture supply for the sugar beet. Both legume and non-legume species may be used. Legumes are usually sown under a cereal crop, while non-legumes may be sown after harvest of the crop grown the year prior to sugar beet. Legumes fix atmospheric nitrogen, which will be available to the sugar beet crop. Experiments testing green manuring in relation to set-aside and rotations are in the next chapter.

Effect of green manures on sugar beet

Legumes ploughed in provide nitrogen and decrease the amount of fertilizer needed for maximum sugar yield. Draycott and Last (1970) found that trefoil contained 60 kg N ha^{-1} and decreased sugar beet requirement from 125 to 65 kg N ha^{-1}. Similarly, Dyke (1965) found that trefoil provided 30 kg N ha^{-1} and Last *et al.* (1981) 50 kg N ha^{-1}. On light sandy soils, green manures produced yields greater than nitrogen fertilizer alone, probably because of some soil physical improvement (Dyke, 1965). However, on loamy soils, maximum yield could be produced with fertilizer alone (Draycott and Last, 1970; Last *et al.*, 1981). Moisture-holding-capacity differences due to green manure on the two textures probably does not account for the increased yield from green manure on the sandy site; for example, Bunting (1963) showed that adding organic manure to sandy soils increased the water supply by only 1–4 days of water use by a sugar beet crop. The improved root development observed following green manure on sand soil is more likely to be the result of a stabilized soil structure.

Draycott (1972) summarized experiments from Poland explaining some aspects of the influence of green manures on sugar beet. The results confirmed that ploughing down field peas, spring vetch and field beans (*Vicia faba*) increased the root yield of sugar beet equivalently to two-thirds of the increase from 26.9 t ha^{-1} of farmyard manure. Leaf yield was greater with green manure than with farmyard manure (which indicates a large supply of nitrogen from the green manure). Sunflower or radish crops were less effective green manures than legumes. In other experiments, red clover used as a green manure gave very good root yields in the following sugar beet, and a mixture of sweet clover and yellow clover gave the best leaf yields. Draycott (1972) concluded that, on average, a good green manure has a similar fertilizing value to an average dressing (17.9 t ha^{-1}) of farmyard manure. However, this depends on the quality of farmyard manure. For example, in Germany, it was found that farmyard manure increased sugar beet yield by 17%, green manure increased it by 36% and the two together by 53%.

Gregg and Harrison (1950) reported the effects of ploughing down stubbles of various forage mixtures as green manures for sugar beet in Michigan, USA. The plots had been established for 5 years before the sugar beet test crops, which were grown for 2 years in succession. Sugar beet established best and gave the largest yield after brome-grass. Fescues and timothy gave the poorest results, while bluegrass and redtop gave intermediate yields. There was a close correlation between the yields of sugar beet and the root habits of the grasses, but not with the aggregation or pore space of the soils.

In more recent work, Halvorson and Hartman (1975) found that the residual effects of ploughed-down lucerne were similar to those of 112 kg N, but slightly less than 22 t of farmyard manure. Biennial sweet clover was more similar to 56 kg N, being less than the effect of farmyard manure or 112 kg N.

The value of green manuring in farming practice

Recent work with green-manure crops has focused on reduction in nitrate leaching (Allison and Armstrong, 1991) and the control of pests. The current terminology used is cover or catch crop; however, the practice is essentially that of a green-manure crop (Allison *et al.*, 1993). These crops may be sown after cereal harvest and then destroyed before sugar beet sowing. A full account of recent work is in the next chapter.

Organic Sugar Beet Production

Guiding principles and conversion

Organic crop-production systems are guided by an overriding philosophy of 'feed the soil to feed the plant' and 'this basic precept is implemented through a series of approved practices designed to increase soil organic matter, biological activity and nutrient availability' (Gaskell *et al.*, 2000). Additions of organic material through green-manure crops, crop residues and composts serve to increase soil humus. As humus increases, there is an increase in the capacity of the soil to hold water and supply nutrients for crop growth. Conversion time from non-organic to organic production is specified by rules of certifying agencies in different countries. Generally, land must be free from non-organic inputs for 2 or more years before crops can be certified as organic in both the EU and the USA.

Land selection and rotational considerations

Sugar beet grown without fertilizers and pesticides are best suited to medium-textured soils. Finer-textured and peat soils are more prone to pests and weeds. Soils with good structure allow roots greater access to water and nutrients, which is essential in organic crop production. In the UK, priming organic seed is seen as a method of improving plant establishment and may help with seedling disease control (Limb and McAughtrie, 2000).

Rotations that are capable of being sustained in the absence of inorganic nutrient inputs are an essential part of organic sugar beet production. Sugar beet has the capability of being compatible with organic cropping. It is deep rooted and extracts moisture and nutrients from deeper zones in the soil. It has the advantage that tops may be left on the surface of the soil and nutrients removed from depth will be added back to the fertility of the soil. This can be appreciable, as shown by Moraghan and Ananth (1985). They found that sugar beet tops returned to the soil contained an average of 171, 208, 245 and 167 kg ha^{-1} total N, Cl, Na and K, respectively. They presented evidence that some of the nutrients came from depths greater than the plough layer.

Practices for providing soil improvement

Rotations including cereal-grain crops, maize and forage legumes add more organic material to soil than vegetable crops and sugar beet. For example, a wheat or maize crop may add as much as 11 t dry material ha^{-1} while the vegetable crops broccoli, tomato, lettuce, onion and garlic add 7.8, 2.8, 1.3, 0.75 and 0.5 t ha^{-1}, respectively (Mitchell *et al.*, 2000).

Green-manure crops are seen as an important part of soil management in organic production systems. Legumes are important for supplying much-needed nitrogen. They all provide a practical and economical means for supplying organic material, suppressing weed growth, attracting beneficial insects and reducing nitrate-leaching losses between periods of regular crop cultivation. In California rye and vetch cover crops planted in October and incorporated in March added 10 and 5.8 t dry matter ha^{-1} respectively.

Organic amendments may include animal manures and composted materials. Genetically modified (GM) plant remains are to be avoided as organic manures in the EU. Organic additions are associated with improved soil tilth and water infiltration. However, few studies have monitored changes in the key indicators or processes that may result from these amendments. Powlson (1998a) reported on a 3-year study of inputs and outputs of major nutrients at Rothamsted, showing how difficult it can be to keep them in balance in an organic farming system, but that it is possible with care and planning.

Tillage

Conservation tillage may be utilized in organic production since tillage of any kind has a negative effect on soil humus. The term 'conservation tillage' describes production systems where at least 30% of the soil surface is covered by residues of previous crops. This creates a dilemma for organic producers. Primary tillage to control weeds is used extensively on organic farms, which incorporates surface residues, excessively aerates the soil and reduces soil humus. The development of high-residue conservation tillage that enables adequate weed control is a challenge facing innovative organic producers.

Nutrition

Conventional fertilizers are not allowed in organic farming systems but producers still need to determine the nutrient requirements of sugar beet prior to sowing. Soil testing is essential (Johnston, 2001) and nutrient recommendations from conventional (non-organic) fertilizer trials are a starting-point for determining nutrient needs for organic sugar beet production. In some cases, this may be supplemented with data from organic-system field experiments. Working in The Netherlands, Voss (1996) described how nutrients for an organic system could contribute to farming in a sustainable way, a topic reviewed recently by Tzilivakis *et al.* (2002) and others at the 65th IIRB Congress.

Management of nutrients, such as phosphorus, potassium, calcium and magnesium should be directed towards increasing these nutrients to optimum levels, as determined by a soil test. Nitrogen and sulphur cannot be increased in the soil except through the addition of organic matter and legumes for nitrogen. The contributions of nutrients from cover crops, animal manures and other organic additions are in an earlier section of this chapter. Tinker (2001) showed how difficult it is to meet the needs of organic crops without conventional fertilizers; hence organic crops are always at risk of nutrient deficiency.

Materials allowed differ from country to country. Those used for fertilizer may include blood meal and mined sodium nitrate for nitrogen and sodium, rock and colloidal phosphate (phosphorus), mined potassium sulphate and potassium magnesium sulphate for potassium and sulphur, and green sand for potassium. The Potash Development Association (1999) provides guidance over which potassium fertilizers are permitted. Calcium sources approved include gypsum and limestone. Magnesium sources include dolomitic limestone and mined potassium magnesium sulphate.

The effectiveness of green-manure crops appears to be related to the amount of nitrogen supplied. Webb *et al.* (1960) made a pot experiment comparing six green manures for sugar beet. Lucerne, smooth brome grass, Ladino clover, cock's-foot grass, hairy vetch and rye were grown in sand cultures with various fertilizer treatments. Weighed amounts of the crops were used to fertilize sugar beet grown in sand. The yield of sugar beet was closely related to the amount of nitrogen added in the green manure. Neither species nor fertilizer treatment to the green manure affected the growth of the sugar beet except when they influenced the nitrogen supplied to the sugar beet.

Organic fertilizers contain a relatively small amount of readily soluble nutrient, with a second fraction gradually available over time. These materials need to be applied 2–4 weeks before the nutrients are needed. The availability of these nutrients will depend on microbial activity. The composition and particle size of the material will be major factors determining the availability. A more concentrated material with finer size will be more available than a coarser-sized mixture. A listing of approved organic fertilizers commercially available is given in Table 8.12. Organic sources of micronutrients (trace elements) often contain more than one element. Some certifying agencies permit the use of synthetic fertilizers to correct these deficiencies.

Table 8.12. Organic fertilizer materials and their approximate analysis on a dry-weight basis (Gaskell *et al.*, 2000).

	N	P_2O_5	K_2O
Material		%	
Fish-meal or powder	10–11	13.7	2.4
Chicken manure	2–3	3.4	1.8
Processed liquid fish residues	4	4.6	2.4
Feather meal	12	0	0
Sea-bird and bat guano	9–12	6.9–18.3	1.2–2.4
Lucerne meal	4	2.3	1.2
Cottonseed meal	6	0.9	1.8
Soybean meal	7	4.6	1.2
Bone-meal	2	11.5	0
Kelp	< 1	0	4.8

Summary

Organic materials applied to soil before sugar beet include animal manures, sewage sludge, green-manure crops and other organic residues. These materials supply significant quantities of nutrients, which have been undervalued in the past. However, with economic and environmental concerns, there is increased pressure to evaluate the availability of these nutrients and to avoid the application of excess fertilizer.

Nutrients present in manure may be accounted for by utilizing tabular values or laboratory analysis. In addition to nutrients, manure enriches the organic fraction in soil. It slowly increases the amount of organic matter; but large amounts of material need to be added to make a noticeable difference. There may be other benefits from animal manure, including growth regulators, vitamins and improvement of the physical structure of soil, including water-holding capacity. Enhanced root growth due to favourable soil structure may improve yield.

Biosolids – the sludge from sewage treatment plants – come under rules promulgated as a result of environmental and health concerns. Nutrients may be accounted for by methods similar to those used for animal manures.

Composting is the controlled biological decomposition and conversion of solid organic material into a stable humus-like substance. Composting decreases the volume of waste, decreases odours, kills weed seeds, makes pathogens impotent and may modify some chemicals, making them less harmful. There are reports of disease and nematode suppression in some crops, but there are few studies showing such benefits for sugar beet. Similarly, there are few trials evaluating the availability of nutrients from composted materials.

Straw (particularly cereal straw) incorporation has shown only small increases in yield. Long-term studies show that soil tilth was not improved by straw incorporation. Any advantage has been attributed to nutrients contained therein. Incorporation of straw does not increase the fertilizer-nitrogen requirement, but does appear to reduce the leaching of nitrate.

Sugar beet is being produced in organic-farming systems in some countries. Challenges include correct tillage, nutrition and weed control. Sugar beet forage deeply for nutrients and this meets one of the objectives of organic crop production where conventional fertilizers are banned.

9
Nutrient Reserves and Crop Rotations

The first eight chapters have covered each nutrient needed by sugar beet and how shortfalls in soil supply can be amended by inorganic and organic applications. Often the amount of nutrient needed to produce the best economic return exceeds the removal in tap roots sent for processing. This is quite usual in crop production, because the fibrous root system explores less than 2% of the total soil volume and nutrients do not move large distances to the root surface. Also some of the applied nutrients may become unavailable through fixation, leaching or volatilization.

This chapter deals with the consequences of applying nutrients before and during sugar beet cropping. Where reserves remain in soil and can be used efficiently by following crops, consequences are beneficial and these are covered first. Of increasing concern since the last major review of this subject (Draycott, 1972) are any negative aspects of residual nutrients carried over in the soil and subsequent losses to the environment.

Future decisions over optimum amounts and timings of each nutrient for sugar beet must not be governed solely by farm economics, as in the past. Of even greater importance is that nutrients be employed such that harmful carry-over effects are, at worst, minimal or, better, maintain the productive capacity of soil in such a way that they have a benign environmental impact.

Sugar Beet in the Crop Rotation

In most countries where sugar beet is produced successfully, it is grown in rotation with other crops. Where monoculture has been attempted, yields have usually declined rapidly, largely due to multiplication of diseases and other pests – particularly sugar beet cyst nematode (*Heterodera schachtii*). Weeds and weed beet also become a problem where the rotation is very short or rotational cropping is not practised. Mainly for these reasons it seems likely that sugar beet will continue to be grown in some form of arable or ley/arable rotation. In farming practice it is therefore important, when considering the amount of fertilizer needed by sugar beet, to take into account the needs of other crops in the rotation. This both maximizes economic returns and minimizes environmental impact.

It is not possible to study any crop in isolation from the rest of those grown in the rotation because plant residues and unused nutrients affect the amounts of fertilizer needed by the following crop. Thus fertilizers applied to crops grown before sugar beet affect the amount required by sugar beet. Previous cropping also affects the amount of fertilizer required, because crops take different quantities of nutrients from the soil and leave different nutrient residues. For example, sugar beet takes up large quantities of some nutrients from the soil, which may deplete soil reserves and affect the following crop. Sugar beet tops are particularly rich in some elements and it is important to take this into account when tops are removed from the field. Many experiments have been made to investigate how these factors affect the nutrient requirement of sugar beet and how nutrition of sugar beet affects the following crop.

Effects of Nutrient Reserves and Crop Residues on Sugar Beet

It is important to distinguish between nitrogen and other nutrients, such as phosphorus and potassium. Mineral nitrogen does not tend to accumulate in soil except in dry climates. In humid areas, where most sugar beet is grown, nitrogen is lost by leaching or denitrification after harvest. The residual effects of nitrogen are often due to mineralization of readily decomposable plant material. In contrast, phosphorus and potassium can accumulate in many soils as plant-available reserves and therefore have beneficial effects on subsequent crops.

The different farming systems in which sugar beet is grown have an impact on the nutrient requirement of the sugar beet, in particular on nitrogen. One of the first detailed accounts was that of Adams (1962), who showed that sugar beet grown after a run of cereals needed much more nitrogen than after one or two cereals. Tinker (1965) followed by showing that the optimum amount of nitrogen increased from about 100 to 140 kg N ha^{-1} as the number of preceding cereals increased from nil to three or more. Boyd *et al.* (1970) examined over 170 field experiments and found that sugar beet following potatoes needed 75 kg N ha^{-1}, whereas after other crops (nearly all cereals) 120 kg N ha^{-1} was best for sugar beet. In the USA Christenson and Butt (2000) found that sugar beet needed an additional 30–50 kg N ha^{-1} following maize compared with following *Phaseolus* field bean. The extra nitrogen needed after maize did not seem to be related to nitrogen fixation by field bean or by nitrogen present in organic forms in the soil. While there is no direct evidence of an allelopathic effect of maize residues on sugar beet, there is some evidence of this effect on other crops.

Hull and Webb (1967) made one of the first large-scale field experiments at Broom's Barn specifically to investigate the effect of previous cropping on the nitrogen requirement of sugar beet. Figure 9.1 summarizes the results. No nitrogen was needed after

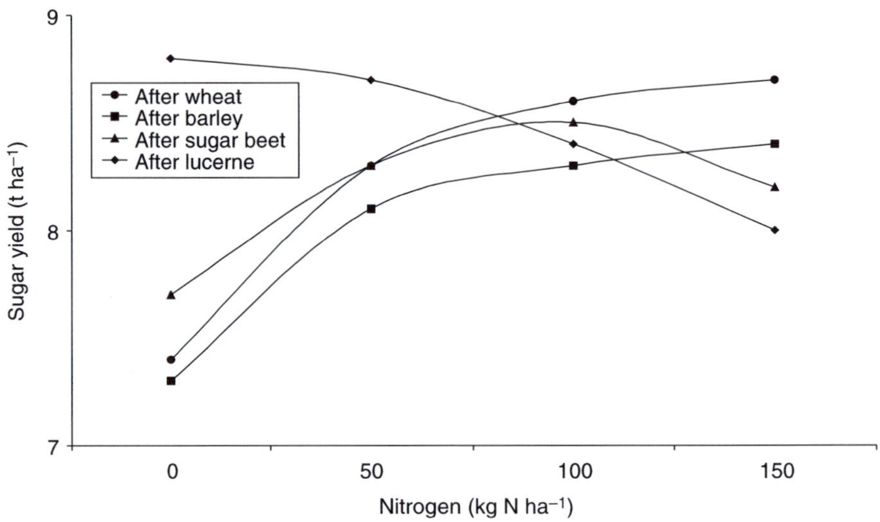

Fig. 9.1. Effect of previous crop on nitrogen requirement of sugar beet.

lucerne, 100 kg N ha^{-1} after sugar beet and 145 kg N ha^{-1} after wheat or barley. Similar experiments were made, but with different crops and soils, at Silsoe and Broom's Barn (Draycott and Last, 1970) and measurements made of nitrogen uptake and residual nitrogen from the crops before the sugar beet. Potatoes left about 70 kg N ha^{-1} and cereals none. Trefoil before sugar beet also returned about 65 kg N ha^{-1}. Thus nitrogen requirements of sugar beet were 125 kg N ha^{-1} after cereals, 70 kg N ha^{-1} after potatoes and trefoil and 125 kg N ha^{-1} after grass ley.

The three winters during the Broom's Barn experiments were relatively dry and there was a linear relationship ($r = -0.86$) between the residual nitrogen from the first-year crops and the amount of fresh fertilizer nitrogen needed for maximum sugar yield. There was some indication that fresh nitrogen was slightly less effective than residual nitrogen. Unpublished survey data from the Red River Valley in the USA suggests that carry-over nitrogen in the soil profile is utilized more efficiently than nitrogen applied in the current season. The reason for this is not clear.

Response to Various Cropping Systems and Fertilizer by Sugar Beet in Long-term Experiments

Much fundamental work in the 'classical' experiments on clay soils has been done at Rothamsted, UK, and at its satellite stations on other soil types (Woburn – light sandy loam; Broom's Barn – sandy loam; Saxmundham – chalky, sandy boulder clay), published over the second half of the 20th century. Some of the main findings are summarized here, followed by details of parallel work in the EU and USA.

Woburn, ley/arable

Boyd (1968) reported this experiment, which was begun in 1937. The object was to compare the effects of 3 years of various leys on yields of two arable crops that followed and with yields of the same two crops in an all-arable rotation. Farmyard manure was also tested, along with fertilizer nitrogen and potassium. At the start of the experiment there were doubts over the sustainability of continuous cash-cropping on such light-texture soils. It is now widely accepted practice, provided weeds, diseases and other pests are controlled and the soil nutrient status and soil organic matter are adequate (Johnston, 1997).

Until 1955, potatoes were the first test crop and barley the second (Mann and Boyd, 1958). The cropping scheme was changed in 1956 because of the build-up of potato cyst nematode, and sugar beet was inserted in place of potatoes as the first test crop (Boyd, 1968). Early results indicated that potassium, nitrogen and possibly other elements, such as magnesium, were being depleted by the arable rotation. When the scheme was changed to include sugar beet, plots were split to test more nitrogen and potassium.

Table 9.1 shows that there were large effects of rotation and large responses to farmyard manure. On the continuous arable sequence, farmyard manure increased yields of sugar by more than 40%, whereas on the ley sequence the increase was only 10%. There was little increase in yield from extra nitrogen and potassium applied – an unexpected result. Soil analyses showed that the crop should have responded to potassium. Warren and Johnston (1961) on a nearby experiment showed that the lack of response was due to the distribution of the potassium in the soil. Where it was dug in, the crop responded; where it was broadcast on the surface, there was little response. Thus the yields of sugar were larger with farmyard manure than with fertilizers because the farmyard manure contained large amounts of potassium that had been ploughed in.

Table 9.1 also shows the mean sugar yields for 1962–1964. During those years, large corrective dressings of fertilizer equalized the potassium status of plots in each rotation. The effect was to decrease the response to farmyard manure by more than half. For all rotations except lucerne, responses to potassium were less than in previous years. Responses to nitrogen differed between rotations but it was not possible to

Table 9.1. Woburn ley/arable. Effect of preparatory cropping, farmyard manure and nitrogen and potassium fertilizer on a sugar beet test crop (after Boyd, 1968).[a]

	Preparatory cropping in previous 3 years			
	Grazed ley	Lucerne for hay	Arable (including seeds and hay)	Arable (roots)
	Sugar yield (t ha^{-1})			
Mean yield in 1956–1961	7.10	6.56	6.10	6.59
Response to:				
Farmyard manure (38 t ha^{-1})	0.85	1.18	1.57	1.83
Extra N[a]	−0.35	−0.38	−0.30	0.01
Extra K[b]	0.34	0.29	0.28	−0.05
Mean yield in 1962–1964	7.61	7.38	6.80	7.66
Response to:				
Farmyard manure	0.11	0.46	0.54	0.72
Extra N[a]	−0.30	−0.23	0.23	0.04
Extra K[b]	0.11	0.29	0.09	−0.14

[a]Basal application, 90 kg N ha^{-1}, 110 kg K_2O ha^{-1}.
[b]Additional 90 kg N ha^{-1}, 110 kg K_2O ha^{-1}.

indicate precisely the optimal nitrogen dressing after each rotation. From 1965 to 1967, four amounts of nitrogen were tested and sugar beet in the arable rotations needed most, the optima depending on whether or not farmyard manure was used.

Thus at Woburn these 12 years of results with sugar beet indicate that the early large effects of farmyard manure can be explained mainly in terms of response to nutrients, but there was still a small effect of farmyard manure not obtainable by fertilizers alone. Another experiment at Woburn also emphasized the importance of thorough mixing of the fertilizer dressing with the soil for maximum response, so that it is not concentrated in the surface soil, where it may be largely unavailable in dry periods.

Woburn, organic-manuring experiment

Mattingly (1974) reported further work on the light sandy loam soil to determine the long-term benefits of organic additions to sugar beet grown in rotation with cereals and field beans (*Vicia faba*). It was a successor to and extension of the green-manuring experiment (1936–1963), the ley/arable experiment, started in 1938, and the market-garden experiment, started in 1941. In the organic-manuring experiment, started in 1964, peat, straw, green manures, farmyard manure and leys were tested in combination with mineral fertilizers.

Mattingly found that organic inputs were necessary to maintain yield, mineral fertilizers alone not producing full yield. Peat and straw, both ploughed in, did not benefit sugar yield much. Green manures and farmyard manure, however, produced more sugar than fertilizers alone. Most of the benefit was from nitrogen released from the organic matter. In the absence of organic additions, 125 kg N ha^{-1} was needed. Green manures provided 60 and farmyard manure 50 kg N ha^{-1}. Farmyard manure produced about 1.5 t sugar ha^{-1} more than fertilizers after adjusting for the available nitrogen released from the farmyard manure.

Rothamsted, continuous sugar beet

Few field experiments have tested continuous sugar beet cultivation over a long period. The Barnfield experiment at Rothamsted probably comes nearest to a sugar beet equivalent of the Broadbalk wheat experiment. Lawes and Gilbert started a manurial experiment on turnips, swedes and mangolds on Barnfield in 1843. Mangolds were grown continuously from 1876 until 1959, except for a few years when the crop failed. From 1946 to 1959 a part of each plot (the same part each year) was cropped with sugar beet. The results up to 1894 were summarized by Lawes and Gilbert in 1895 and reported in detail by Hall (1902). Watson and Russell (1943) examined the results up to 1940, and Warren and Johnston (1962b) up to 1959. Cooke (1967) extracted much interesting information about mangolds up to 1945 and Draycott (1972) produced an overview of the sugar beet from 1946 to 1959.

The experiment compared $(NH_4)_2SO_4$ with $NaNO_3$ and tested phosphorus, potassium, sodium and magnesium, rape-cake and farmyard manure. Ammonium and nitrate nitrogen gave the same increase in sugar beet root yield in the presence of phosphorus and potassium or phosphorus and sodium. Potassium and sodium each increased yield where the nitrogen was given as ammonium sulphate, but neither had any effect where sodium nitrate was used. Fertilizers and farmyard manure were both needed for maximum yield.

The Barnfield study also provides unique data relating to nutrient uptake by mangolds and sugar beet, and corresponding changes in the quantities of nutrients in soil brought about by long, continuous cropping, although detailed chemical analyses of crops and soils were only made in the later years of the experiment. Many of the data have been referred to in previous chapters dealing with nitrogen, phosphorus and potassium.

Broom's Barn, continuous sugar beet and crop rotations

An experiment was started in 1965 to compare five contrasting rotations of crops given nutrients in amounts considered optimal for each crop. Continuous sugar beet was tested, three three-course rotations consisting of one sugar beet and two other crops and one six-course rotation of one sugar beet and five barley. Four amounts of nitrogen were tested each time sugar beet was grown.

Results of the first 6 and second 6 years of the experiment were published by Draycott *et al.* (1972a, 1978b). Tops of sugar beet were ploughed in but cereal straw removed. During those first 12 years, diseases and other pests had little effect on sugar beet in any of the rotations and yields were similar in all five rotations. Response to nitrogen by sugar beet varied greatly from year to year but not much between rotations, provided adequate potassium was supplied. To achieve a satisfactory potassium balance, the amount applied had to be revised upward several times.

The performance of the continuous sugar beet changed dramatically from the 13th (!) year of the experiment onwards. This resulted from the appearance of a heavy *H. schachtii* infestation discovered when the crop was harvested, together with a light incidence of violet root rot (*Helicobasidium purpureum*). The latter had been seen on a few plants on continuous sugar beet in previous years. The experiment in its original form was therefore abandoned and the area used for work on the relationship between *H. schachtii* and sugar beet yield (Webb *et al.*, 1997).

Saxmundham, Norfolk, four-course rotations (RI and RII)

A long-term rotation experiment was begun at Saxmundham (RI) on sandy clay loam in 1899 with the Norfolk four-course system of roots, spring barley, legume and winter wheat. In 1965 sugar beet was grown in place of mangolds and the four-course system continued until 1969. From 1956 to 1964 sugar beet and mangolds were grown side by side on half plots. From 1899 to 1965 eight fertilizer treatments of nitrogen, phosphorus and potassium, in small quantities by current standards, were tested for each crop. Farmyard manure and bone-meal were also tested. In 1966 the fertilizer treatments were changed to test dressings near to current commercial practice. The experiment was described by Trist and Boyd (1966) for the period 1899–1961 and by Williams and Cooke (1971) for the period 1964–1969.

Response to phosphorus was outstanding and yields were doubled. Response to nitrogen was less than to phosphorus and very small unless phosphorus was also given. Yields were a little less with farmyard manure than with a full fertilizer application.

The same rotation nearby tested farmyard manure and supplementary nitrogen and phosphorus, seeking the best financial return from the small amounts of fertilizer available at the time (RII). Johnston *et al.* (2001) have drawn interesting conclusions from sugar beet grown since 1969 on these plots, which previously received widely differing amounts of phosphorus and potassium over many years. In respect of phosphorus, previous treatments produced a wide range of soil phosphorus concentrations not normally encountered in commercial practice. Table 9.2 shows yields of sugar beet averaged over the 6-year period 1969–1976. In the absence of fresh phosphorus fertilizer a soil concentration between 20 and 30 mg P kg^{-1} was needed for optimum yield.

Table 9.2. Yields of sugar beet grown at Saxmundham in 1969–1976 on soils with four concentrations of soil phosphorus (after Johnston *et al.*, 1986).

	Available soil phosphorus (mg P kg^{-1})			
	7	17	30	48
Sugar (t ha^{-1})	4.3	6.0	6.5	6.6

In 2000 on the RI experiment, first- and second-order interactions between nitrogen, phosphorus and potassium were investigated on plots that had received the same phosphorus and potassium applications since 1965. The plots had been 'mothballed' and from 1985 to 1998 grass was grown and was cut and removed annually. For the crops grown in 1999 and 2000, phosphorus and potassium were applied to those plots receiving phosphorus and/or potassium between 1899 and 1965. In 2000 there were differences in soil phosphorus, which were related to the large amount of phosphorus applied recently. There were, however, no differences in exchangeable potassium probably because the grass had removed much potassium and most of any potassium balance from the application in 1999 had been fixed. Yields of sugar from sugar beet in 2000 were similar on soils with least phosphorus at all levels of nitrogen and there was no response to potassium (Table 9.3). There was a large response to phosphorus, especially when most nitrogen was applied. In the presence of phosphorus, there was a response to potassium with the smaller amounts of applied nitrogen but not at the largest. This suggests that, with too little nitrogen, sugar beet did not fully exploit the soil mass to find sufficient potassium to achieve optimum yield. This response to applied potassium may have occurred because both soils had about 150 mg K kg^{-1} soil, which is near the critical value for sugar beet on this soil type (Chapter 4).

Table 9.3. Yields of sugar from sugar beet given four amounts of nitrogen at Saxmundham 2000 (after Johnston *et al.*, 2001).

Treatment	Soil data (mg kg^{-1})		kg N ha^{-1}			
			40	80	120	180
2000[a]	P	K	Yield of sugar (t ha^{-1})			
Nil	16	189	5.6	6.6	8.1	8.4
K	20	143	5.5	7.1	7.9	8.5
P	32	150	6.0	7.8	8.5	10.3
PK	32	148	8.1	8.6	9.3	10.2

[a]P and K treatments in 2000 continued those from 1965 (for earlier treatments see Williams and Cooke, 1971).

Woburn and Rothamsted, six-course rotation experiments

Nitrogen, phosphorus and potassium fertilizers were tested on a six-course rotation of crops at Woburn and Rothamsted. Results for the period 1931–1955 were reported by Yates and Patterson (1958). The 15 fertilizer treatments were applied so that each plot received every treatment once in 15 years and there was no build-up of phosphorus and potassium in soil. The six crops grown were wheat, barley, rye, potatoes, sugar beet and clover. Responses were estimated by the method of Crowther and Yates (1941) to standard amounts of the three major elements by all crops. In the case of sugar beet there were large responses to nitrogen at both sites. Phosphorus had little effect but potassium increased yield, particularly at Woburn. One of the most interesting results was the evidence that nitrogen improved yield most at both sites in wet summers and least in dry summers, a result confirmed elsewhere (Chapters 2 and 12).

Rothamsted and Woburn, exhaustion land and depleted plots – value of potassium reserves

Warren and Johnston (1960) reported experiments at Rothamsted on the exhaustion land, where some plots had received no potassium fertilizer since 1856. The exhaustion-land site is a strip of 1 ha of arable land at the north end of Hoosfield, and derived its name from the unmanured cereal cropping, which was begun in 1902 to measure the residual effects of manures that had been applied in previous experiments between 1856 and 1901. Residual potassium was 'worth' 75 kg K_2O ha^{-1} of new potassium fertilizer and, in a somewhat similar experiment at Woburn, the potassium was worth about the same on one site but only 20 kg K_2O ha^{-1} on another. Both at Woburn and on the exhaustion land the response curves to fresh potassium did not show a maximum with any amount of potassium tested, and yield on the enriched soil always exceeded that on the depleted soil, a result that cannot be seen in annual experiments.

In a later report, Johnston *et al.* (1970) found that residual potassium increased sugar yield by 1 t ha^{-1} on the exhaustion land and at Woburn by between 0.5 and 1 t ha^{-1}, but the yield of sugar was the same on depleted and enriched soils provided a large dressing of new potassium was given. The residual potassium on the exhaustion land could not be valued in terms of a new dressing of potassium, but at Woburn they were worth between 84 and 96 kg K_2O ha^{-1} as fresh potassium fertilizer.

Woburn, reference-plot experiment

An experiment at Woburn was begun in 1960 on a field where arable crops, mainly in a four-course rotation, had been grown since at least 1876. The first 5 years of cropping were described by Widdowson and Penny (1967). Nitrogen, phosphorus and potassium fertilizers and farmyard manure were tested during a five-course rotation of barley, ley, potatoes, oats and sugar beet. Draycott (1972) summarized the sugar beet yield data, which emphasized increasing response to potassium on this light sandy loam soil. Uptake data showed that sugar beet recovered 67% of the nitrogen applied, the most of all crops grown.

The average yield of sugar in 1965–1969 on plots with and without potassium since 1960 is shown in Table 9.4. By 1965 there were large differences in exchangeable potassium in the soil. There was a large difference in yield as soil potassium increased from 36 to over 300 mg kg^{-1} (Chapter 4).

Table 9.4. Yield of sugar beet on the reference-plot experiment at Woburn (adapted from Johnston *et al.*, 2001).

	Soil potassium (mg K kg^{-1})		
	36	131	311
Sugar (t ha^{-1})	2.8	5.4	7.3

Broom's Barn, long-term fertilizer experiment

Shortly after the UK sugar beet research station was set up at Broom's Barn, a long-term fertilizer experiment was begun to investigate the use of nitrogen, phosphorus, potassium and sodium fertilizers and farmyard manure. A rotation of sugar beet followed by two cereals was started in 1965 and continues today. Nil, standard and double amounts of fertilizer were tested in factorial combinations. Draycott *et al.* (1972b, 1977) reported on the first and second 6-year periods and Last *et al.* (1981) on the third 6-year period. Sugar beet tops were always ploughed in (but sampled and weighed) and straw was removed.

Nil plots are now very impoverished in contrast to double-application plots, where soil nutrient concentrations have increased. Yields and nutrient offtakes were measured for every crop every year and soil was sampled every third year. Some of the main findings after six rotations (18 years) are summarized as follows and in Table 9.5.

Nitrogen (with phosphorus and potassium) increased yield greatly, on average by 0.8 t sugar ha^{-1} and in the 18th year by nearly 50%. The standard (or recommended amount) of 100 kg N ha^{-1} gave best results throughout. The double amount (200 kg N ha^{-1}) was always harmful.

Phosphorus (with nitrogen and potassium) had little effect on sugar yields in the early years of the experiment because the soil was able to supply sufficient for full yield. Towards the end of the 18 years, responses of about 0.5 t sugar ha^{-1} were the norm from 50 kg P_2O_5 ha^{-1} because soil phosphorus on nil plots had declined, but only by about 5 mg P l^{-1} soil. The experiment showed that this sandy loam soil had a large reserve of phosphorus, which became available for crop uptake.

Table 9.5. Long-term effects of fertilizers and farmyard manure (FYM) on sugar yield at Broom's Barn. Averages of 18 years.

	t sugar ha^{-1}
No fertilizers or FYM for 18 years	4.5
Response to:	
NPKNa[a] at recommended amounts for soil and crop	+2.3
NPKNa × 2 – NPKNa	+0.1
FYM – NPKNa	+0.7
N (PKNa given)	
N0	5.3
N100–N0	+0.8
N200–N100	−0.3
P (NKNa given)	
P0	6.0
P50–P0	+0.1
P100–P50	0.0
K without Na (NP given)	
K0	4.8
K1–K0	+1.2
K2–K1	+0.5
K with Na (NP given)	
K0	6.5
K1–K0	+0.3

[a]100 N, 50 P_2O_5, 100 K_2O, 150 Na, kg ha^{-1}, for sugar beet and recommended amounts for cereals.

Potassium (with nitrogen and phosphorus) increased yield greatly (even more than nitrogen) throughout the 18 years. In the absence of sodium, 100 kg K_2O ha^{-1} (the amount recommended) increased yield by about 30% and 200 kg K_2O ha^{-1} an additional 10%. Soil analysis confirmed the low reserves of potassium in the soil, resulting in serious decline in yield. In the presence of the standard amount of sodium (150 kg Na ha^{-1}), response to standard potassium was decreased from 1.2 to 0.3 t sugar ha^{-1}.

The farmyard manure plots (30 t farmyard manure ha^{-1} applied every 3 years before sugar beet) regularly produced the largest yields – even with double amounts of mineral fertilizers. Soil analysis showed that farmyard manure greatly increased soil potassium, which explained part, but not all, of its benefit. Subsoil analysis showed that potassium from farmyard manure had moved into the subsoil and was probably beneficial in dry years. Farmyard manure also improved plant establishment.

Overall the experiment proved that the recommended amounts of nitrogen (100 kg N ha^{-1}), phosphorus (50 kg P_2O_5 ha^{-1}), potassium (100 kg K_2O ha^{-1}) and sodium (150 kg Na ha^{-1}) were correct for the soil type and rotation over 18 years of testing. Double amounts were of no benefit except for potassium in the absence of sodium. Farmyard manure improved yield still further, partly, perhaps, through small increases in soil organic matter and subsoil enrichment with nutrients leached out of the plough layer.

Swedish rotation experiment

Agerberg (1969) reported on a long-term experiment in which three four-course rotation systems were tested: A – sugar beet (given 20 t ha^{-1} farmyard manure), barley, ley and winter wheat; B and C – sugar beet, barley, winter rape and winter wheat. In B, the straw and sugar beet tops were ploughed in, and in C the straw was burned and the tops carted off. Four amounts of compound fertilizer were given during the first rotation and the plots were split for nitrogen dressings during the second rotation.

Yields during the first 8 years showed no marked advantage from the ley system, although organic matter in the soil was affected. In system A it increased but in B and, more especially, in C the organic matter decreased. There was a tendency for greatest losses where least fertilizer was given. Table 9.6 shows the yields of roots for the two sugar beet crops. The rotation systems affected the amount of nitrogen fertilizer required for maximum yield and the magnitude of the response to nitrogen, but, given the correct nitrogen dressing, the yields were similar from all three systems.

Belgian long-term experiment

Frankinet *et al.* (1987) reported on work at Gembloux from 1959 to 1983. Mineral fertilizers were tested in conjunction with organic-matter additions in the form of straw, beet tops, farmyard manure, slurry

Table 9.6. Effect of three rotation systems and of nitrogen fertilizer on yields of sugar beet in Sweden (after Agerberg, 1969).

		Rotation		
		A	B	C
	N application (kg ha^{-1})	Yield of roots ($t\ ha^{-1}$)		
1st crop		56.4	57.9	59.4
2nd crop	0	41.2	35.9	36.3
	40	42.5	40.6	43.9
	80	46.1	45.6	43.4
	120	45.2	43.9	46.4

and sugar-factory lime. Soil was analysed annually for major elements and organic matter.

In common with the Broom's Barn experiment described above, changes in available soil phosphorus and exchangeable soil potassium were small compared with the cumulative input–offtake balances of the elements. Large amounts of both elements moved from the 'unavailable' to the 'available' form where none was applied. The authors recommended that, for soils with good reserves of phosphorus and potassium, the amount given in fertilizer should be decreased by the amount in any organic matter applied.

Ferden Farm long-term experiment, USA

This experiment was initiated in 1941 to evaluate the need for forage legumes and the effect of the previous crop on sugar beet yield with two fertility levels (Cook *et al.*, 1945). The fertility treatments were 220 and 560 kg 2–16–8 ha^{-1} of N–P_2O_5–K_2O to each sugar beet crop during the first 10 years of the study. A similar amount was split among the cereals and row crops in the rotation. During the next 7 years, the amount of fertilizer was adjusted to equalize the total amount applied between the two fertility treatments. Two nitrogen treatments (0 and 45 kg ha^{-1}) were superimposed on each of the two fertility treatments.

Additional fertilizer increased the yield of beets by an average of 4.4 t ha^{-1} across the different rotations over the course of the first 10 years of the study (Table 9.7), essentially no difference in the response across the six rotations. A differential response occurred from additional nitrogen in the various rotations in the next seven seasons (Table 9.7). When sugar beet followed lucerne–brome, there was a 1.3 t ha^{-1} reduction in yield compared with the nil treatment. If *Phaseolus* field bean was inserted the reduction declined to 0.9 and nitrogen had no effect when maize was included. When the forage legume was removed from the rotation, the response to applied nitrogen increased, showing the importance of additional nitrogen under these conditions.

The yield of sugar beet following lucerne-brome was 3.1 t ha^{-1} less than when field bean was inserted between the forage and sugar beet. Robertson *et al.* (1952) commented that, in wetter years, sugar beet did not yield well after lucerne. This was attributed to lack of air in the soil, resulting in accumulation of toxic decomposition products where fresh lucerne was ploughed under. However, a later report (Schneider and Robertson, 1975) suggested that *Aphanomyces* infection in sugar beet was significantly greater following lucerne than following other crops. Complete evaluation of nutrition of sugar beet should include such observations.

Table 9.7. Response to additional fertilizer in years 1–10 and to additional nitrogen in years 11–17 for various rotations in the Ferden farm study (from Cook *et al.*, 1945; Robertson *et al.*, 1952; Guttay *et al.*, 1958).

Rotation	Response to additional fertilizer (t ha^{-1})	Response to additional N	
		Roots (t ha^{-1})	Sugar (kg ha^{-1})
C, Ba, A/Br, A/Br, SB	4.5	−1.3	−343
Ba, A/Br, A/Br, FB, SB	4.4	−0.9	−295
Ba, A/Br, A/Br, C, SB	4.2	0.0	−128
Ba, O, A/Br, C, SB	4.8	1.3	39
Ba/Cl, FB, W/Cl, C, SB	4.2	1.6	180
Ba, FB, W, C, SB	4.3	2.0	286

A/Br, lucerne–brome grass; Ba, barley; C, corn (maize); Cl, red clover; FB, field bean (*Phaseolus*); O, oats; SB, sugar beet; W, wheat.

Fertilizer Value of Sugar Beet Tops

Sugar beet tops are particularly rich in nitrogen and cations, so several field experiments have been made to measure their value to following crops. Pizer (1954) drew attention to the organic matter they provide, particularly on infertile, sandy soils. In many stockless rotations without leys and where cereal straw was burnt, sugar beet tops and cereal straw and stubble were the only crop residues that returned any quantity of organic matter. He estimated that sugar beet left little organic-matter residue in the soil as roots but up to 8 t ha^{-1} dry matter as leaves and crowns.

In France, Crohain and Rixhon (1967) made a thorough examination of the value of sugar beet tops in terms of nitrogen fertilizer on three succeeding cereal crops. In the first year after ploughing down, tops were equivalent to 35–45 kg N ha^{-1}. In the second cereal crop, tops increased yield equivalent to 10–13 kg N ha^{-1}, but in the third crop it was difficult to find any residual effect. It is interesting to compare these values with the amount of nitrogen in the tops as determined by analysis before ploughing down; the total nitrogen value in this case was 115 kg N ha^{-1}.

Widdowson (1974) made similar experiments at Broom's Barn. Sugar beet tops were either ploughed in or removed and spring barley grown in the following year with four amounts of nitrogen. Table 9.8 summarizes the results. Where no nitrogen was given to the barley, tops increased the yield by 0.4 t ha^{-1} grain. However, only a small amount of nitrogen fertilizer was needed to give the same effect. Ploughed-in tops were 'worth' about 20 kg N ha^{-1} to the barley, and this value was remarkably similar in all three experiments.

Webb *et al.* (1997) used spring wheat as a test crop in new experiments in the 1990s, which were similar to those of Widdowson with spring barley in the 1970s. They were interested in the fate of the nitrogen left by sugar beet (in tops ploughed in and in soil) on sandy soil. About 60 kg N ha^{-1} was present in early winter and by spring this had decreased to about 50 kg N ha^{-1}, a greater amount than was reported on heavy soils by Sylvester-Bradley and Shepherd (1997).

Thomsen and Christensen (1996) in Denmark incorporated sugar beet tops labelled with ^{15}N into a light sandy loam soil in autumn. Leaching and availability of this nitrogen to 2 years of subsequent crops were measured. About 15% was harvested in the following barley and ryegrass and a similar amount lost by nitrate leaching. The remaining nitrogen was mostly in the 0–20 cm soil layer, including roots. Denitrification losses were thought to be less than 15% over the 2-year period. It was calculated that, if the tops had contained 100 kg N ha^{-1} (about the usual amount – see Table 2.1), there would be 20 kg N ha^{-1} available to the spring barley, replacing this amount of mineral fertilizer.

Moraghan *et al.* (1997) in the Red River Valley of North Dakota/Minnesota, USA, and Manitoba, Canada, also determined the amount of nitrogen in sugar beet tops and their residual value for spring wheat. Three categories of autumn growth and canopy colour were recognized: crops that were obviously very green and vigorous prior to harvest, those that were 'average' and those that looked impoverished due to yellow leaves. The mean amounts of nitrogen present in the tops were 270, 125 and 80 kg N ha^{-1}, respectively.

Table 9.8. Yield of barley grain after sugar beet. Means of three experiments (after Widdowson, 1974).

	Barley grain (t ha^{-1})			
	kg N ha^{-1} applied for barley			
Sugar beet tops	0	41	83	126
Removed	3.31	4.41	4.79	4.72
Ploughed in	+0.44	+0.11	+0.08	−0.04

Aerial photographs, satellite positioning and variable-rate fertilizer applications were used in the following crop to account for residues. Soils in the states of North Dakota, Minnesota and Manitoba freeze up just after harvest, preserving the nitrate in sugar beet tops and thus preventing loss. The authors found that, in general, 1 kg N ha^{-1} in the very green tops was equivalent to 0.5 kg N ha^{-1} as urea given to following spring wheat. Even for the yellow impoverished crops, 1 kg N ha^{-1} in tops was equivalent to 0.25 kg N ha^{-1} for wheat. Shepherd *et al.* (1997) did parallel work recently in the UK, where autumn and winter are usually both mild and wet, so mineralization of soil organic matter, followed by leaching of nitrate, can be considerable. They found that average sugar beet crops returned 100 kg N ha^{-1} or more to soil in ploughed-in tops. They stressed the need to synchronize the nitrogen released with the following cereal requirement.

Value of Cover (or Catch) Crops before Sugar Beet

Serious concerns exist over the leaching of nitrate into aquifers or by run-off into surface-water sources. Field experiments were initiated in several countries to determine whether short-term cropping in late summer/autumn/early winter could alleviate the problem. The hypothesis was that such a crop might absorb much of the nitrate present in soil after cereals, oil-seed rape and early-harvested potatoes and vegetables. Often at this time of year nitrification of soil organic nitrogen goes on apace, adding to the quantity of nitrate potentially leachable, especially from bare soil in wet winters before spring-sown sugar beet.

Allison *et al.* (1998a,b) described 17 detailed experiments on sandy loams in eastern England after cereals and oil-seed rape. In that area, 90% of sugar beet follows cereal, so there is much scope for the practice, particularly as autumn-sown cereals are harvested in July or August and the following spring-sown sugar beet only starts to take up large amounts of nitrate in late May or early June. In the UK, cover crops grown over winter are a requirement in nitrate-sensitive areas (MAFF, 1998) and are advised in *The Soil Code* (MAFF, 1998) and the new *Arable Cropping and the Environment – a Guide* (DEFRA, 2001b).

A number of cover crops were tested with different sowing dates and times of destruction. Average dry-matter yield was 1.6 t ha^{-1} and nitrogen uptake 35 kg N ha^{-1}. Soil mineral nitrogen was decreased on average from 46 to 32 kg N ha^{-1}. There were large differences between years and sites. The carbon : nitrogen ratio of all the cover crops tested was wide and the authors thought that little nitrate would be released from them during winter after ploughing (no legumes were tested). They concluded that, if costs were taken into account, the preferable cover crop was simply volunteer cereals and weeds, allowed to grow after cereal harvest.

In France, Duval (2000) reported on a long-term study in the Champagne region on a typical calcareous loam over a 10-year period. The aim was to compare several types of crop management for decreasing nitrate leaching and their influence on yields in a sugar beet, pea and wheat rotation. Cover crops were wheat or radish after sugar beet and grass, peas and radish after wheat. Reduced nitrogen applications to the sugar beet were also tested.

Nitrate in soil water at a metre depth was measured by means of porous cups on one site and by lysimeters on another site. On the former site, over the period of the experiment, nitrate in soil before winter drainage started was least after sugar beet (up to 60 kg N ha^{-1}). After peas and wheat it was similar (up to 140 kg N ha^{-1}). Cover-cropping decreased the amount slightly (by about 10–20 kg N ha^{-1}).

On the other site, measurements on the water flowing out of the lysimeters showed larger variations. Much of the water contained more than the threshold 50 mg nitrate l^{-1} required for drinking-water, e.g. between peas and wheat, water concentration exceeded 140 mg nitrate l^{-1} and 116 mg nitrate l^{-1} with reduced nitrogen fertilization. However, cover-cropping between wheat and sugar beet decreased it from 133 to 42 and between sugar beet and peas from 59 to 33 mg nitrate l^{-1}. Table 9.9 summarizes the results of the lysimeter work, which suggests benefits from cover-cropping, greater than those indicated by Allison *et al.* (1998a,b) and Duval's (2000) porous-cup measurements.

Effects of Set-aside and Cover Crops

All EU member countries agreed on reform measures to the Common Agricultural Policy (CAP) in 1992. An important element was the introduction of the Integrated Administration and Control System (IACS), requiring detailed information about use of land from farmers for the first time. The rules of IACS are set out in EU legislation; relevant here is that set-aside was introduced as a prerequisite for receipt of CAP payments to arable farmers, the aim being to regulate surpluses. Since then set-aside in arable rotations has varied from year to year from 5 to 15% of eligible land.

Sugar beet may be grown before or after set-aside. The new rules encouraged research to discover how best to grow sugar beet in conjunction with set-aside. The use of cover crops grown on set-aside land before sugar beet is reasonable because there is ample time available for establishment. These crops are seen both as a means of improving soil for sugar beet (nutritional and physical) and of decreasing leaching, particularly of nitrate. Several authors reviewed their work at the IIRB Congress in Brussels in 1998.

Working in Italy, Amaducci *et al.* (1998) grew radish, *Phacelia*, pigeon bean, ryegrass and wheat on set-aside and ploughed these cover crops in before sugar beet. The amount of biomass produced varied greatly, as did nitrogen uptake of the cover crop and its carbon : nitrogen ratio. Ryegrass absorbed nitrate well and decreased leaching, but released little to following sugar beet. The legume provided most nitrogen for sugar beet and *Phacelia* was intermediate.

Couvreur *et al.* (1998) did similar work in Belgium on a silt soil. Before ploughing in the cover crops, soil to 1.5 m contained from 6 to 84 kg N ha^{-1}, largely due to the growth of the cover crop and its depth of rooting. They also found that ryegrass released its nitrogen slowly and actually decreased sugar yield (presumably the sugar beet was starved of nitrogen, because an additional 50 kg N ha^{-1} for the sugar beet resolved the problem). On the other hand, clover as a cover crop permitted a reduction of at least 50 kg N ha^{-1} for sugar beet.

Koch (1998) measured the uptake and release of nitrogen by several cover crops and their influence on nitrate leaching and sugar beet yield. Cover-crop biomass and nitrogen uptake were always greatest with lucerne followed by oil radish, *Phacelia*, winter wheat and ryegrass. The risk of nitrate leaching was greatest after *Phacelia*, which decomposed quickly after ploughing in, followed by lucerne. To give most protection to groundwater from leached nitrate, he suggested that cover crops should grow as long into the winter as possible. Also, nitrogen fertilizer for sugar beet must be matched carefully to that provided from the ploughed-in cover crop.

Table 9.9. Effect of cover cropping and nitrogen fertilization on nitrate in water at 1.10 m depth and quantity of nitrogen leached. Averages over eight winters in eastern France (from Duval, 2000).

	No cover Standard N		Cover Standard N		No cover Reduced N	
	Nitrate in water (mg l^{-1})	Nitrogen leached (kg ha^{-1})	Nitrate in water (mg l^{-1})	Nitrogen leached (kg ha^{-1})	Nitrate in water (mg l^{-1})	Nitrogen leached (kg ha^{-1})
Peas – wheat	143	50	93	31	116	43
Wheat – sugar beet	133	48	42	11	113	42
Sugar beet – peas	59	11	33	5	45	9

Rotational Application of Phosphorus, Potassium and Magnesium Fertilizers

Nutrients, such as nitrogen, which can leach easily must be applied to coincide with crop requirement and uptake, as described in Chapter 2; otherwise they may be lost. In contrast, other, less mobile major nutrients, such as phosphorus, accumulate in the plough layer. For phosphorus, only in excessively enriched soils is there a risk of leaching but, if such soils are eroded to water, then there is a risk of eutrophication. Cations, such as potassium and magnesium, move downwards very slowly over a period of years and then generally only where soil contains little clay and organic matter (Chapters 3 and 4).

Thus the task of applying phosphorus, potassium and magnesium every year for every crop is generally unnecessary. It is possible to calculate expected offtake of the three elements by crops over the rotation (often 3 or 4 years) where sugar beet is grown. If soil concentrations at the beginning of the rotation are optimum for crops being grown, an amount equivalent to the offtake can be applied at any suitable time.

Table 9.10 shows optimum ranges of nutrient concentrations for sugar beet rotations, highlighted in Chapters 3 and 4 and by MAFF (2000). These are regarded (MAFF, 2000) as the 'targets' and soils with concentrations greater or less than these may need adjustment by smaller or larger applications of fertilizer. In most sugar beet soils where the crop has been grown for many years, only maintenance amounts of the major nutrients are needed. In others, concentrations may need to be increased to these targets and then maintained there by replacing offtakes.

Fate of Nitrogen, Phosphorus and Potassium Applied in Fertilizer for Sugar Beet

Several papers in the proceedings of the International Sugar Beet Research Congress in 1997 reported on the environmental impact of sugar beet production in the EU. Draycott *et al.* (1997) produced a budget for the input of the three major nutrients as fertilizer and their eventual fate. Annually the total application for sugar beet in the EU was estimated from survey data as 292 t N, 204 t P_2O_5 and 331 t K_2O (excluding that in organic manures).

The average yield of fresh, clean, sugar beet roots in the UK was then about 51 t ha^{-1}, equivalent to dry-matter yields of 12.4 t roots ha^{-1} and 4.6 t tops ha^{-1}. Using this ratio of roots to tops, total EU beet production and typical nutrient concentrations, described in Chapters 2, 3 and 4, the amounts of nutrients taken up by tops and roots for the EU crop can be calculated (Table 9.11).

Generally, tops are left on fields to be ploughed in or consumed by grazing animals, leaving much of the nutrient in the soil. Removing roots plus tops has a dramatic effect on the nutrient balance (Table 9.11), both nitrogen and potassium being in deficit.

Some 157,000 t N, 66,000 t P_2O_5 and 159,000 t K_2O are delivered to EU sugar factories in roots. The sugar and other soluble components of the roots are extracted by diffusion and the resultant 'juice' purified with lime. For the EU crop it is estimated that about 2.5 Mt dry matter of factory lime is produced annually. Most of the lime is returned to arable fields or grassland. Assuming an average analysis of 0.6% N,

Table 9.10. Soil concentrations of PKMg considered optimal for rotational fertilizer application (from Chapters 3 and 4, and MAFF, 2000).

	Optimum soil concentration	
	Arable and forage crops and grassland	Vegetables
mg P kg^{-1}	16–25	26–45
mg K kg^{-1}	120–180	181–240
mg Mg kg^{-1}	50–100	50–100

Table 9.11. Estimates of the amounts of the three major nutrients in the EU crop compared with those applied in fertilizer.

	N	P_2O_5	K_2O
		1000 t	
Quantity of nutrients in:			
Tops	264	77	372
Roots	157	66	159
Tops plus roots	421	143	531
Amount applied in fertilizer	226	185	292
Balance if roots only removed	+69	+119	+133
Balance if roots and tops removed	−195	+42	−239

1.6% P_2O_5 and 0.2% K_2O on a dry-matter basis, then the quantities of nutrient leaving the factories in lime are 15,000 t N, 40,000 t P_2O_5 and 5000 t K_2O. Thus only a small amount of nitrogen and potassium may be accounted for in the factory lime. Most of the remaining phosphorus, with some nitrogen, is within the insoluble part of the root (the marc), which is returned to agricultural production as animal feed. After sugar crystallization, the soluble components constitute molasses, which contains the remaining nitrogen, phosphorus and potassium. Molasses has a variety of uses, but much returns to the farm as animal feed. Small amounts of nitrogen are lost to the atmosphere as ammonia or other gaseous emissions during processing. Losses of all three nutrients in waste water are small, due to strict legal limits.

Detecting Nutrient Reserves using Global Positioning Systems

The availability of global positioning systems (GPS) has increased over the past 30 years, allowing maps of characteristics such as pH and available nutrients to be constructed. This was the beginning of 'precision farming'. The next step in this new technology is the use of remote sensing of differences in soil and/or crop growth, dealt with below.

Maps for a few experimental farms showing contours of pH, phosphorus, potassium and magnesium were prepared before GPS. Fields were marked out laboriously, as on the Broom's Barn farm, so that soil samples could be taken from every 100 m × 100 m square. This was done in 1960, when the farm was purchased, and again in 1975 to examine the changes resulting from 15 years of cropping. Such information is valuable both for siting field experiments and for spreading lime and fertilizer to even out fields for future experiments (Draycott *et al.*, 1977; Cooke *et al.*, 1982).

With the introduction of GPS, commercial farms could be mapped easily, allowing differential application of lime and fertilizer within a field. Areas with large reserves receive little or no fertilizer and deficient areas more than they would from an even application over the whole field. In theory, fertilizer is better used and yields are improved, but few experiments have been made to explore the effects on crops and profitability.

To produce yield maps of fields, harvester-machinery manufacturers have also adopted GPS and it has become common practice, especially on cereal combine harvesters. Jaggard *et al.* (1997a) wondered if such cereal yield maps of fields coming into sugar beet could be useful in predicting sugar beet requirements and yield. In a theoretical study they examined the relationship between cereal yields and nutrient availability, and between cereal yields and following sugar beet yields. The long-term fertilizer experiment at Broom's Barn was used for this study. They found that the relationship between one sugar beet crop and the next sugar beet crop were significant and positive, but, between cereal and subsequent sugar beet, yield relationships were seldom significant and sometimes negative.

The implication is that cereal yield maps could not be used to predict sugar beet-yield variability because the two species appear to react in different ways to variations in soil. While the rooting pattern of the two crops is very different and probably contributes to the situation, other factors may include diseases, other pests and variations in soil water availability. Further work is in progress to determine how cereal yield maps could be used to adjust inputs to sugar beet, but on the evidence produced so far it will not be simple.

Detecting nutrient reserves by remote sensing

Instruments on land-based machines, aeroplanes and satellites are being introduced into agricultural applications to sense differences in soil and/or crop growth. This is with the object of utilizing past or current information to vary inputs to suit within-field variation. Most measure the sun's reflected radiation but some are beginning to use radar, which is not obstructed by cloud.

Generally, the instruments used so far are cameras or radiometers. Both make use of variations in the wavelength of the reflected light from the field and are calibrated to pick up differences, e.g. in the ratio of soil to leaf cover, species, status of leaf for nitrogen concentration, diseases and water stress. Steven *et al.* (1986) made considerable progress with radiometry in forecasting the yield of sugar beet in the UK and Wood (2002) has successfully used digital photography to improve nitrogen applications to wheat. K. Duthoit (Broom's Barn, 2002, personal communication) is currently investigating interrelationships between soil nitrogen, the uptake of nitrogen by sugar beet and optimum nitrogen fertilizer application for sugar production, by means of ground-based radiometry.

Summary

Too few field experiments have been made to check the nutrient requirements of sugar beet on a long-term basis. Applying fertilizers in autumn, winter or spring and then testing the response in one crop often gives confusing results. This is because all the relatively immobile nutrients, such as phosphorus, potassium, calcium and magnesium need years to be fully mixed and incorporated with soil. Only then is the full picture seen.

Of necessity, sugar beet is grown as one crop in a rotation and the aim should be to use fertilizers to establish optimum soil concentrations. It is misleading (particularly with phosphorus and potassium) to study any crop in isolation from the rest. Generally these remarks do not apply to nitrogen, due to its mobility, and each crop should receive just the amount needed with minimum surplus as described in Chapter 2. Much recent work shows the benefits of cover-cropping for decreasing the leaching of nitrate.

From studies reviewed in this and previous chapters, soil phosphorus concentration should ideally be above 25 mg P l^{-1} soil (Olsen's method), potassium above 180 mg K l^{-1} and magnesium above 50 mg Mg l^{-1} (ammonium nitrate extraction). Fertilizers can be used gradually to reach these background soil concentrations and then to maintain them by balancing inputs and offtakes. More work of a long-term nature is needed to verify these concentrations on other soil types and in other growing conditions.

10
Soil Physical Conditions

In addition to being present in sufficient quantity, nutrients and water must be accessible to plant roots, so soil has to provide a suitable environment for roots to grow in. The sugar beet crop, particularly just after sowing, is very sensitive to soil physical conditions. To establish and yield well, sugar beet roots must be able to extend rapidly.

Soil structure and tilth are terms used to describe the physical state of soil in respect of its suitability for seed germination, seedling emergence and root growth through to harvest. Russell, E.W. (1971) stated, 'good tilth is something a farmer can recognize with his boot, but scientists cannot describe'. Indeed, the literature contains many different measurements. There does not seem to be a single analysis or even several sufficiently sensitive analyses that adequately describe conditions for ideal crop performance. The following addresses aspects of some of the commonly used evaluations of soil physical properties and factors affecting nutrient requirement.

Plant roots must grow through the many pore spaces in well-structured soils. Consequently, any factor that affects pore space in soil affects plant growth. The 'ideal' soil contains approximately 50% solid matter and 50% pore space. That pore space contains 50% water-holding micropores and 50% air-holding macropores. Compaction of soil removes the macropores first, forcing water into the medium and smaller soil pores and thus reducing the supply of air for root growth. Limited air supply restricts nutrient uptake by roots because oxygen is required for metabolic processes involved in active nutrient uptake.

Soil texture and structure both influence percentage pore space. Soil texture is defined as the percentage of sand, silt and clay in a specific soil type. There is very little opportunity through common farming practices to change texture. Soil structure, however, is the aggregation of textural and organic components into larger and more dynamic particles, easily compacted by traffic and tillage, which are very much under farmers' control.

Physical conditions are also affected by rainfall, temperature variation, wind, crop rotation, field operations and tillage. Bulk density, pore-size distribution, water-holding capacity, penetrometer resistance, sheer strength, hydraulic conductivity and aggregate size and stability have been used to describe some of the physical properties of soil.

Optimum Conditions

Bulk density

Bulk density is commonly used as an indicator of the relative degree of compaction in soils. General values range from 1.0 to 1.5 g ml^{-1} for fine- and medium-textured soils to 1.3 to 1.8 g ml^{-1} for coarse-textured soils. Miller and Donahue (1990) suggested that bulk densities should not exceed 1.4 g ml^{-1} for fine-textured soils and 1.6 g ml^{-1} for coarse-textured soils. These values correspond to 48 and 40% pore space, respectively. Work by Jaggard (1984) confirms this suggestion when he reported that sugar-beet yields declined as bulk densities of medium-textured soils increased above 1.45 g ml^{-1} (Fig. 10.1). Pabin *et al.* (1991), working in Poland, showed a marked reduction in yields on a loamy sand soil with bulk densities above 1.51 g ml^{-1}. Håkansson (1990) proposed a 'degree of compactness' measurement, which is the ratio (%) of the dry bulk density of the soil and the bulk density of the same soil in a compacted state. On mineral soils, the maximum crop yield of barley was obtained at the same degree of compactness, irrespective of soil type. Unfortunately, no data were presented for sugar beet.

In summary, the maximum bulk density a sugar beet can withstand without limiting growth is about 1.50 g ml^{-1}. Generally, the bulk-density measurement is not sufficiently sensitive to detect changes in soil structure affecting plant growth.

Pore size and distribution

Pore size and its distribution govern air and water movement, as well as the capacity of soil to hold both components. Kuipers (1955) felt that the soil structure of marine clays in The Netherlands could best be described by the amount of large pores. Independently, Baver and Farnsworth (1940) and Baver (1949) showed that, on a fine-textured soil, beet yield decreased sharply as the non-capillary pore space dropped below 10% on US soils. Pendleton (1950) found unrestricted root growth of sugar beet at 14 and 18% non-capillary pore space on a sandy loam and a silt loam soil, respectively. Conversely, root growth was restricted at 6.5 and 11.7% macropore space for the two soils. He also reported improvement in the shape of roots as non-capillary pore space increased in field studies. Later work confirmed these values for sugar beet growth (Blake *et al.*, 1960).

Thus, to grow sugar beet where pore space is a non-limiting factor, a total pore space of at least 45% and a non-capillary pore space of at least 12% are necessary.

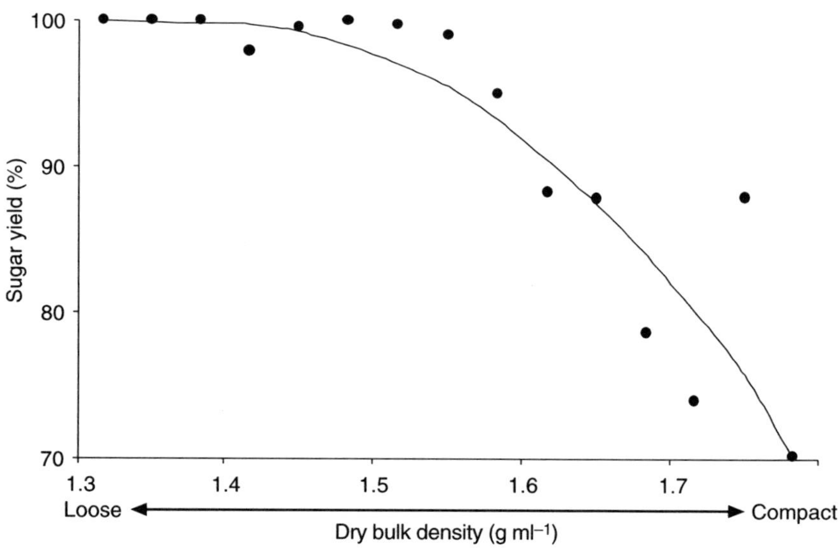

Fig. 10.1. Effect of bulk density of a medium-textured soil on sugar yield expressed as a percentage of the highest yield (from Jaggard, 1977).

Aggregate stability and size distribution

Aggregate stability and size distribution are important considerations for evaluation of the physical conditions of soil. Sieving techniques are usually used to evaluate size distribution. Van Bavel (1949) introduced a mean weight diameter (MWD) concept that quantifies aggregate size. De Boodt *et al.* (1961) advanced the methodology further, suggesting a 'change in mean weight diameter' (CMWD). A bulk sample of soil is air-dried under controlled humidity. The MWD is measured for the dry aggregates. The aggregates are then moistened and MWD is measured again. The result gives an indication of the water stability of the soil aggregates.

Measurement on one soil was related to the yield of mangolds, as shown in Fig. 10.2. The parabolic nature of this plot is surprising. One might expect a curvilinear relationship, where a maximum yield could be determined, or possibly a linear relationship, showing a steady-state effect. De Boodt *et al.* (1961) explained the parabolic relationship as follows:

> At lower values of CMWD, the soil aggregates are sufficiently stable to act as gravel reducing the water holding capacity of the soil and making it difficult for mangold roots to penetrate. At the higher values of CMWD, the soil is not stable causing the soil to slake with associated aeration problems and low yields. At the intermediate values of CMWD, the stability of aggregates approached the optimum with appropriate yields.

Gummerson (1989) reported the importance of fine aggregates for good seedling emergence. He showed an increase in emergence of sugar beet as the percentage of aggregates less than 5 mm increased from 40 to 60. Dürr and Aubertot (2000) found that the per cent emergence decreased exponentially with aggregate size over 10 mm. They also showed that the number of seedlings impeded increased markedly when the aggregate maximum length exceeded 25 mm. When the weight of the aggregate exceeded the force exerted by the sugar beet, emergence was decreased. The force exerted by sugar beet was in the range of 0.10 to 0.15 newtons (N). For comparison, wheat exerts about 0.30 N. Dürr *et al.* (1992) found that seedling size was greatly affected by rate of emergence, the largest seedlings being those which emerged first.

Whether aggregates are too small or too large, emergence is reduced. Small aggregates increase the strength of the surface

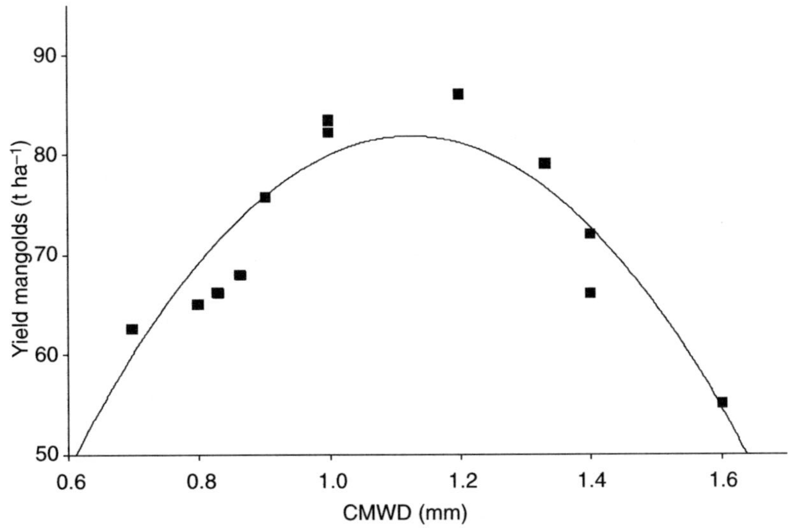

Fig. 10.2. Comparison of yield of mangolds with change in mean weight diameter (CMWD) on a sand soil (from de Boodt *et al.*, 1961).

while large aggregates are too heavy for the sugar beet seedling to force aside. Aggregates in the range of 0.5–5 mm seem to be the most favourable but there is a need for additional work describing the pore-size distribution needed for optimum sugar beet emergence and growth.

Saturated hydraulic conductivity

Saturated hydraulic conductivity (K_S) is an indirect measurement of soil structural stability. Undisturbed soil cores are placed in water and saturated for up to 48 h. Then saturated flow is measured by applying a small head of water above the surface of the soil contained by the core. Darcey's law is applied to calculate K_S. Relatively greater aggregate stabilities from different treatments can be compared for increasing K_S values. However, there are no data indicating an ideal value or range for sugar beet production. Probably the best use of this measurement is for comparing treatments within an experiment.

Soil strength

Strength of soil has an impact not only on root growth, but also on emergence. Robertson (1952) used penetrometer resistance to show greater crust strength in plots without soil-improving crops. Smucker and Leep (1975) reported that sugar beet emergence could be doubled when soil-crust strengths exceeding 1.0 MPa were reduced to 0.57 MPa, when rows were banded with anti-crusting agents. Sugar beet root yield was increased by 630 kg ha^{-1} for each 0.1 MPa reduction in crust strength. Taylor and Bruce (1968) presented the root yield of sugar beet in respect of increasing penetrometer resistance. There was a near-linear decrease in yield with increasing soil strength between 0.1 and 2 MPa resistance. Individual tap root weight declined from 1.1 to 0.6 kg over this range. Pabin *et al.* (1991) found an increase in yield of 603 kg ha^{-1} for each 0.1 MPa reduction in resistance between 1.75 and 4.65 MPa.

Nutrition of Sugar Beet on Compacted Soils

Oxygen

Oxygen is not usually considered a limiting nutrient in the growth of crops since there is an abundant supply of the element in the atmosphere. Oxygen supply may be limiting for adequate growth of plant roots, particularly where there are dense subsoil layers. In some early work, Bertrand and Kohnke (1957) observed that oxygen diffusion was slower in compacted subsoil layers than in the looser counterpart. About the same time, Gill and Miller (1956) found that normal root growth is negatively affected when the oxygen content in soil air is reduced to 10%. However, they noted that roots continued to grow with as little as 1% oxygen, provided they were not subjected to a mechanical barrier.

Lemon and Erickson (1952) suggested that the oxygen diffusion rate through soil is the important factor and not necessarily the absolute amount in the soil air. Stolzy *et al.* (1961) found that root growth was reduced when the oxygen diffusion rate was less than 38 μg O_2 cm^{-2} s^{-1}. Wiersma and Mortland (1953) reported that oxygen supply to roots is a limiting factor in the growth of sugar beet. Scott and Erickson (1964) showed that oxygen supply to sugar beet roots limited their penetration of a dense soil layer. Supplying the root with an oxygen source (calcium peroxide) promoted penetration of roots through the layer, which was compacted to a bulk density of 1.9 g ml^{-1}. Barley and oats are able to survive lower soil oxygen diffusion rates from 8 to 25 μg O_2 cm^{-2} s^{-1}, while sugar beet requires from 13 to 50 μg O_2 cm^{-2} s^{-1}. Thus, while oxygen supply is important, adequate diffusion is necessary and may be greater for sugar beet than for other crops.

Nitrogen

Most of the research work on the effect of soil compaction on nitrogen centres on the effect of additional fertilizer required for sugar beet under compacted conditions. Kuipers (1955) pointed out that sugar beet responded to higher nitrogen rates on compacted soil than on looser conditions of soil. Draycott *et al.* (1970b) reported that increased fertilizer nitrogen was needed for sugar beet grown under compaction, which was confirmed by Jaggard (1977). The former showed an additional need of 75 kg N ha^{-1} on compacted soils compared with looser soils. Jaggard's results, averaged over 3 years, showed that soils with less compaction did not respond to amounts above 75 kg N ha^{-1}, while on more compacted plots there was a response to 150 kg N ha^{-1}, thereby confirming earlier work.

Wiersum (1962) found that nitrate uptake by plants was independent of rooting density, undoubtedly due to the mobility of the ion. The difference in the fertilizer nitrogen needed appears to be due to restriction of the amount of mineralization under the reduced aeration of compacted soils. Whisler *et al.* (1965) and Clement and Williams (1962) found less mineral nitrogen in compacted than in uncompacted soils after incubation. The latter's work showed 16.1 mg N kg^{-1} mineralized on compacted soils with 3% air-filled pores. Less than 10% of the mineralized nitrogen was in the nitrate form. On the uncompacted soil, there was 41.7 mg N kg^{-1} mineralized, with nearly all of it as nitrate. In a study in Norway, Bakken *et al.* (1987) found from five to seven times as much nitrogen loss through denitrification in compacted compared with non-compacted soil. These factors may explain the need for additional fertilizer in compacted soils compared with their less compacted counterparts.

Phosphorus

Decreased phosphorus uptake by crops on soils with poor structure has been reported (Wiersum, 1962). Wiersum demonstrated the influence of soil structure on root growth and that uptake of nutrients depended on the mobility of the nutrient. More intensive rooting in finer soil improved the utilization of phosphorus, but phosphorus uptake diminished with coarse aggregates, when only few roots were developed. Lawton (1945), Flocker *et al.* (1959) and Flocker and Nielsen (1962) also found that soil compaction affected plant growth and decreased phosphorus uptake.

Shierlaw and Alston (1984), working with maize and annual ryegrass, found that compaction decreased root length in compacted layers, but increased root length in the overlying soil. They also showed that phosphorus uptake per unit length of root was generally decreased with increasing compaction. Draycott *et al.* (1970b) suggested that the amount of phosphorus fertilizer needed for sugar beet was affected little by compaction. Jaggard (1977) also reported that sugar beet did not respond to additional phosphate fertilizer under compacted conditions. These crops were grown on soils with high residual phosphorus concentrations and so would not be expected to respond much to applied fertilizer.

Prummel (1975) described an interaction between soil phosphorus concentration and soil compaction in respect of dry-matter yields of sugar beet. On compacted soils there was a response where soil concentration was more than 24 mg P l^{-1} extracted in water. However, on uncompacted soils there was no response with 13 mg P l^{-1} in the extract. Fried and Broeshart (1967) suggest that uptake is diminished by restricted root growth. Barber (1995) clearly demonstrated that uptake was influenced most by root surface area, which also increased with greater root length. Voorhees *et al.* (1975) reported that the root elongation rate increased by nearly 80% as soil air-filled pores increase from 1 to 30%. Therefore, as mechanical impedance of soil increases, causing root thickening and stunted roots, greater supplies of soluble phosphorus are needed to supply the uptake demands of plants having smaller root systems (Silverbush and Barber, 1983).

Adjustment of phosphorus application for sugar beet on compacted soils does not seem to be needed in most conditions. Under conditions of high residual phosphorus concentration in soil, no additional phosphorus should be needed. With low residual phosphorus, it appears that normal amounts of phosphorus fertilizer should be adequate. Recommendations based on soil analysis may already account for some degree of compaction, since the correlation work has been done over a wide range of conditions, including varying degrees of compaction.

Potassium

Lawton (1945) showed that reduced aeration in water-retentive soils reduced the potassium concentration in maize. Merely forcing air through the soil increased uptake. Philips and Kirkham (1962) reported a reduction in total potassium in maize leaves due to tractor traffic, whether potassium was applied or not. Poor aeration was an important factor in potassium nutrition of sugar beet in experiments conducted in Montana, USA (Larson, 1954a). The results showed 31% less potassium in petioles when grown on soils containing a small number of large pores.

Potassium uptake seems to be affected to a greater extent by poor aeration than other nutrients. Lawton (1945) showed the following ratios of uptake on non-aerated to aerated cultures of a silt loam containing 50% water: K, 0.3; N, 0.7; Mg, 0.8; Ca, 0.9; P, 1.3.

General considerations

Smucker *et al.* (1978) studied the interaction of soil compaction and sugar beet variety on nutrient concentration in beet leaves. Averaged across three varieties, concentrations of nitrogen, phosphorus and calcium were decreased with increasing compaction. Conversely, there was little effect on potassium and magnesium concentration due to soil compaction. Yield on the compacted plots was 11% lower than on the looser soil. Russell and Goss (1974) provide an excellent discussion concerning physical aspects of uptake.

While restricted root growth generally reduces uptake of nutrients from soil reserves, spatial distribution of compacted regions within the soil profile greatly affects the functional efficiency of nutrient uptake by roots. For example, in fields where wheel tracks compact only portions of the root zone, the vertical distribution of roots is most noticeably affected. Compensatory root growth into less compacted areas of soil that are more hospitable to root growth and function often results in greater vertical root growth. Consequently, some compacted conditions may not decrease the amount of nutrient taken up, especially if nutrients are uniformly distributed within the soil where the zone explored by fibrous roots has a sufficient concentration of the desired nutrient. Then no additional fertilizer nutrient would be needed. However, at lower concentrations in soil or when compaction excludes root growth in proximity to fertilizer bands, additional nutrients probably need to be added. These spatial considerations affecting soil compaction need to be included in conclusions on the nutritional needs of sugar beet grown on compacted soils. Consequently, a better description of rooting patterns, localized soil compaction, types of fertilizer applications and the nutrient concentrations of the soil is needed. Such data could be incorporated into a model proposed by Aubertot *et al.* (1999).

Previous Cropping

Improvement of soil structure by inclusion of forage legumes and/or grasses in a rotation is well documented. R.H. Eliot, as quoted by Low (1955), said: 'four to six years good turf on old arable land would restore it to a condition comparable with old pastures'. Following up on this, Low conducted a study to evaluate the time taken by a ley to change the physical state of an old arable soil to that of an old grassland. The process was found to be slow, taking possibly 50 years on some clay soils, but only 5–10 years on sandy soils. Barber (1959) measured the effect of lucerne, lucerne–brome grass, brome grass and maize on aggregation of a silty clay loam soil in Indiana, USA. The rank order of formation of aggregates was brome grass > lucerne–brome grass > lucerne > maize. Maize did not affect aggregation and yields were not affected by the improved aggregation. Lucerne promotes aggregation on coarse-textured as well as fine-textured soils (Miller and Kemper, 1962).

Robertson (1952) reported the effects of several crop rotations on sugar beet production on a sandy clay loam. Growing a lucerne–brome grass sward in the rotation significantly increased sugar beet yields. One year was as beneficial as 2. Both total and non-capillary pore space was increased by production of this forage. Differences in pore space were greater during the latter part of the season than at the beginning. Crusts formed on plots that did not have legumes, but were not present on plots with a legume in the rotation, as measured by a penetrometer. Since greater yield and improved soil structure were the result of rotation, Robertson concluded that some of the increase in yield was due to improved structure.

In Michigan, cropping patterns changed after this work by Robertson. A study was initiated to evaluate the use of non-forage crops for reducing the decline of soil organic matter due to row crop practices. The study was described in detail, including yields (Christenson *et al.*, 1991b). Momen (1985) evaluated the soil structure after 10 years of cropping. MWD was significantly affected by the amount of maize present in the rotation. There was a steady progression of increased MWD with increasing amount of crop residue returned to the soil (Table 10.1). All systems lost carbon over the course of the study, but those systems with more crop residues lost less. Other measurements including bulk density, K_S, total porosity, air porosity and aggregate size distribution, were not affected by the amount of residues returned to the soil.

Table 10.1. Mean weight diameter (MWD) and aggregate size as affected by amount of crop residue returned over 10 years of cropping (Momen, 1985; Christenson, 1997).

Crop residue (t ha^{-1})	MWD (mm)	Aggregate size range	
		< 1 mm (%)	1–5 mm (%)
33	0.49	82	18
50	0.50	81	20
68	0.55	78	22
90	0.67	77	23

In this same study, the effect of residues on nutrient concentration in sugar beet leaves was striking (Table 10.2). Concentrations of nitrogen, phosphorus, magnesium and iron were all reduced by returns of greater than 5 t ha^{-1} year^{-1} of crop residue, while the concentration of potassium increased. Where maize was included in rotations, it returned large residues.

Christenson and Butt (2000) reported that some of the differences could be accounted for with increased nitrogen on sugar beet following maize. However, there appear to be other factors when maize is included in the rotation, and their specific nature is not clear. Crookston and Kurle (1989) suggest that response was not due to the beneficial effects of decomposing above-ground residue. Some evidence suggests that maize residues might produce phytotoxic compounds, causing negative effects on subsequent crops (Guenzi and McCalla, 1966; Guenzi *et al.*, 1967; Yackle and Cruse, 1983). While the effects have been measured on maize and soybean, there is no reason to suggest that similar effect would not occur on sugar beet following maize. This has not been investigated because very little sugar beet production includes maize in the rotation.

Table 10.2. Crop-residue effects on nutrient concentration in sugar beet leaves (Christenson *et al.*, 1979; Christenson, 1997).

Nutrient		Crop residue (t ha^{-1} year^{-1})	
		< 3.5	> 5.0
N	%	4.0	3.7
P	%	0.36	0.31
K	%	4.6	5.1
Ca	%	0.85	0.83
Mg	%	0.74	0.66
B	mg kg^{-1}	58	53
Fe	mg kg^{-1}	205	122
Mn	mg kg^{-1}	24	26
Zn	mg kg^{-1}	36	37

Soil Compaction

Soil compaction results from both natural and anthropogenic (human-made) processes (Horn, 1998). Natural compaction may be the result of processes during soil formation, including glaciation, or because of natural settling due to rainfall and associated water movement, wind and other forces outside the scope of this book. Mechanized farming caused the loaded wheel to appear as a 'tillage tool', which on average 'treats' every point, on average, several times a year (Håkansson *et al.*, 1988). The effects of traffic and tillage compaction both affect sugar beet production and a recent symposium covered many aspects of the anthropogenic effects (IIRB, 1998). Here these are divided into the effects of wheeled traffic on ploughed soil and the effects of heavy axle loads on deep-soil compaction.

Wheeled traffic on ploughed soil

The structure of freshly ploughed soils is fragile with a very low load-bearing capacity. Autumn-ploughed medium- and fine-textured soils and spring-ploughed coarse-textured soils are subject to compaction by spring field operations. Tractors and other farm implements, regardless of size, tyre configuration or tyre pressure, often exceed the load-bearing capacity of these ploughed soils and cause damage (Werner *et al.*, 1998).

Consequently, compaction as a result of wheel tracks frequently creates problems for crop production. It has been estimated that up to nine field operations may be carried out for seedbed preparation, spraying, fertilizing and sowing, cultivating and harvest in conventionally tilled sugar beet production. The total track area from tractor wheels and other implements may be more than the field area (Jaggard, 1984).

Henriksson and Håkansson (1993) pointed out that the extent of compaction of the plough layer is determined by the soil moisture content, wheel track distribution, the number of passes, wheel positioning in the furrow or on land, the load and slippage of the wheels, the wheel tracking arrangements and wheel lugs, tyre pressure and other characteristics of the wheels. Therefore, it appears that tyre pressures affect wheel traction, causing greater soil compaction, since the soil cannot support the weight of the implement.

Hebblethwaite and McGowan (1980) found that sugar beet was very sensitive to compaction during the emergence and establishment phases. They found a 45% yield reduction from compaction, due mainly to the decreased population and uneven distribution of the plants. Gemtos and Lellis (1997) also reported that sugar beet is more sensitive to compaction by increasing pressure on the soil than is cotton. Compaction reduced the amount of root dry matter, but the aerial portion was affected less. At the lowest compaction pressure the root-to-shoot dry matter ratio was 1.3, and with 400 kPa the ratio declined to 1.0.

In a survey of farms, Cook *et al.* (1959) found a reduction in the yield of sugar beet with increased numbers of passes for secondary tillage. When the soils were worked once or twice, the yield was the same. With three workings, the yield was reduced by 9%, with four 12% and with five or six 15%. Even with the lower axle loads of that time, it is apparent that additional wheelings decreased the yield of sugar beet. Allmaras *et al.* (1988) demonstrated that iso-stress lines exceeding 0.1 MPa penetrate to depths below 30 cm during one pass of a moderate-sized tractor. Subsequent and repeated passes continue to compact soils to greater bulk densities, reducing soil porosity and aeration to depths of at least 50 cm.

Heavy axle loads

Farm machinery, such as sugar beet harvesters and attendant trailers, is an important cause of compaction (Håkansson *et al.*, 1988). With heavy axle loads, compaction may even extend into subsoils and the effect can persist or become permanent. Wheeled traffic studies show that compaction on a silty clay loam from normal farming operations is to a depth of 45 cm. Autumn tillage essentially alleviated compaction in the 0–15 cm layer (Voorhees *et al.*, 1978). Ploughing was more effective than discing or chiselling in decreasing compaction in the 15–30 cm layer. Compaction below tillage depth was not completely ameliorated by annual freezing and thawing. The strength and density of wheel track clods were greater and the average aggregate diameter was larger than with clods outside tracks, a difference that persisted over winter.

High axle loads on a dry soil caused little subsoil compaction. On wet soils, high axle loads compacted soil to a depth of 60 cm or more. Surface-layer compaction from annual inter-row wheel traffic did not cause a significant yield response consistently at any site (Voorhees *et al.*, 1989). Recent experiments have examined the effect of sugar beet harvesters up to 38 t in weight (Arvidsson, 1998). Soil was damaged to 50 cm depth.

Controlled Traffic

In conventional tillage systems

Sugar beet is universally a row crop, so it is feasible to arrange wheelings between rows. Soil is then not compacted in the zone below the plant. Trouse (1985) suggested that two zones are needed for crop production – a loose-soil zone in which to grow crops and a firm-soil zone to provide support and mobility for machinery.

Robertson *et al.* (1978) presented a graph, reproduced here as Fig. 10.3, showing the effect of subsequent passes on soil compaction. Over 70% of the compaction from wheel traffic on ploughed soils is with the first pass. Since the soil after one pass is too dense for plant roots to enter, it seems reasonable to reduce compaction by following the same track for subsequent passes across the field.

Spoor (1979) earlier suggested that maximum use of natural weathering and drying should be made to reduce the number of tillage operations and wheelings prior to sowing. He suggested that 90–95% of the preparations for the final seedbed should be made during the autumn. He also suggested that consideration should be given to establishing the bed and tramline systems after primary tillage, rather than after sowing.

Lamers *et al.* (1986) conducted a comprehensive analysis of controlled traffic in a 9-year study in The Netherlands. A controlled traffic system of crop production was compared with conventional tillage at two sites. Soils in the beds had 4% greater pore space, double the saturated water permeability and oxygen diffusion, and smaller aggregates than conventional tillage. Using established lanes reduced tool resistance by 25% and energy used in seedbed preparation by 50%. However, the yield of sugar from the two systems was not different, averaging 11.0

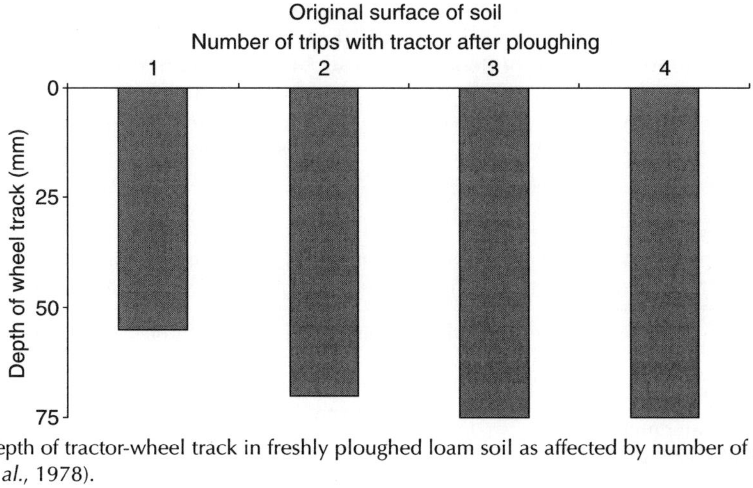

Fig. 10.3. Depth of tractor-wheel track in freshly ploughed loam soil as affected by number of passes (from Robertson *et al.*, 1978).

and 11.1 t ha^{-1}, respectively, for the conventional and controlled systems. The extra cost incurred by modifying tractors and acquiring machinery was estimated to be 30% of annual costs. They concluded that controlled traffic could not be recommended for arable farming in Dutch polders, due to the limited benefits in comparison with increased costs.

Henriksson and Håkansson (1993) summarized a number of studies related to controlled traffic. They suggested that controlled traffic systems have resulted in energy saving at subsequent tillage. However, crop yields have been improved only on soils most sensitive to compaction.

Ridge tillage

In some countries ridges are formed either in autumn, in spring or at inter-row cultivation time. The major advantage, particularly if ridges are established in the autumn, is that soil on the ridges is drier and warmer at sowing. A second advantage is that a smaller area of the field is compacted by traffic, since all wheels are in the same tracks. Robertson and Erickson (1983) estimated a reduction of 38% for fuel consumption and 32% for labour associated with ridge tillage compared with conventional tillage for maize production. However, ridge tillage is not suited to all soils. Sandy and organic soils are subject to wind erosion and fields with slopes greater than 3% are subject to water erosion if the ridges run parallel with the slope. Soils with greater than 40% clay are not suitable for ridge tillage (Robertson and Erickson, 1983).

Liebig *et al.* (1993) made an intensive study of soil properties on a silty clay loam soil after 10 years of continuous ridge tillage. Traffic lanes were installed and used throughout the study. Bulk density, soil strength, aggregate MWD, water content at field capacity and wilting point were all greater in the trafficked inter-row than in the row (Table 10.3).

Friessleben *et al.* (1988), in Germany, described a system of building ridges in autumn for sugar beet seeding in spring. Xu (1991) compared ridge tillage with conventional tillage for 2 years on a loam soil. Early season growth and yield were similar on both tillage systems when averaged across 2 years. Brutlag *et al.* (1989) reported that the ridge-tillage system gave a greater yield of both roots and sugar than conventional tillage. In addition, they estimated a 42% reduction in fuel and 18% reduction in labour. Yield increases on ridges over conventional tillage were 13.7% and 22.0% for roots and sugar, respectively, in a 1-year study in North Dakota on a silt loam soil (Giles *et al.*, 1993).

Insufficient studies have been conducted on this tillage system to evaluate its widespread viability for sugar beet production. Effects on nutrition are completely lacking in all studies.

Tillage

The objectives of tillage may include the production of suitable conditions for sowing, seedling establishment and plant growth by: (i) loosening soil; (ii) controlling weeds; and (iii) burying plant residues and incorporating amendments, such as fertilizer, manure, pesticides and lime (Henriksson and Håkansson, 1993).

Table 10.3. Comparison of several physical measurements of soil taken from the trafficked inter-row, non-trafficked inter-row and the ridge in a long-term experiment on a silty clay loam (Liebig *et al.*, 1993).

Measurement	Trafficked inter-row	Non-trafficked inter-row	Ridge
Bulk density (g ml^{-1})	1.5	1.4	1.3
Soil strength (kPa)	23	15	11
Mean weight diameter (mm)	3.1	2.9	2.8
Water content at wilting point (v v^{-1})	0.29	0.28	0.25

Primary tillage

Mouldboard ploughing is the most common form of primary tillage used for sugar beet production. In northern Europe and the USA, autumn ploughing is usually done on medium- and fine-textured soils. Over-winter weathering of soil creates a finer soil structure, since the freezing and thawing ameliorates large clods. Coarse-textured soils are usually mouldboard-ploughed in the spring, with subsequent secondary tillage. In Mediterranean climates, autumn sowing of sugar beet means that ploughing is in late summer, giving soil time to weather only as a result of rain or irrigation and wetting and drying cycles.

Chisel and disc ploughs are also used for primary tillage. The main advantage is that residues are left on the soil surface and the energy needed is less than for mouldboard ploughing. Robertson *et al.* (1979c) estimated that a chisel plough system used 38% and a disc plough system 57% less fuel per unit of land compared with conventional tillage. Hao *et al.* (2001) found no difference in sugar beet yield between conventional tillage and a chisel plough minimum-tillage system. They felt that implementation of the minimum-tillage system would be highly feasible in irrigated crops in Alberta, Canada.

Secondary tillage

Secondary tillage is the process of seedbed preparation after primary tillage. One or two harrowings are all that are needed for seedbed preparation on autumn-ploughed soil with a smooth surface and a good frost mould or on spring-ploughed soil that has been pressed after ploughing. Work in Sweden (Henriksson and Håkansson, 1993) has shown that a harrow in which each section is supported by rollers prepares a superior seedbed to a standard S-tine harrow. Stout *et al.* (1956) concluded from a study concerning the effect of moisture and compaction on sugar beet emergence that the use of minimum tillage would conserve moisture so that seeds may be placed in more optimal conditions.

Reduced tillage

Mouldboard ploughing is expensive in terms of time taken, fuel and machinery costs. Consequently, there have been many studies in various countries to determine whether sugar beet could be produced successfully without conventional ploughing. Results have been variable. Deibert *et al.* (1984) in North Dakota, USA, found that yields were depressed in reduced-tillage systems, whereas Miller and Dexter (1982) found the opposite, provided weeds were controlled, which is more difficult without the plough. Halvorson and Hartman (1984) found that the potential advantages of reduced crusting, better soil water conditions in the seedbed for germination and reduced energy requirements and production costs made reduced tillage a viable option for irrigated sugar beet in eastern Montana.

Reduced tillage had no effects on row and furrow bulk density, water infiltration rate, weed populations, emergence percentage, final stand, yield, quality or recoverable sugar as compared with conventional tillage (Glenn and Dotzenko, 1978). Reduced tillage produced a sugar beet crop of comparable tonnage and quality to that of conventional tillage, while reducing the number of field operations by 40%. In Germany, Hoffman (1998) reported that reduced-tillage sugar beet yielded less than with conventional tillage, even though there was better growth with the reduced tillage early in the season. The reduction in growth rate was always when the taproot swelling started. Causes such as emergence, response to applied nitrogen and mineralization of nitrogen were ruled out. Root length density measurements were made on both treatments, showing more roots per unit of soil in the reduced-tillage system. It was suggested that greater root resistance in the reduced tillage system promoted formation of abscisic acid in the root. Translocation to the leaves may have induced growth changes later in the season. Sanchez *et al.* (2001) reported a similar yield of sugar beet with reduced tillage compared with conventional ploughing. However, the net return per unit of land was 30% greater with reduced tillage compared with conventional ploughing and harrowing.

Deep ploughing and subsoiling

All the operations so far discussed are confined to the plough layer of approximately 25 cm depth. Sugar beet roots use nutrients and water to 2 m and more. Many experiments have examined whether tillage below conventional ploughing might increase production.

Russell (1956) cites work from the 18th century, where the yield of turnips was increased 20% by ploughing to a depth of 30 cm compared with shallower ploughing. He further reported on his work in the 1940s and 1950s that ploughing to a depth of 30 cm or subsoiling to a depth of 45 cm increased crop yields on about half the fields in which an experiment was carried out. He further states:

> Unfortunately, it was not possible to recognise what was the difference in soil properties between the 50% of the fields that responded appreciably to deep tillage and the remainder whose yields were either unaffected or sometimes reduced by deep tillage.

It is not surprising, then, that there are mixed results in the literature concerning these two practices. Robertson *et al.* (1979b) stated: 'Deep tillage is likely to improve crop yields only on problem soils and under average conditions will have little effect.' Similarly, in Switzerland, Furrer (1973) found no yield benefit from deep ploughing but increases from subsoiling on six out of eight fields.

These observations suggest that appraisal of subsoil conditions is essential to the success of subsoiling, making an examination of the soil profile necessary. Subsoiling or deep ploughing without adequate evaluation of conditions would be analogous to application of fertilizer without first assessing the needs through soil analysis.

On sandy soils in North Carolina, Vepraskas and Miner (1986), Vepraskas *et al.* (1986) and Vepraskas (1988) found that the bulk density of the EB horizon (transitional layer between an E and a B horizon), the sand content of the B horizon and cumulative rainfall accounted for 96% of the observed variation in subsoil root development for the non-subsoiled treatment. Subsoiling reduced the resistance of penetrometer readings and significantly increased root penetration into the subsoil. These results support the need for adequate evaluation of conditions prior to such deep tillage.

Soane *et al.* (1986) demonstrated the need to consider wheel traffic after subsoiling. They compared deep loosening followed by mouldboard ploughing, mouldboard ploughing followed by deep loosening and simultaneous deep loosening and mouldboard ploughing in a single operation. Their results showed that deep loosening followed by mouldboard ploughing and subsequent random surface wheeling can cause significant recompaction of the subsoil zone, regardless of whether the wheels operate on the surface or in the furrow. Three possible methods of ensuring that the effect of deep loosening remains to benefit the next crop were suggested. One would be to adopt a one-pass system incorporating deep loosening, surface cultivation and sowing; a second would use controlled wheeling in a bed-management system; and a third would use lightly loaded tractors equipped with low-ground-pressure tyres on simultaneously ploughed and deep-loosened surfaces.

Hull and Webb (1967) at Broom's Barn in the UK showed that subsoiling increased the yield of sugar beet roots by 1.8 and sugar by 0.26 t ha^{-1}. Also McEwen and Johnston (1979) found that subsoiling increased yield by 11% on average over 4 years. In the Saginaw Valley of Michigan, the response of two varieties of sugar beet was evident each year and was attributed in part to improved soil aeration by subsoiling and where wheel traffic was absent (Johnson *et al.*, 1989; Johnson and Erickson, 1991). Subsoiling improved root yield by up to 10.6 t ha^{-1} and was attributed to improved soil aeration in the root zone. Elimination of preplant wheel traffic increased root yield by an average of 8.3 t ha^{-1}. Others point out the need to minimize wheel traffic in sugar beet production systems, regardless of the primary tillage practices employed (Weatherly and Dane, 1979; Winter, 1983).

Where land has been ploughed regularly at about the same depth, a plough pan (or dense layer) often forms just below this depth. Commonly this is 2–6 cm in thickness, and it may prevent free passage of sugar beet roots, water and air. If it is a thin layer, slightly deeper ploughing may remove most of the problem at least cost. A compacted layer of greater thickness or at greater depth requires subsoiling. Spoor (1979) demonstrated the importance of close-spaced winged tines; working in dry conditions, just below the compaction, these gave uplift and removed barriers to root growth most effectively for sugar beet. We have much field experience that supports Spoor's experimental work, having seen large improvements in sugar beet yield where barriers to root growth have been removed.

Freezing and Thawing

The action of freezing and thawing has been recognized as improving soil for a long time. The creation of 'frost mould' on the surface of autumn-ploughed soils is considered essential to the early sowing of many crops, particularly sugar beet, in northern Europe and the USA. The issue concerning freezing and thawing is not whether there are advantages to the action, but rather the depth to which it will ameliorate soil compaction.

Although there are some reports indicating that freezing and thawing ameliorates soil compaction to greater depths than the plough layer, the majority of evidence is to the contrary. Van Ouwerkerk (1968) found no alleviation of compaction in a sandy loam subsoil during a 6-year period. After compacting a clay loam soil in Minnesota, USA, Blake *et al.* (1976) found no change in subsoil bulk density over a 9-year period, in spite of freezing to a depth of 1 m in the winter. Following artificial compaction of a sandy loam forest soil in Minnesota, the bulk density gradually decreased in the 0–7 cm layer and after 9 years was the same as in uncompacted plots (Thorud and Frissell, 1976). There was no change in the bulk density of the 15–22 cm layer during this time. In Idaho, USA, on a loamy soil, the main part of the increase in bulk density at 5 cm had disappeared after 23 years, but at 15 and 30 cm a large part of the compaction remained.

Voorhees (1983) reported only slight amelioration of surface-layer compaction by one year's freeze/thaw activity in a clay loam in Minnesota not subjected to tillage. Soils under zero tillage accumulated more snow over winter and did not freeze as deeply as soils ploughed in autumn. The greater depth of frost penetration resulted in significantly greater surface displacement on autumn-ploughed soil. However, soils under both tillage treatments quickly reconsolidated on thawing and returned to near-prefreezing bulk densities prior to spring planting. The formation of ice lenses appears to create pores that are inherently unstable and collapse as the ice melts and the soil drains. The unstable nature of the pores appears to be in contrast to pores created by tillage, which, on account of their formation by vertical and horizontal displacement of peds, are more stable (Kay *et al.*, 1985). Thus it appears that, even in a climate with deep annual freezing, compaction persists for many years, except in layers near the surface.

Soil Conditioners

Numerous synthetic products have been tested over the past 50 years with a variety of claimed benefits to soil structure. Work with soil conditioners for sugar beet centred on a product marketed under the name of Krilium™. Results were mixed. Baird *et al.* (1954) found that a heavy application (2.2 t ha^{-1}) improved soil structure, but yield between treated and non-treated was not significantly different. Haise *et al.* (1955) found an improvement in soil structure, but a linear decrease in yield with increasing quantity of conditioner. Martin *et al.* (1952) reported that there was no increase in sugar beet yield due to conditioner. Smith (1954) also reported reduced yield from the use of Krilium as a soil conditioner for sugar beet production, but Bolton and Aylesworth (1968) showed that sugar beet yield was improved and the effects of conditioner plus fertilizers were additive.

The product is not used much now, but interest continues (Smucker and Leep, 1975; Wallace and Wallace, 1986; Nelson, 1998) in finding benefits. During the 1970s there was widespread use of Vinamul, a spray that bound together surface particles to prevent wind erosion, but this has largely been replaced by other less expensive techniques.

More recently, other products have been studied with the aim of improving yield by altering the physical structure of soil. Research has centred on the following compounds: sulphonated polystyrene, including a low-density and a high-density formulation, polystyrene, polyacrylonitrile, hydrolysed polyacrylonitrile, hydrolysed sulphonated styrene acrylonitrole, polyvinyl alcohol and polyacrylamide gel. Research has shown improved soil aggregates and increased water-holding capacity and water-use efficiency from use of these materials. There is also evidence of improved emergence, reduced crusting, better seedling vigour and, in some cases, improved yields. It is not the scope of this book to review all of the work concerning these materials. Wallace and Terry (1998) have recently reviewed the subject.

Summary

Manipulation of soil physical conditions is an important component of soil management for crop production in general and sugar beet in particular. Development of good soil structure allows roots to explore unimpeded for water and nutrients, leading to maximum yield potential. Soil structure can be improved by cropping and tillage systems and the effects of both are reviewed above.

Proper timing and/or reducing traffic on ploughed soil are the best means of reducing effects of wheelings on soil structure. On autumn-ploughed soils most of the necessary smoothing should be done in the autumn so that winter frosting ameliorates the soil surface. Limited smoothing should be done in the spring. Harrows with tines that do not bring up unweathered soil, supported by rolling baskets, appear to be best for secondary tillage.

Reduced-tillage systems have been researched with mixed results. Systems wherein traffic is limited to specific roadways show promise from a soil compaction point of view. However, detailed studies have shown that yield increases do not offset the cost. Ridge tillage appears to save fuel, but evaluation of its effect on yield and nutrition is still incomplete. Neither system has been adopted widely.

Oxygen supply is usually non-limiting for leaves in sugar beet production but it may be limiting for roots when soils are compacted. The overall effect is that diffusion is slowed, depriving roots of an adequate supply. This limits growth and other nutrient uptake. A diffusion rate approximately twice that for cereals seems to be required by sugar beet.

Nitrogen and potassium need to be supplied in greater quantities on compacted soils. In the case of nitrogen, reduced mineralization and nitrification, coupled with increased denitrification, increase the fertilizer need of sugar beet. In the case of potassium, the need for additional fertilizer is related to diffusion rates in compacted soil and the quantity of potassium accessible to roots. In contrast, additional phosphorus fertilizer is not needed on compacted soil for sugar beet production, probably because crop needs are small and most soils are well supplied with available phosphorus.

11
Time, Form and Method of Fertilizer Application

The goal of a fertility programme is to provide adequate, but not excessive, quantities of plant nutrients for optimum growth. Not only is the amount supplied important, but also the timing and form of all nutrients. Nitrogen timing is critical for obtaining the optimum production of both roots and sugar. Early-season supply is needed to promote the development of a full canopy, but excessive availability later in the season causes a decline in sugar production. Early-season availability of phosphorus is important for the development of a vigorous root system, but the rate of uptake over the course of the growing season is fairly constant. Calcium and magnesium are also needed at a fairly constant rate over the season. Potassium and copper are taken up rapidly in the first 2 months but the rate declines later in the season. Micronutrients are needed early, but the rate of uptake seems to be fairly constant. Since such small quantities of material are used to provide for the needs of sugar beet crops, care must be exercised so that reactions in the soil do not render the nutrient unavailable. Weather may also influence the availability of the nutrient and therefore the timing, form and position in the soil in respect of the growing crop.

TIME OF APPLICATION

Nitrogen

Growers have several options with the timing of the fertilizer, depending on the climatic region: autumn, late winter, early spring, planting time and side dressing or top dressing after plants have emerged. The choice is governed by rainfall pattern, the price of the fertilizer, availability when market supplies are limited or workload in the field. Where several row crops with different planting dates are grown on a single farm, field operations for later-planted crops and/or weather may interfere with timely post-emergence treatment of sugar beet.

Autumn compared with spring application in the USA

It may be desirable to apply nitrogen in the autumn because the cost of fertilizer is favourable or to reduce spring field operations. It is recognized that this application has the risk of greater losses. However, in late autumn, applications of nitrogen in the NH_4 form when soil temperature is below 10°C reduce the risk of losses associated with nitrification. Two studies outlined here compared the effects of autumn and spring application.

Smith (1982) found no difference in sugar beet yield or sugar content between autumn- and spring-applied nitrogen when averaged over 3 years and four forms of nitrogen in the Red River Valley of North Dakota, USA. Contrasting results were obtained in Michigan, where autumn application resulted in a lower yield than spring application (Christenson, 1986). Data in Table 11.1 show that an additional 39 kg N ha^{-1} was needed in the autumn to give the same yield as in spring. It is apparent that autumn application is less efficient than spring application in Michigan. With increasing concerns over nitrate in the water supply, there is little to recommend the practice of autumn application of nitrogen where leaching takes place.

Top dressing experiments

Comparisons between seedbed and top-dressed nitrogen have been made in the UK and USA. A study with ten experiments in Scotland showed that, although top dressings of 50 and 75 kg N ha^{-1} increased the yield of tops, there was no effect on the yield of sugar (Edinburgh and East of Scotland College of Agriculture, 1957). Williams and Cooke (1971) reported that top dressings increased the yield of sugar beet after a wet spring on a clay soil at Saxmundham, presumably by making good losses probably due to denitrification, since leaching is slow on these soils.

Draycott (1972) summarized the results of two large groups of experiments in the UK, comparing various seedbed and top dressing treatments. The data are reproduced in part here (Table 11.2). Adams (1960) made 28 experiments comparing 75 and 150 kg N ha^{-1} applied in the sugar beet seedbed or as a top dressing at the end of June. The smaller amount of nitrogen increased the yield of sugar most when applied in the seedbed. When 150 kg N ha^{-1} was given in the seedbed, the yield of sugar was increased slightly, and giving half this amount in the seedbed and half as a top dressing increased it no further, but there was an increase in the yield of tops. This split dressing decreased juice purity greatly, but affected sugar percentage no more than when 150 kg N ha^{-1} was applied in the seedbed.

In 34 experiments from 1959 to 1962, Last and Draycott (1972) found no advantage in splitting the application between seedbed and top dressing (Table 11.2). In every year

Table 11.1. Yield of roots and sugar as affected by time and amount of nitrogen application (Christenson, 1986).[a]

	Amount (kg N ha^{-1})			
	0	73	112	151
Time of application	Yield (t ha^{-1})			
Autumn				
Roots	56.2	+5.2	+6.9	+9.7
Sugar	8.49	+0.63	+0.68	+0.96
Spring				
Roots	56.2	+5.5	+9.7	+8.8
Sugar	8.49	+0.67	+1.02	+0.85
June				
Roots	56.2	+5.3	+7.6	+6.6
Sugar	8.49	+0.57	+0.66	+0.24

[a]The autumn application was ploughed down and the spring broadcast and tilled into the soil and a top dressing was applied on the soil surface approximately 50 days after sowing.

Table 11.2. Comparison of nitrogen in the seedbed, as a top dressing or as a split application.

	No nitrogen	In seedbed	In seedbed	As top dressing	Split application
			kg N ha^{-1}		
	0	75	150	75	150
Average of 28 experiments, 1956–1958 (Adams, 1960)					
Sugar (t ha^{-1})	5.23	+0.79	+0.85	+0.60	+0.89
Tops (t ha^{-1})	25.1	+8.5	+14.5	+9.0	+16.1
Sugar (%)	16.0	−0.3	−0.7	−0.3	−0.6
Juice purity (%)	88.1	−0.5	−0.3	−0.5	−1.2
Average of 34 experiments, 1959–1962 (Last and Draycott, 1972)					
Sugar (t ha^{-1})	5.76	+1.17	+1.29	–	+1.02

there was a smaller response to the split application than to giving all of the fertilizer in the seedbed. Furthermore, there was a greater chance of depressing yield by splitting the application than from all of it being given in the seedbed. No crop on soil containing more than 1.7% organic carbon responded to top dressing. Rainfall and response were not clearly correlated. A regression combining spring and early summer rainfall and percentage organic carbon accounted for a small part of the variation in yield response to top dressing. Even on crops damaged by ectoparasitic nematodes (Docking disorder), giving all of the nitrogen in the seedbed was as good as giving part as a top dressing (Cooke and Draycott, 1971).

Similarly, in the USA, Christenson (1986) reported a lower yield with top dressing in June than when the entire application was made in the seedbed before planting. Carter and Traveler (1981) also found that nitrogen supplied 12 weeks after sowing promoted top growth at the expense of sugar and root growth. Interestingly, sugar beet from both treatments took up the same amount of nitrogen. The root : total dry-weight ratio at harvest was similar for early and late application. Carter and Traveler suggested that there was insufficient time for the sugar beet plant to utilize fertilizer supplied late under conditions in central Idaho, USA. Experiments in France (Christmann, 1963) and Ireland (Gallagher, 1967) add further evidence that late application of nitrogen is detrimental to sugar production.

The concentration of NO_3-N in leaves was smaller with split dressing early in the season but larger later, which probably accounted for the decreased juice purity of the roots at harvest (Last and Tinker, 1968). Top dressings on some fields had little effect on NO_3-N concentration of petioles in August, because dry weather made fertilizer unavailable in the dry surface layer of soil. Where plots were irrigated, a top dressing of nitrogen caused a large increase in NO_3-N concentration and decreased sugar percentage at harvest.

In summary, experiments in the USA and EU show no advantage from giving some of the nitrogen required by spring-sown sugar beet as a top dressing at the conventional time of early June. It is more likely to do harm rather than good at this late stage, being used to produce more top growth but no more sugar yield. However, where a large amount of fertilizer nitrogen is required, which is likely to decrease plant population if given in one application at sowing (see Fig. 2.10), part should be given in the seedbed and the remainder after emergence (Chapter 2).

Phosphorus, Potassium and Other Major Nutrients

Table 11.3 summarizes early work with phosphorus fertilizers. Although carried out some 40 years ago, when soils probably contained lower concentration of phosphorus than

Table 11.3. Comparison of autumn and spring broadcast application of phosphorus for sugar beet.

	No phosphorus	Autumn	Spring	Reference
		Yield (t ha^{-1})		
Sugar	5.42	+0.03	+0.11	Adams, 1961b
Roots	25.3	+15.9	+13.5	Larson, 1954b
Roots	26.9	+4.7	+5.1	Schmehl *et al.*, 1954

today, no important differences were found between the various times of application. Attention was often drawn to the danger of large fertilizer applications just prior to sowing, as mentioned above with nitrogen. Brummer (1966) in Finland found that seedling emergence was drastically decreased by superphosphate and potassium chloride in the seedbed and advocated application after ploughing in autumn as did Holmes *et al.* (1973).

Increasingly in the UK over the past 30 years, there has been a trend towards autumn or winter application of the four major nutrients phosphorus, potassium, magnesium and sodium. The four fertilizers are blended in amounts dictated by soil analysis of individual fields or groups of fields. Fresh studies were made about 10 years ago in two papers to recheck optimum timing (Armstrong and Jaggard, 1990; Allison *et al.*, 1994). Armstrong and Jaggard (1990) showed the damage caused to soil physical properties by inopportune timing of spreading traffic. In both papers (where compaction was avoided) there was no difference in yield between times of application (Table 11.4). In the case of sodium, a nutrient that leaches quite quickly on sandy soils (Chapter 4), clearly this should be taken into account when planning the timing of blended fertilizers containing sodium.

Micronutrients

Boron

Berger (1950) showed that boron deficiency could be corrected equally well by broadcast and by placement to the side and below the seed at planting. Initially, recommendations in Michigan suggested broadcasting 12 kg borax ha^{-1} prior to sowing sugar beet (Cook, 1948). Based on Berger's work and other studies, recommendations in Michigan were modified to place a fertilizer containing 0.25% B to the side and below the seed (Cook *et al.*, 1957). This supplied sufficient boron for sugar beet when 650–850 kg ha^{-1} fertilizer was applied.

Foliar application is another option and Hamence and Oram (1964) found that a foliar spray, applied in June or after singling, was just as effective as fertilizer broadcast in the seedbed. Gupta and Cutcliffe (1978) found that a band application of 1.12 kg, a broadcast application of 2.24 kg or two foliar sprays each with 1.12 kg B ha^{-1} eliminated deficiency (brown heart) on rutabaga (*Brassica napobrassica*).

In summary, an application of boron broadcast in the seedbed, side-placed in a band or on foliage as a spray can all be used effectively to correct deficiency.

Table 11.4. Effect of time of application of phosphorus, potassium, sodium and magnesium on sugar yield in 14 experiments in the UK (Allison *et al.*, 1994).

	Autumn	Early spring	Late spring	Autumn/spring
Sugar yield (t ha^{-1})	9.75	9.78	9.86	9.85

Manganese

Broadcasting prior to planting is an attractive method, particularly when the nutrient can be mixed with fertilizer, but success has been limited due to the large amounts needed. Murphy and Walsh (1972) and Martens and Westermann (1991) summarized the results of broadcast manganese for a number of crops, including sugar beet. The optimum amount ranged from 14 to 134 kg Mn ha^{-1}. Since there is very little residual effect over the short term, this practice is expensive in that applications need to be made every year. Henkens and Smilde (1967) and Smilde (1968) suggested that broadcasting manganese was far too costly since the practice would have to be repeated annually. Most crops that need manganese in Europe are treated with foliar sprays as described in Chapter 6. However, in the USA, fertilizer containing manganese applied to the side and below the seed is successfully used to correct manganese deficiencies on sugar beet.

Iron

Soil applications of inorganic forms of iron are generally ineffective unless very large amounts are used (Martens and Westermann, 1991). Anderson (1982) stated that most sources of iron were effective in alleviating chlorosis when soil-applied but few were economically feasible. Consequently, in areas of the USA where chlorosis occurs, broadcast application of $FeSO_4$ is not recommended (Blaylock *et al.*, 1996; Rehm *et al.*, 1998). Since application timing is closely tied to the source, the topic is dealt with below.

FORMS OF FERTILIZER

Nitrogen

Much of the research comparing different forms of nitrogen was prior to 1990 and the interest in conducting source studies seems to have waned in recent years. Most of the research has shown little difference in the effectiveness of the various sources, provided an advantage was not given to one source over another. Late-season or emergency application seems to favour the nitrate form of nitrogen; however, cases where this would be desired are expected to be few in number.

Positional availability may be a factor. In a 4-year study, Smith (1982) found that anhydrous ammonia yielded 7.4 t ha^{-1} more roots than urea, urea–ammonium nitrate (UAN) or ammonium nitrate (Table 11.5). It was felt that the reason that anhydrous ammonia yielded more was due to the positional avail-

Table 11.5. Comparison of form of nitrogen and method of application on yield of sugar beet in the Red River Valley of North Dakota, USA (Smith, 1982).

Nitrogen source	Method of application	Root yield (t ha^{-1})
Average of 4 years		
Urea	Surface	43.9
Urea–ammonium nitrate	Surface	+2.2
Ammonium nitrate	Surface	+4.0
Ammonia	Injected	+7.4
Average of 2 years		
Urea	Surface	48.6
Urea	Injected	+7.2
Urea–ammonium nitrate	Surface	+0.7
Urea–ammonium nitrate	Injected	+5.4
Ammonia	Injected	+7.4

ability of the ammonia. In a companion 2-year trial, where urea and UAN were injected, the yield of roots was equivalent to the yield from ammonia (Table 11.5).

While positional availability may be a factor in these results, when surface-applied, both urea and UAN can lose nitrogen through volatilization of ammonia. In the 4-year study, the difference in yield between surface-applied urea and injected ammonia declined with increasing rainfall between May and September. When there was 260 mm precipitation, urea yielded 11.2 t ha^{-1} less than with ammonia. With 470 mm, the difference was 3.8 t ha^{-1}. The average difference between urea and ammonia was 7.4 t ha^{-1} while with ammonium nitrate it was 3.4 t ha^{-1}. The estimated yield loss due to ammonia volatilization from urea in this case was 4 t ha^{-1}.

Several studies conducted across the sugar beet-growing area of the USA show little difference between several nitrogen-fertilizer sources (Table 11.6). In Michigan, where fertilizers were broadcast and tilled into the soil before sowing, there was a smaller increase in yield over the control from UAN than from urea or urea–urea phosphate. In California, Loomis *et al.* (1960) found a slightly better yield from ammonium sulphate than from the other sources. On a soil with high residual nitrogen, Winter (1975) found ammonium sulphate slightly inferior to ammonium nitrate and calcium nitrate. Other workers (Hills and Axtell, 1950; Boawn *et al.*, 1960b) have reported no important differences.

The main economic advantage of anhydrous ammonia is that it is the least expensive source of nitrogen. In the USA, approximately 42% of all nitrogen used as fertilizer is anhydrous ammonia. It is often delivered to farm communities via pipeline, which reduces the cost. The main disadvantage is that it must be injected into the soil. This can be problematic on wet or stony soils, where there are problems with a good seal on the slit opened by the tine. Even though anhydrous ammonia is the least expensive source, the greatest financial benefit is for crops that required more added nitrogen than sugar beet.

In The Netherlands, when anhydrous ammonia injected at 9–5 weeks before sowing was compared with injection 3–1 week before sowing, the earlier application was inferior to the later (van Burg *et al.*, 1967). Draycott and Holliday (1970) and Draycott (1971) con-

Table 11.6. Comparison of nitrogen sources in respect of root and sugar yield in the USA.

Nitrogen source	Yields (t ha^{-1}) Roots	Sugar
Michigan (Christenson, 1986)		
Control	54.2	8.08
Urea	+9.1	+1.14
Urea–ammonium nitrate	+8.3	+0.87
Urea–urea phosphate	+9.2	+0.95
California (Loomis *et al.*, 1960)		
Control	46.6	6.78
Ammonium sulphate	+13.0	+1.72
Ammonium carbonate	+11.9	+1.76
Ammonium nitrate	+12.6	+1.44
Calcium nitrate	+11.4	+1.63
Texas (Winter, 1975)		
Ammonium sulphate	63.3	–
Ammonium nitrate	+0.9	–
Calcium nitrate	+1.5	–

firmed these results under conditions in the UK. Ammonia injected 6–4 weeks prior to sowing yielded less sugar than ammonia injected into the seedbed. Ammonia in the seedbed gave similar yields to calcium ammonium nitrate (Nitro-Chalk). They found that delaying application until singling time decreased sugar yield, even though the crop contained as much nitrogen at harvest as when applied in the seedbed. Apparently with later application there was insufficient time for the sugar beet plant to metabolize the nitrogen (Carter and Traveler, 1981).

Jameson (1959) compared ammonium sulphate with anhydrous ammonia in one field experiment and found no difference in yield of sugar beet grown with two forms of nitrogen. Where the gas escaped, it scorched the growing plants, but the effects were not serious. On the other hand, Hera *et al.* (1961) found that 80 kg N ha^{-1} as anhydrous ammonia or ammonia in aqueous solution was more effective than ammonium nitrate. Ammonia increased root yields by up to 29% more than ammonium nitrate. However, in experiments in Belgium, Roussel *et al.* (1966) found that anhydrous ammonia gave the same yield in one year and less in another year than ammonium nitrate. Dutch experiments (van Burg *et al.*, 1967) indicated that ammonia gas gave larger yields than calcium ammonium nitrate, whereas in England Draycott and Holliday (1970) found that 125 kg N ha^{-1} as anhydrous ammonia gave the same yield of sugar beet roots, sugar and tops as Nitro-Chalk, although the amount of nitrogen recovered in the crop at harvest was slightly less from ammonia than from Nitro-Chalk.

Over the course of the past 50 years, urea has become a significant source of nitrogen fertilizer. Favourable economics of manufacturing, storage, transportation and handling make urea a competitive source of nitrogen fertilizer. It is also used in foliar sprays because it is un-ionized but concentrated, which gives solutions with a low osmotic pressure, minimizing risk of foliar scorch. Urea hydrolyses in soil to NH_4^+, which then nitrifies to NO_3^-, the form generally taken up by plants like sugar beet. In alkaline soils nitrogen may be lost to the atmosphere as ammonia gas, so incorporation in soil is important. Urea also sometimes contains biuret ($(CONH_2)_2NH$) as an impurity, which is toxic to many plants. For foliar application the acceptable level of the impurity is about 0.3%, but for broadcast soil applications up to 2% of biuret does not have any adverse effect. Few recent reports of problems with biuret have been reported.

Adams (1960) compared urea with ammonium sulphate and calcium nitrate; all were applied to the sugar beet seedbed before sowing or as a top dressing at the end of June. All three forms were equally effective in increasing sugar yield and there was no damage to germination. Calcium nitrate produced sugar beet with the largest yield of tops. Devine and Holmes (1963) compared urea (containing less than 1% biuret) with ammonium nitrate (33.5% N). They made 19 experiments, mostly on alkaline soils in eastern England, and detected no difference between the mean increase in yield of sugar or tops. However, in other experiments with barley, potatoes and grass, urea gave smaller mean increases in yield than ammonium nitrate or sulphate. The conclusion from these experiments and from many made abroad is that urea gives about the same increase in yield of sugar beet as ammonium and nitrate fertilizers.

Potential benefits from slow-release nitrogen fertilizers include: (i) more efficient use of nitrogen by the crop; (ii) less leaching of nitrogen; (iii) lower toxicity; (iv) longer-lasting nitrogen supply; (v) reduced volatilization losses of nitrogen; and (vi) lower application cost (Allen, 1984). In an attempt to minimize leaching losses on a loamy sand soil, Cooke and Draycott (1971) conducted tests to determine if sugar beet would respond favourably on such soils to forms of fertilizer that released nitrogen more slowly. On some soils, isobutylidene diurea showed marked growth and yield increases over calcium ammonium nitrate, but the advantage was not consistent from field to field. Allen (1984) cites several studies on tomatoes, potatoes, watermelon and oats, where slow-release fertilizers gave greater yields than ammonium nitrate and/or urea. This was particularly true in years when leaching losses were greatest.

The relative cost of urea, sulphur-coated urea, Formolene® (urea plus methylolurea), and isobutylidene diurea or urea formaldehyde was 1.0, 2.0, 2.5 and 3.0 per kg of N, respectively. In conclusion, Allen (1984) says, 'Cost has severely limited acceptance of slow release nitrogen fertilizers for most field crops.'

In summary, differences between nitrogen sources for sugar beet production are generally small and the reports above do not consistently show that any sources are superior. Comparisons between forms of nitrogen, in terms of their effect on sugar yield, are only valid where the application was below the optimum, where a range of dosages were applied or where nitrogen uptake by the crop was measured. It seems safe to assume that yields of sugar beet grown with any nitrogen source will not differ greatly, provided there is no damage to the crop and no loss through leaching or to the atmosphere.

Phosphorus

In much of the older work, calcium phosphates with varying water-solubility characteristics were evaluated. For example, Schmehl *et al.* (1952) compared several sources and found that concentrated superphosphate (0–47–0) and ammoniated concentrated superphosphate were equally available. On the other hand, calcium metaphosphate was satisfactory only when the particle size was small, not larger than 0.40 mm. In another study, Olsen *et al.* (1950) found that the absorption was highest from ordinary superphosphate and calcium metaphosphate, followed by dicalcium phosphate, and least from tricalcium phosphate. Generally, phosphorus was more available from materials with greater water solubility.

Lachowski (1959) compared powdered with granulated superphosphate, both placed and broadcast, in fertilizing sugar beet in Poland. There was little difference between the two forms in the yield of tops or sugar or in the quality of the roots. Where the powdered form was used, 7.5 kg P_2O_5 ha^{-1} placed gave the same yield of roots and sugar as 30 kg ha^{-1} broadcast. Where both forms were broadcast, the sugar beet seedlings took up more phosphorus from the granulated superphosphate.

Numerous experiments comparing ammonium phosphates with calcium phosphates have been reported for a wide range of crops. Results summarized by Terman (1971) and Englestad and Terman (1980) suggest a stimulatory effect of NH_4 contained in ammonium phosphates. This has been attributed to lower pH in the rhizosphere, stimulation of root growth and increased metabolic activity of the crop, among other effects. The net effect on deficient soils is appreciably greater, with an early growth response and occasionally higher crop yields from ammonium phosphates than from calcium phosphates with similar water solubility (Terman, 1971). These benefits are not as pronounced on soils more adequately supplied with available phosphorus. With calcium phosphates, dicalcium phosphate (a less soluble compound) is formed within days of application, reducing the availability from calcium phosphates.

Diammonium phosphate (DAP) temporarily increases pH in the zone of application. In some soils this will inhibit complete oxidation of the NH_4^+-N, causing accumulation of nitrite, which is toxic to plants. In addition, DAP releases NH_3 in an alkaline environment, resulting in loss of nitrogen to the atmosphere and/or damage to plants. Consequently, in alkaline soils DAP may give inferior results to mono-ammonium phosphate (MAP) or calcium phosphates.

In 1971 a significant improvement in the manufacturing of ammonium polyphosphates (APPs), known as the pipe reactor, was introduced (Young and Davis, 1980). This allowed the production of polyphosphate fertilizer at local plants on an 'as needed' basis. Anhydrous ammonia is combined with wet process phosphoric acid, forming liquid fertilizers with grades of 10-34-0 and 11-37-0. Up to 70% of the phosphate present in these formulations exists as polyphosphate. Murphy (1979) reviewed the literature comparing ammonium ortho- and polyphosphate for crop production. He concluded that the availability of phosphorus from these materials was essentially equal. Even though no studies were found evaluating sugar beet, it would seem that these results would apply to sugar beet.

In Switzerland, Siegenthaler (1987) compared mineral phosphorus fertilizers (basic slag and superphosphate) with phosphorus in sewage sludge. On soils that were slightly acid or near neutral, all forms were of similar efficiency in terms of both yield and phosphorus uptake (the latter ranged from 52–112 kg P_2O_5 ha^{-1}).

Potassium

In one of the few reports on potassium sources, Gascho *et al.* (1969) compared four potassium fertilizers – KCl, KNO_3, K_2SO_4 and $K.MgSO_4$ – in respect of the yield and quality of sugar beet. The experiments were on loam, clay loam and organic soil. Yields and quality were the same with all four potassium fertilizers, but giving 225 kg K_2O ha^{-1} decreased the quality of sugar beet compared with 112 kg ha^{-1}, which had little effect. Yield with no potassium averaged 48.1 t ha^{-1} with marginal increases of +0.9, +0.9, +1.0 and +0.9 for KCl, KNO_3, K_2SO_4 and $K.MgSO_4$, respectively.

Magnesium

Comparisons between sulphate, carbonate and oxide have been discussed in detail in Chapter 5.

Manganese

Foliar

Foliar application is used to correct manganese deficiency symptoms when soil application is not feasible or when deficiencies occur even though there has been soil application. Generally, $MnSO_4$ and manganese chelate are the products used. However, $MnCl_2$ has also been used, but it is not commonly available. Most work has shown that $MnSO_4$ is the most effective source for foliar application. Murphy and Walsh (1972) and Martens and Westermann (1991) both summarized work done with foliar application over a number of crops. The rates range from as little as 0.3 kg Mn ha^{-1} applied twice to as much as 5.4 kg ha^{-1} applied in three applications. The volume of solution ranged from 31 to 230 l ha^{-1}. When manganese chelate is used, less manganese may be needed than with $MnSO_4$. For example, Randall *et al.* (1975) showed optimum yield from $MnSO_4$ with one spray of 0.56 kg Mn ha^{-1}, compared with 0.17 kg ha^{-1} with manganese chelate.

The most commonly used and widely tested material is manganese sulphate ($MnSO_4.H_2O$ – 33% Mn) throughout the sugar beet areas of the world. Allison (1996) reviewed this work and concluded that 10 kg manganese sulphate ha^{-1} was the most cost-effective way of eliminating deficiency, provided a suitable wetter or adjuvant was included. For severely deficient crops several applications are needed.

Last and Bean (1991) also found that manganese sulphate (15 kg ha^{-1}), plus an adjuvant, rapidly cured deficiency symptoms and was highly cost-effective, yield being increased from 8.8 to 9.6 t sugar ha^{-1}. They also investigated the use of both chelated and cutonic manganese sprays, with and without adjuvant (Last and Bean, 1990). These experiments confirmed the need for an adjuvant and manganese sulphate as the superior source. Only manganese sulphate was able to increase rapidly and maintain manganese concentrations in dried leaves above 30 mg Mn kg^{-1}, the threshold above which the crop is not deficient in late spring/summer.

Working on sugar beet, Voth (1977) showed that sulphate was slightly more effective than chelate, but it took three applications of each for optimum production. The optimum rate for the sulphate was 2.24 kg ha^{-1} total applied in three applications. For the chelate, the rate was 1.12 total in three applications. However, the increase in yield for the chelate was only 60% of that for $MnSO_4$. There was some indication of toxicity at 2.24 kg Mn as chelate. General recommendations suggest 1.25–2.50 kg ha^{-1} of Mn as a foliar spray (Vitosh *et al.*, 1998), applied when deficiency symptoms first appear and repeated at 7–10 day intervals until new growth no longer shows symptoms.

Soil

Voth and Christenson (1980b) reported that manganese deficiency affected sugar beet crops on high-pH soils in Michigan. It was observed that NP fertilizer banded near the seed decreased symptoms. Laboratory, greenhouse and field experiments showed that this was due to slight acidification in the band increasing the availability of soil manganese. Several forms of manganese were also tested. Manganese sulphate and finely ground manganese oxide (MnO) increased manganese concentration equally in sugar beet leaf tissue, from about 30 to 45 mg Mn kg^{-1} dry matter. They concluded that a combination of banded acidic NP fertilizer plus sulphate or oxide would eliminate early-season manganese deficiency.

Other work with banding manganese with fertilizer has been investigated to determine which fertilizers aid uptake of manganese (Table 11.7). Hossner and Richards (1968) reported that manganese uptake was greatest with phosphorus fertilizers with an initial pH of 2–4. Their rank order of phosphorus fertilizers in increasing manganese uptake was MAP = APP >> monocalcium phosphate (MCP) = DAP. Mortvedt and Giordano (1975) gave an order of MAP > DAP > APP > fluid polyphosphates. The results suggest that when manganese deficiencies occur early in the season, use of an acidic fertilizer in combination with $MnSO_4$ or finely divided MnO would be helpful in elimination of the problem.

A broadcast soil application of 56 kg Mn sulphate ha^{-1} (18 kg Mn ha^{-1}) greatly decreases deficiency symptoms and improves yield (Draycott and Farley, 1973). However, a foliar spray of a tenth of this amount is much more cost-effective, so spraying of manganese is preferred in most countries.

Other soil applications have been tested in field experiments using MnO and silicate frits (Farley and Draycott, 1973). Where large amounts are used, deficiency is greatly decreased and the effects can be long-lasting (Henkens and Smilde, 1967), as with very large application of $MnSO_4$. Unfortunately,

Table 11.7. Relative ranking of manganese sources as measured by uptake and ranking of fertilizer sources in respect of improving uptake by sugar beet and other crops.

	Crop(s)	Reference
Relative ranking of manganese sources as measured by uptake		
$MnSO_4$ = MnO (100 mesh) > oxysulphate = MnEDTA	Sugar beet	Voth and Christenson, 1980b
$MnSO_4$ > MnO (60 mesh) > $MnCO_3$ > MnEDTA	Field bean Oat Pangolagrass Tomato	Fitts *et al*., 1967
$MnSO_4$ > MnO > MnO_2 = MnEDTA	Tomato	Fiskell and Mourkedes, 1955
$MnSO_4$ = MnO	Onions	Shephard *et al*., 1960
Fused S plus $MnCO_3$ > fused S plus MnO	Ryegrass	Ludwick *et al*., 1968
Relative ranking of fertilizer sources to improve manganese uptake		
MAP = APP >> MCP = DAP	Soybean	Hossner and Richards, 1968
MAP > DAP > APP > fluid polyphosphates	Oat	Mortvedt and Giordano, 1975
Acid fertilizer > neutral fertilizer	Oat	Hammes and Berger, 1960
NH_4NO_3–MCP–KCl acidic > urea–DAP–KCl alkaline	Sugar beet	Voth and Christenson, 1980b

MAP, mono-ammonium phosphate; APP, ammonium polyphosphate (solid); MCP, monocalcium phosphate; DAP, diammonium phosphate.

the cost of these materials for sugar beet has limited the broadcast application of manganese as a means of correcting deficiencies.

Seed

Correcting manganese deficiency on seedlings by foliar spray is not effective because plants have little leaf area to receive spray. In addition, they are also growing rapidly, causing dilution of the application, and manganese is not translocated much between leaves. Thus several attempts have been made to apply some manganese to seed before sowing, or a little in the seed furrow near the seed.

Sugar beet seed in much of the EU and all of the UK has been pelleted since the 1970s to aid sowing to a stand. Farley and Draycott (1978), working with a seed company (Germain's), experimented in the greenhouse and field with manganese incorporated in seed pellets. At the time, a clay-based pellet was used. MnO prevented symptoms of deficiency and replaced a foliar spray, but plants developed deficiency later in the spring.

Farley (1980) later showed that, in certain conditions, MnO could improve plant establishment, even when plants were not likely to be deficient, probably by accelerating seed germination and emergence. In field trials under conditions of severe manganese deficiency, plants weighed more and contained more manganese on plots sown with seed pelleted with material containing 50% MnO than on plots sown with ordinary seed. Applying a foliar spray of $MnSO_4$ in the third week of June in addition to pelleting the seed with MnO gave higher yields than either the seed pellet treatment or foliar spraying alone.

Manganese-pelleted seed was introduced in the 1980s and used widely in commercial practice on fields known to be severely deficient. Since then the main constituents of the pellet have changed from clay to lighter (wood-based) materials. Unfortunately, this greatly decreased the availability of manganese to the seedlings and the supply of seed including manganese in the pellets has stopped.

Iron

Iron chlorosis is one of the most difficult deficiencies to control. As mentioned earlier, soil application is not consistently effective in correcting the problem. Rapid precipitation of insoluble $Fe(OH)_3$ makes inorganic salts unattractive. Most chelates are not sufficiently stable in the soil for the correction of deficiencies. Norvell (1991) showed that Ca^{2+} competes with the iron in the ligand, decreasing the availability of iron when the soil pH exceeds 7. Iron chelate is the most stable in soil but cost prohibits its use in the soil for crops such as sugar beet. Murphy and Walsh (1972) and Martens and Westermann (1991), reviewing the literature for general crop production, concluded that foliar application seemed to be the best means of correction of the problem.

METHOD OF APPLICATION

Placement Compared with Broadcasting

Yields of many crops are reputedly less from broadcast fertilizers than from the same amount of fertilizers localized near the seed and there are many reviews on this subject (Miles, 1947; Cooke, 1954). The usual method of broadcasting solid fertilizer for sugar beet has been compared with less conventional methods of application, in many experiments in most sugar beet-growing countries. Over the last 60 years the subject has probably had more attention than it deserves, in the opinion of the authors.

Contact with Seed

Fertilizer in close proximity with seed of various crops has been shown to promote early season growth and/or yield. When fertilizer is in contact with seed, there may also be increased risk of crop damage. Yerokun and Christenson (1989) worked with several fertilizers in contact with maize seed. They found that stand reductions were related to the amount of rainfall in the first 10 days after sowing. When precipitation was greater than 18 mm, neither stand nor yield was affected by up to 44 kg N ha^{-1}. However, when less precipitation was received, even as little as 11 kg N ha^{-1} reduced stands and yield. With 44 kg N ha^{-1} in contact with the seed, urea reduced yield by 155 kg ha^{-1} and ammonium nitrate by 80 kg ha^{-1} for each kg N ha^{-1} applied.

Early experiments in England with sugar beet were made by McMillan and Hanley (1936). They mixed fertilizer and seed and sowed them together. This 'combine drilling' decreased germination and emergence. Lewis (1941) found that sowing seed and fertilizer in the same row sometimes gave larger yields than broadcasting, especially on soil deficient in one or more nutrients, but he also found that the technique involved risks to germination. Applying fertilizer in contact with sugar beet seed is risky at best. Results show that free ammonia from urea-based fertilizers is detrimental to germination and emergence. Other fertilizers also have a negative effect, primarily due to osmotic effects.

Band Placement

Lewis (1941) also placed fertilizers in a band below and to the side of the seed, and recommended that the bands should be 4 cm to the side and 2 cm below the seed. Cooke (1949) found that contact with a compound fertilizer (9% N, 7.5% P_2O_5, 4.5% K_2O) damaged seed; there was also some damage when fertilizer bands were 5 cm below and 2 cm to the side of the seed, but fertilizer 8 cm to the side was safe. Yields were reduced by methods of applying fertilizer that decreased plant population. There was little difference between mean yields for all experiments given by broadcast fertilizer and fertilizer placed in safe positions near the seed. Placed fertilizer promoted much more vigorous growth of tops early in the year but by harvest the advantage had disappeared. This is somewhat unexpected as most cultural treatments that increase vigour of sugar beet early in the season usually improve yield.

In further experiments (Cooke, 1951), a phosphorus/potassium fertilizer (16% P_2O_5, 13.4% K_2O) was applied in different ways. There was no damage to germination or plant establishment by fertilizer placed in bands 5 cm to the side and below the seed. There were no significant differences between the yields of sugar given by placed and broadcast fertilizer. Similar yields were given by broadcast applications applied in early spring and worked into the seedbed, and by dressings on the seedbed that were worked in shallowly. In most of the experiments, placement again gave more vigorous growth during late spring and early summer than broadcasting the fertilizer, but by harvest the superiority had disappeared. Shotton (1962) made similar experiments to compare a compound fertilizer (average analysis: 10% N, 10% P_2O_5, 15% K_2O) broadcast with the same fertilizer placed by commercial drills. Placement of the fertilizer produced larger yields than broadcasting on 11 fields, gave the same yields on three fields and was inferior to broadcasting on two fields where the broadcast fertilizer was deeply incorporated in the soil. The average increase in yield from placement was 1 t roots ha^{-1} or 0.2 t sugar ha^{-1}. Germination was not impaired by placement and final plant populations were larger when the fertilizer was placed.

The annual and residual effects of fertilizer placement were investigated in a long-term experiment by Hanley and Ridgman (1963). Sugar beet was given a relatively small dressing of fertilizer. The effect of placement was variable and sometimes reduced yields in dry springs. Analysis of soil showed that placing a small quantity of fertilizer decreased reserves of phosphorus compared with broadcasting twice as much fertilizer – an obvious result but easily overlooked.

Parallel work has been done in the USA, particularly with placement of phosphorus. The practice of applying mixed fertilizers in bands to the side and below the seed was well established in the USA by the early 1950s for a number of crops (Lill *et al.*, 1938, 1940; Millar *et al.*, 1938, 1940, 1945, 1948; Mellor *et al.*, 1950). This method of fertilizer application resulted in the highest yields of sugar beets per unit of phosphate fertilizer.

Radioactive tracer (^{32}P) studies show where and when sugar beet absorbs phosphorus from soil. Olsen *et al.* (1950) showed that placement near the seed increased uptake at thinning time compared with broadcasting. In June, the uptake was greatest from band-placed fertilizer, but later in the season uptake was the same whether banded or mixed with the soil, confirming the Olsen *et al.* (1950) results (Schmehl *et al.*, 1954). Similarly, Lawton *et al.* (1954) found that, in the early stages of growth, sugar beet tops and roots obtained the largest percentage of phosphorus from applied fertilizer in a band.

Table 11.8 gives the effects of banded phosphorus on yield. Over a large number of studies, the preponderance of evidence indicates that fertilizer phosphorus broadcast and tilled into ploughed soil in the spring produces more root and sugar yield than when banded to the side and below the seed (Table 11.8). The average across all studies show that broadcast (seedbed) and applied phosphorus averaged 1.4 and 0.2 t ha^{-1} more roots and sugar, respectively.

Work in Michigan separated the effects of banding on soils with little available phosphorus from those where the soils were well supplied with soil phosphorus. Table 11.9 shows that sugar beet responded most to banding when soil phosphorus was low but was of small benefit compared with broadcasting when soils were well supplied.

During the past decade there has been renewed interest in placement of fertilizer for sugar beet in the EU, probably generated by new machinery capable of accurate placement and used for other row crops. Part of the IIRB Winter Congress in 1991 covered the subject and Dunham (1991b) reviewed the papers. At the IIRB Summer Congress in 1997 the topic of nitrogen placement was discussed in a session on reducing the environmental impact of sugar beet growing (Vandergeten *et al.*, 1997).

Table 11.9. Effect of fertilizer placement on soils with small and large amounts of available phosphorus (Christenson *et al.*, 1975).

	Broadcast	Broadcast plus banded
	Yield (t ha^{-1})	
Low soil phosphorus (4 experiments)		
Roots	54.2	+3.4
Sugar	8.69	+0.45
High soil phosphorus (17 experiments)		
Roots	65.7	+0.3
Sugar	8.54	+0.14

Table 11.8. Effect of phosphorus fertiilizer placed in a band to the side and below the seed compared with fertilizer broadcast and disc-ploughed into soil prior to sowing.

	No phosphorus	Band-placed	Broadcast	
		Yield (t ha^{-1})		Reference
Roots	25.3	+11.6	+13.5	Larson, 1954b
Roots	29.8	+3.7	+7.2	Lawton *et al.*, 1954
Roots	30.2	+10.2	+10.5	Romsdal and Schmehl, 1963
Sugar	6.11	+1.85	+2.01	
Roots	32.5	+5.8	+6.6	Schmehl *et al.*, 1954
Roots	48.6	+3.1	+4.8	Christenson *et al.*, 1975
Sugar	6.53	+0.45	+0.80	
Roots	34.0	+1.1	+2.2	Moraghan and Etchevers, 1981
Sugar	5.86	+0.24	+0.47	
Roots	48.6	+1.6	+2.0	Papanicolaou *et al.*, 1982
Sugar	7.57	+0.23	+0.23	
Average				
Roots	35.6	+5.3	+6.7	
Sugar	6.52	+0.69	+0.88	

Dunham (1991b) concluded from his own and other papers that, in fertile arable soils, placement did not reliably increase yield. In infertile soil, there was a small benefit; also the amount of fertilizer needed to achieve a full yield might be less. Vandergeten *et al.* (1997) confirmed that less nitrogen fertilizer was needed when banded and residues were less. They listed all the benefits and disadvantages of placement.

Fluid Fertilizers

Fluid fertilizers are available as clear liquids and as suspension fertilizers. Clear liquids contain all nutrients in solution while suspension fertilizers contain suspended solids. Liquid fertilizers generally have lower analysis than suspension and dry fertilizers. The principal components of liquid fertilizers are urea, ammonium nitrate, ammonium ortho- and polyphosphate and finely ground soluble potassium chloride. Similar products are used in suspension fertilizers, except that lower-quality materials may be used, thus lowering the cost. The downside of suspension fertilizers is that special equipment is needed and commercial applicators are usually employed for the application.

Liquids can be sprayed on the soil surface (broadcast), or injected accurately into the soil through tines in a band at the required distance and depth below the seed or young plant (Draycott *et al.*, 1967). Few experiments have been made testing liquid fertilizers on sugar beet crops, but with other crops it has been shown that yields from sprayed liquids and from broadcast solids are the same provided that the crop is not damaged and that both fertilizers are applied at the same time.

Devine (1962) compared solid fertilizer with liquid fertilizer of similar composition and nutrient ratio for sugar beet. The solutions containing ammonium nitrate, ammonium phosphate and urea were applied by means of a knapsack sprayer adjusted to give large droplets. There were no significant differences in the yield of the sugar beet between the liquids and solids. Draycott and Holliday (1970) reported similar findings with liquid compound fertilizer (average analysis: 17% N, 8% P_2O_5, 15% K_2O) compared with the solid granular compound of the same analysis. The liquid was sprayed on the seedbed and the compound was broadcast by hand. Both were worked into the soil and both gave the same yield of sugar. They also compared liquid sprayed on the soil with the same liquid injected to give a band of fertilizer, either 5 cm to the side and 5 cm below the seed or 5 cm to the side and 15 cm below the seed. Yields from sugar beet grown with fertilizer applied by the three methods were not significantly different, but deep placement gave a small consistent increase in yield over shallow placement, probably due to the increased availability of the deep-placed fertilizer as it was in soil which did not dry as readily as the surface soil. This was substantiated by greenhouse pot experiments with sugar beet.

Foliar Applications of Nitrogen, Phosphorus and Potassium

Desprez (1963) compared the effects of urea solutions sprayed on leaves of sugar beet in France at 20 kg N ha^{-1} with the same amount of nitrogen broadcast as solid sodium nitrate. Both were applied 3 weeks after sowing, in addition to the conventional dressing of 150 kg N ha^{-1} applied in the seedbed, and late sprays of urea were included in later experiments. Results indicated that, provided urea was applied at least 3 months before harvest, it did not decrease sugar percentage. Unfortunately, it increased root yield little. Whereas the plants rapidly assimilated nitrogen from urea, nitrogen from the broadcast sodium nitrate was taken up later, thus decreasing the sugar percentage.

Thorne and Watson (1956) found that, when leaves of sugar beet crops were sprayed in September and October with ammonium nitrate or urea solution, only 0.7% of the nitrogen was recovered in the plants, compared with 40% from soil applications. Spraying slightly increased the dry-matter yield of the tops, but not of the roots. It decreased the sugar percentage of roots by 1%. Kozera and Lachowski (1959) in Poland sprayed nitrogen, phosphorus and potassium solution on the foliage of sugar beet several weeks before harvesting. The solution increased the yield of tops slightly but did not affect the yield of roots or the sugar percentage. James *et al.* (1968) found that foliar application of potassium late in the season had no effect on sugar beet yield and sugar production. Hills and Ulrich (1971) summarized data from California, where a basal application of 90 kg N ha^{-1} was given to sugar beet. A supplemental application of 45 kg N ha^{-1} was made either to the soil or as a foliar spray. The soil-applied treatment yielded 64 t roots ha^{-1}, while the foliar spray yielded 59.2 t ha^{-1}.

In general, it seems that foliar application of major (or macro) nutrients is not an effective way of providing the large amount required by sugar beet. Work with magnesium substantiates this principle, a soil supply being much more effective than foliar sprays (Chapter 5). However, for trace (or micro) nutrients, when the crop requires only a few g ha^{-1}, foliar application is highly effective (Chapter 6).

SUMMARY

Nitrogen fertilizers may be applied to the seedbed before, at or after sowing. Application prior to sowing works well if there is sufficient rainfall/time between application and sowing to allow the 'salt effects' to be minimized, so as not to affect germination. In the UK, nearly all the nitrogen is broadcast – a third immediately after sowing and two-thirds after emergence. This approach removes the danger of toxicity and minimizes leaching.

Several different forms of nitrogen may be used satisfactorily. In comparing them, care is needed to make sure that timing and placement do not give one source an advantage over another. Urea, ammonium nitrate, ammonium sulphate, urea–ammonium nitrate solution, calcium ammonium nitrate and anhydrous ammonia are commonly used. Essentially there is no advantage to any of these sources with regard to utilization by sugar beet. Cost and convenience should be the primary considerations in determining which to use.

Evidence suggests that phosphorus should be distributed throughout the plough layer for optimum utilization by the sugar beet crop. Potassium likewise only works well when distributed throughout the plough layer (Chapters 4 and 9).

Boron fertilizer may be broadcast, banded or foliar-applied. Generally, manganese is applied as a foliar application in Europe while soil application is often used in the USA. Broadcast application of manganese fertilizers is not cost-effective. Banding to the side and below the seed is effective if fertilizer phosphorus is also included. On moderately to severely manganese-deficient soils, a combination of banded and foliar application shows much promise. Banded manganese in the soil provides early-season supply when the plants are too small to benefit from foliar application. Foliar application is used later in the season, particularly with recurring symptoms, which need repeated applications.

In the few cases where iron is needed, foliar application is recommended because the element is quickly immobilized in soil. Zinc may be either sprayed on the foliage, banded to the side and below the seed or broadcast on the soil (Vitosh *et al.*, 1998). Deficiencies of other micronutrients are not common and little work has been done with sources and times of application.

12
Interactions between Nutrients, Water Supply, Pests and Diseases and Agronomic Practices

Availability of nutrients to plants is often affected by the quantity of water present in soil. In dry soil conditions, irrigation increases yield directly by supplying water needed for transpiration and indirectly by improving the availability of nutrients. Thus the quantity of fertilizer required may be influenced both by increased yield of the irrigated crop and by changes in the availability of nutrients present in soil. Similarly, changing plant density greatly affects the growth and yield of individual plants and the depth and distribution of their roots in the soil. This also affects the uptake of nutrients and amount of fertilizer needed for maximum yield. Diseases and other pests slow growth and more fertilizer is often needed by affected crops than by healthy ones, especially if roots are damaged.

Soil/Water Relationships in Sugar Beet

Irrigation increases the leaf area of sugar beet early in the growth cycle and later helps to maintain it (Carter *et al.*, 1980a; Dragović *et al.*, 1996). This improves leaf cover and light interception, which results in more photosynthate being produced. In some ways nitrogen fertilizer has similar effects, and increases dry matter production mainly by increasing radiation capture. Other plant nutrients, such as sodium and potassium and, to a lesser extent, phosphorus and magnesium, increase or help maintain leaf area. Where these elements have been tested both with and without irrigation, interactions are small but the effects of the nutrient and irrigation are additive (Draycott and Farley, 1971).

There is also evidence that phosphorus and potassium availability is increased by irrigation, which may be explained by the findings of Brown *et al.* (1987) that many fibrous roots die during periods of water stress. In another study, Brown and Biscoe (1985) showed that nitrogen fertilizer decreased fibrous root development, particularly below 50 cm early in the season.

Water from both rain in the growing period and that stored in soil from the previous winter plays a crucial role in the growth and nutrition of sugar beet. Generally, the crop is grown on the best land in a region, often with good water-holding capacity. It also has a deep and spreading root system and so makes good use of stored soil water. Despite these attributes, irrigation is often needed in dry conditions to satisfy the evapotranspiration demand and ensure continuous nutrient uptake.

Inside the plant, water is the medium in which chemical transformations take place and is necessary for the reduction of carbon dioxide during photosynthesis. Nutrient ions, carbohydrates and other chemicals are transported both by diffusion and in the moving stream of water. When plants come under water stress, all these processes are disrupted (Dunham, 1993).

Climatic conditions affect the growth of sugar beet. Early work suggested that temperature was the primary climatic factor influencing sugar beet yield. Dubetz and Oosterveld (1976) analysed 50 years of sugar beet yields in Alberta, Canada (latitude 49°N). They concluded that under irrigation, temperature was the only climatic variable that significantly affected sugar beet yields. However, Scott *et al.* (1973) found that beet yields were closely correlated with the amount of solar energy intercepted by the leaf canopy.

Water stress

As soon as stomata open and evaporation begins, leaf water potential starts to fall, as does leaf relative water content. How far they fall depends on the availability of water to the root system and the conductance of the pathway from roots to leaves. As the soil dries, a point is reached at which further decrease in water availability or increase in evaporation reduces growth relative to conditions where water is more plentiful. Beyond this point, the plant experiences water stress. Initially plants are stressed for only part of the day, but the duration of the stress increases daily without rain or irrigation.

Exactly when stress starts is difficult to determine since the first effects of transient stress on growth are small. Similarly the point at which water stress begins to affect nutrient transport to and within sugar beet plants is not clearly defined. It has been discovered that the rate of appearance of individual leaves is only slightly reduced but their productive life before senescence is considerably shortened (Milford *et al.*, 1985; see also Chapter 2). In this respect a shortage of water bears a resemblance to a shortage of nitrogen. This has led to many experiments looking for interactions between these two very important inputs for sugar beet production.

The sugar beet root system

By harvest, the fibrous root system of sugar beet is deep but increasingly sparse at depth, except in the immediate vicinity of the storage roots. Seedling roots initially grow downwards at about 10 mm day^{-1}. Subsequently, for much of the growing season the depth of the root system increases at about 15 mm day^{-1} (Brown and Dunham, 1989). The considerable depth of the final root system (shown in Fig. 12.1) – at least 2.0

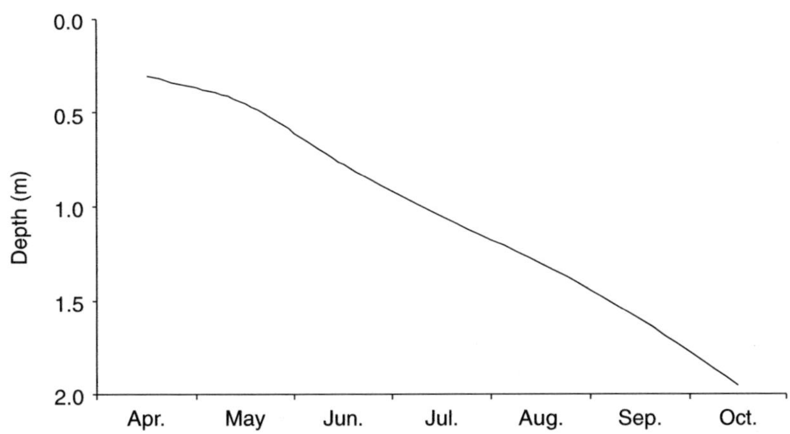

Fig. 12.1. Depth of active sugar beet roots as indicated by the drying front measured by neutron probe over many years at Broom's Barn.

m (Weaver, 1926; Draycott and Messem, 1977) – allows the crop to utilize water and nutrients from deep layers in the soil more efficiently than most other annual crops. Deep extraction of nutrients and water is confirmed by Moraghan (1972) and Christenson and Zinati (1991). Moraghan showed water extraction to a depth of 1.8 m, while Christenson and Zinati showed that sugar beet removed ^{15}N from 1.5 m within 90 days after sowing. Nitrogen fertilizer may influence fibrous root development and activity, so interacting with water supply (Brown and Biscoe, 1985).

Irrigation

Worldwide, about one-fifth of the 7 Mha of sugar beet is irrigated, so most of the production is without supplementary water. In the UK and similar climates, the crop is spring-sown. It performs well in most years on most soil types with a combination of rainfall and water stored in the soil from the previous winter. In the Mediterranean basin, the crop is either sown in autumn and grows on winter rainfall or sown in spring and irrigated. In California, sugar beet sown in autumn needs much irrigation, since winter precipitation is sparse. In other arid regions, sugar beet may be spring-sown and irrigated over the course of the growing season. Very often the seed is placed in dry soil and the land is irrigated to provide moisture for emergence. Furrow irrigation seems to work well for this practice because the soil is wetted from the bottom, reducing surface crusting.

When Penman (1952, 1962, 1970) did his classical transpiration studies on field-grown crops, sugar beet was one of his test crops, first at Rothamsted and then at Woburn. He encouraged a thorough assessment of the value of irrigation for sugar beet at Broom's Barn particularly, measuring water consumption by the crop at intervals throughout the growing period, using neutron probes.

Measurements were made in soil to 150 cm deep under sugar beet at weekly or fortnightly intervals from sowing time to harvest. This study continued for many years to determine the effect of weather patterns, nutrient supply and different agronomic practices, described below (Dunham *et al.*, 1993).

Penman had used sugar beet in weighable volumes of soil at Rothamsted when testing his formulae, comparing evaporation from an open water surface with transpiration by crops. The Broom's Barn studies, however, showed how much water was needed to support sugar beet growth month by month (Fig. 12.2). This proved to be an invaluable aid to both understanding the effect of water shortage on yield and the effective use of irrigation and nutrients.

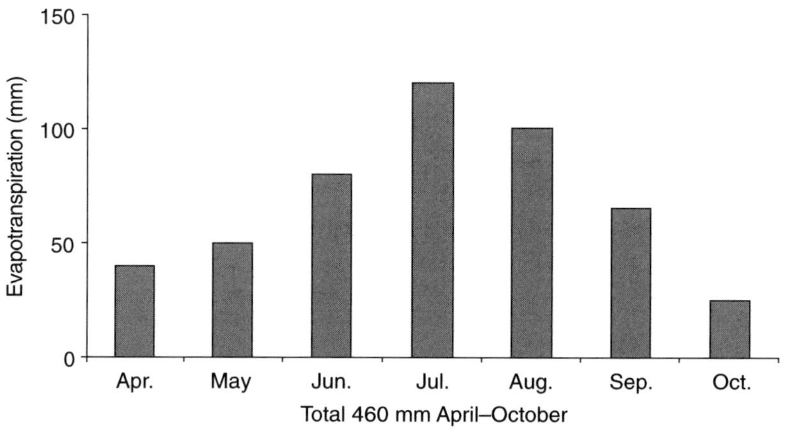

Fig. 12.2. The water needed each month of the growing season for a crop to grow at the potential rate in average weather conditions in the UK.

In average weather conditions in East Anglia, about 450 mm water are needed by sugar beet for the crop to grow at the potential rate from April to October. Peak demand was identified in July and August (over 200 mm). This has encouraged irrigation in this crucial period. It was then discovered that sugar yield per unit area was directly related to water uptake in UK studies (Draycott and Messem, 1977) and in California (Ghariani, 1981), which has further increased interest in irrigation. Measurements of water use by sugar beet have been made in other countries, these ranging from 400 mm in the cool climate of Finland to 1500 mm in the hot, dry desert-like conditions of California.

Evapotranspiration demands by the sugar beet crop reach a peak during the most rapid growth period and warmest temperatures. Daily demand ranges from 6 to 9 mm day^{-1} between the Northern Plains and Columbia basin and Arizona and California in the southern USA (Jensen and Erie, 1971).

Interactions between Water Supply and Nutrition

Europe

Early experiments were described by Garner (1950) and Penman (1952), who grew the crop with variable amounts of nitrogen fertilizer, with and without irrigation. Later work was reported by Price and Harvey (1962) at Gleadthorpe on a very light soil and also by Penman (1962) at Woburn, and on a loamy soil at Reading by Harris (1970). Common conclusions were that irrigation did not improve response to nitrogen, which was disappointing, but that irrigation greatly improved the colour of the leaves with the smaller amounts of nitrogen. In dry seasons, irrigation greatly increased sugar yield by increasing root yield, with little effect on sugar percentage.

Information gained from experiments at Broom's Barn benefited from measurements of water use and plant and soil analysis (Draycott and Durrant, 1971a,b; Draycott and Webb, 1971). On the deep sandy loam soils over marl or clayey glacial drift the average response to irrigation was small, but it was larger on the sandy soils. Figure 12.3 shows that there was a large response to nitrogen fertilizer but that the two factors, irrigation and nitrogen, interacted little. Plant analysis confirmed that irrigating improved the uptake of nitrogen, which explained improvements in leaf colour (Holmes and Whitear, 1976).

Last *et al.* (1983) analysed the growth of sugar beet from seedling emergence to harvest. Plots were given a wide range of amounts of nitrogen fertilizer, with and without irrigation. For the first time there

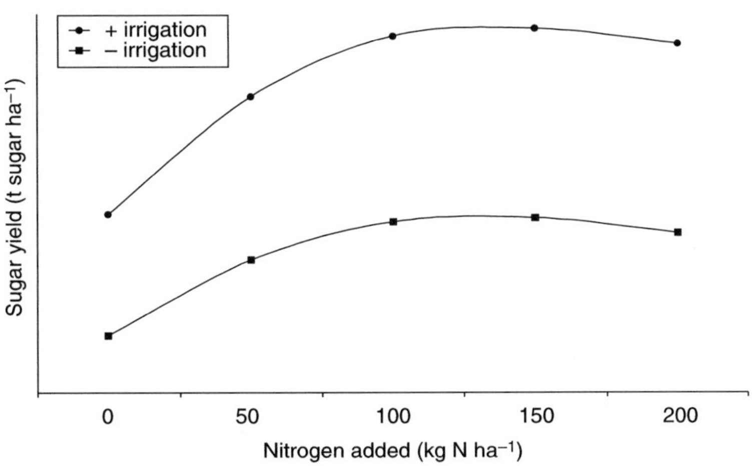

Fig. 12.3. Average effect of nitrogen fertilizer without and with irrigation on sugar yield.

was clear evidence that the uptake of nitrogen for maximum sugar yield was about 200 kg N ha^{-1}. It was also shown that irrigation improved root quality by ensuring that nitrogen was used for growth in dry periods in midsummer. In contrast, without irrigation there were occasions when unused nitrogen spoilt quality if it was taken up too late following autumn rain.

Metochis and Ophanos (1988) tested a much wider range of nitrogen fertilizer (0–600 kg N ha^{-1}) under three water regimes in Cyprus. Very large yields indeed were produced in the Mediterranean climate, up to 165 t roots ha^{-1} or 20 t sugar ha^{-1}. With full water application, such yields were obtained with no more than 150 kg N ha^{-1}. As in the UK, irrigation ensured the optimum use of the fertilizer but did not increase the amount required for maximum yield.

Work in a very dry area of Morocco by Ezekari *et al.* (1993) with nitrogen and various irrigation frequencies showed the additive effect of the two factors. Both greatly increased yield. Best yields resulted from full irrigation and giving nitrogen mostly at the start of the growing cycle; the two together increased yield by nearly 5 t sugar ha^{-1}. In droughty conditions slightly more nitrogen was justified.

Mompin *et al.* (1993) conducted very similar experiments in Spain but their objective was to discover how to make the most efficient use of the limited amount of water available for irrigating sugar beet (aquifers were showing signs of depletion). The crop was spring-sown and irrigated by sprinklers. Optimum nitrogen and irrigation trebled sugar yield and there was a highly significant interaction. For maximum water efficiency in terms of kg sugar per unit of water, about 160 kg N ha^{-1} was needed, coupled with about 600 mm.

A very similar result was reported from a detailed study in Italy's Po Valley by Mannini and Venturi (1993). They produced data from 8 years of trials with irrigation, nitrogen and phosphorus. For best returns (11.5 t sugar ha^{-1}), 160 kg N ha^{-1} was needed, together with water given up to 95% of evapotranspiration. Phosphorus fertilizer had no effect.

USA

As in Europe, most experiments in the USA have investigated interactions between nitrogen and irrigation but a few have included additional plots with other major nutrients. The American experiments have been made where sugar beet is irrigated in commercial practice. Consequently, in most experiments, all plots have been given some irrigation and treatments either have been additions to this or have tested different methods of application of water. The majority of results substantiate the finding that irrigation does not affect nitrogen-fertilizer requirement greatly.

Working in Utah, Haddock (1959) grew sugar beet on a deep, well-drained loam and compared sprinkler and furrow irrigation. Plots were subdivided and given additional nitrogen and phosphorus. Nitrogen did not increase yield but phosphorus gave a large increase, particularly with sprinkler irrigation, suggesting that the availability of phosphorus was increased by irrigation. Henderson *et al.* (1968) substantiated this by comparing subirrigation with subirrigation plus sprinkler irrigation in California. With subirrigation only, the surface soil was relatively dry for the major part of the growing season but the crop was well supplied with water for transpiration. Sprinkler irrigation nevertheless increased nitrogen and phosphorus concentrations in the sugar beet petioles and increased sugar yield. They concluded that the response to sprinkler irrigation was partly due to improved water supply and partly to more favourable conditions for phosphorus and nitrogen uptake.

On soil where nitrogen greatly increased yield, Woolley and Bennett (1962) investigated the amount of nitrogen needed for maximum yield when grown with four amounts of irrigation. Compared with giving none, 85 kg N ha^{-1} increased root and sugar yield with all the moisture treatments. However, giving 250 kg N ha^{-1} increased yield further only with the largest amount of water and decreased yield on plots given little water.

Robins *et al.* (1956) tested the effect of excessive irrigation on the nitrogen requirement of sugar beet in the Columbia basin, Washington. Supplying more water in the early season than was needed to bring the soil to field capacity decreased yield and increased the amount of nitrogen fertilizer needed by the crop; they concluded that this was mainly due to leaching, though denitrification may have explained some of the effect. The concentration of plant nitrate was decreased and the sugar percentage increased by water. When excess water was given in autumn, yields were unaffected but the status of the plants was decreased.

Loomis and Worker (1963) in California showed that moisture stress and nitrogen deficiency affected sugar beet growth in similar ways. Both decreased vegetative growth and increased sugar percentage in roots; purity and sugar yield were increased by decreasing the nitrogen dressing, though not by increasing moisture stress. The effects of the two factors were independent and additive. In California, where the crop is irrigated, sugar beet is often allowed to wilt for a few weeks before harvest, supposedly to concentrate the sugar in the roots. Loomis and Worker found that the practice did not increase sugar yield or quality in their experiments, although in commerce there might still be some advantage through lower haulage costs.

More recently, work by Carter and coworkers in Idaho cast some additional light on the effects of irrigation and nutrients on sugar beet (Carter *et al.*, 1972, 1980a,b, 1982). They found no indication of an interaction between applied nitrogen and the amount of water supplied in respect of sugar yield. Carter (1982) summarized all of the research as follows:

> Optimum nitrogen applied pre sowing or during the early plant growth stages, maintains sugar concentration to a near maximum level for the season. Excessive or late season application and plant nitrogen uptake from fertilizer or residual nitrogen sources caused an increasing proportion of the photosynthate to be used for top growth. Increased sugar concentration caused by irrigation water deficit results from dehydration and does not increase sugar yield.

Reducing the amount of water applied through irrigation has many advantages, provided no loss of yield occurs. Carter *et al.* (1980b) showed that sugar yield is reduced very little when irrigation is discontinued in early August, provided the soil profile contained at least 200 mm of available water in the surface 150 cm.

Irrigation Technology

Furrow irrigation is an age-old method of irrigation. Irrigation with overhead systems (gun-type applicator or sprinkler) is mentioned earlier in the chapter. Both methods have been successful, but they tend to lead to inefficient water use. In addition, there is potential for leaching and denitrification of nitrates and leaching of agrochemicals.

Low-volume irrigation technology offers potential cost and environmental advantages over furrow or sprinkler irrigation. Drip irrigation is used extensively for the production of many crops. A reduction of leaching of nitrate and pesticides is the major advantage of this low-volume irrigation technology (Gregory, 1990; Bihery and Lachmar, 1994; Gelata *et al.*, 1994). Some have shown improved agronomic efficiencies and yields with sugar beet using drip irrigation (Mambelli *et al.*, 1992; Urbano *et al.*, 1992).

A recent study in Wyoming suggests that drip irrigation reduced water use and the quantity of nitrogen required for the production of sugar beet (Cassel Sharmasarkar *et al.*, 2001b). These workers also made a detailed economic analysis of drip irrigation for sugar beet production (Cassel Sharmasarkar *et al.*, 2001a). As has been shown in other studies, there was no significant interaction between irrigation and amount of nitrogen applied. They found that root and sugar yield was greater under drip than under furrow irrigation and costs were lower for drip irrigation. The economic return for drip irrigation was 11% greater than for furrow irrigation under the conditions of the study. Obviously, there is a need for additional studies under different conditions before drip irrigation can be widely recommended for sugar beet production.

Diseases and Other Pests

Wherever sugar beet is grown, it is rarely free from attack by a range of diseases and other pests. If a shortage of a plant nutrient decreases vigour and yield, crops are often more susceptible to attack. Where pathogens damage the root system, nutrient uptake is affected and fertilizer requirement increased. Examples of some of these interactions between diseases and other pests and the nutrition of the crop are discussed below.

Effect of fertilizers on inoculum in soil

A widespread root disease of sugar beet, which appears to be increasing in incidence, is violet root rot, caused by the soil-inhabiting fungus *Helicobasidium purpureum*. Hull and Wilson (1946) showed that 100 kg N ha^{-1} as ammonium sulphate greatly decreased the number of roots infected by the disease. Phosphorus and potassium fertilizers had no effect but both sodium fertilizer and organic manure were slightly beneficial. More important was to grow the crop in a wide rotation, avoiding other hosts for the fungus, particularly carrots and potatoes. *Sclerotium* root rot is a serious disease of sugar beet in some countries, being caused by the fungus *Sclerotium rolfsii*, the white mycelium of which covers the surface of affected roots. Leach and Davey (1942) in California showed that nitrogen fertilizers decreased the percentage of plants affected by this fungus. Ammonium sulphate, anhydrous ammonia and calcium nitrate were equally effective when applied in equivalent amounts of nitrogen. The percentage of roots affected was halved by 125 kg N ha^{-1} and the fertilizer greatly increased yield. Laboratory trials showed that ammonium solutions were toxic to the mycelium and sclerotia but failed to explain the effect of the fertilizers in the field, for calcium nitrate was not toxic in the laboratory tests. The partial control in the field may have been due to changes in metabolism of the causal fungus that decreased its growth, to increased resistance by the sugar beet or to changes in the balance of microorganisms in the soil.

Effect of fertilizers on virus diseases

Several virus diseases infect sugar beet, usually causing yellowing of leaves, which results in poor photosynthesis and seriously depressed yield. Hull and Watson (1947) did some of the earliest work to determine the losses from beet yellowing viruses when the sugar beet was grown with different amounts of fertilizers and farmyard manure. The fertilizers increased the root and sugar yields of both infected and healthy plants but the losses caused by infection increased proportionally as the mean yields increased and there were no significant interactions on average. The fertilizers had little effect on the symptoms of the disease or on the rate of spread of the infection; nitrogen, phosphorus and potassium occasionally increased the rate of spread and agricultural salt consistently decreased it. In a large series of nitrogen trials Adams (1962) recorded the proportion of plants with virus yellows at the end of August. There was little relation between yellows incidence and response to nitrogen, indicating that the disease was not a major factor affecting response. Similarly, in experiments described by Tinker (1965), the mean response to nitrogen fertilizer was unaffected by different amounts of virus infection.

The effect of nitrogen fertilizer on the spread of yellowing viruses – beet mild yellowing virus (BMYV) or beet yellowing virus (BYV) – and the appearance of symptoms on the leaves of sugar beet have been studied by Heathcote (1970, 1972, 1974). Nitrogen applied up to 225 kg N ha^{-1} had a large effect on the LAI and the height of the plants. Nitrogen tended to increase the number of aphids, which are vectors of the viruses, but this effect was not entirely consistent. Heathcote suggested that the increased numbers may have been due to the increased plant height or leaf area, or to increased productiveness of the aphids caused by the increased concentration of nitrogen in the plant sap. However, all plants were later sprayed with aphicide (common practice on commercial crops in the UK at the time), so most of the virus spread was by winged aphids.

Hull (1965) reported that large amounts of nitrogen fertilizer masked symptoms of BMYV. In experiments with trace elements, Russell, G.E. (1971) found that aphids settled more readily on plants given lithium, zinc and nickel but less readily on those treated with boron. Aphids transmitted BYV more efficiently to plants treated with lithium and boron than to those treated with copper, zinc and tin.

Defoliation

The growth stage of the sugar beet may be divided into two periods in regard to the effect of defoliation damage. The first period is from emergence until plants have five or six leaves. During this period removal of leaves often leads to the death of the plant since there are insufficient energy reserves to regenerate the top growth. After the plant has seven or eight leaves, the root systems are well developed and defoliation does not cause death. However, since leaf surface area is reduced, reduction in yield may be expected.

Leaf-feeding insects, such as silver Y moth caterpillars, cause sporadic loss of leaf area (Sands, 2002). However, sugar beet may be defoliated at any time during the growing season by hail, by soil particles during wind erosion in spring, by weed cutting in summer or by topping before harvesting. Dunning and Winder (1972) artificially defoliated field-grown plants completely each month from May to September on plots with and without 150 kg N ha^{-1} fertilizer. At harvest in mid-November, minimum root weights followed defoliation in July/August and later defoliation gave minimum sugar percentages. With nitrogen, sugar yields were smallest after August defoliation but, in the absence of nitrogen, they were smallest after July defoliation. Up to 40% of the sugar yield was lost by defoliation in these 2 months but yields and recovery from defoliation were greater with nitrogen than without.

Hail damage to sugar beet is particularly devastating in the Northern Plains and intermountain regions of the USA, but also occurs in other sugar beet regions of the world. Afanasiev (1964) made a detailed study of the effect of simulated hail over a 6-year period in Montana. The treatments included 0, 25, 50, 75 and 100% defoliation at 2-week intervals from about 48 to 132 days after sowing. Beets were harvested 150–160 days after sowing. Similar results were obtained with 25, 50 and 75% defoliation so those values were averaged (Fig. 12.4). For these three treatments, the most significant depression of sugar yield occurred at 104

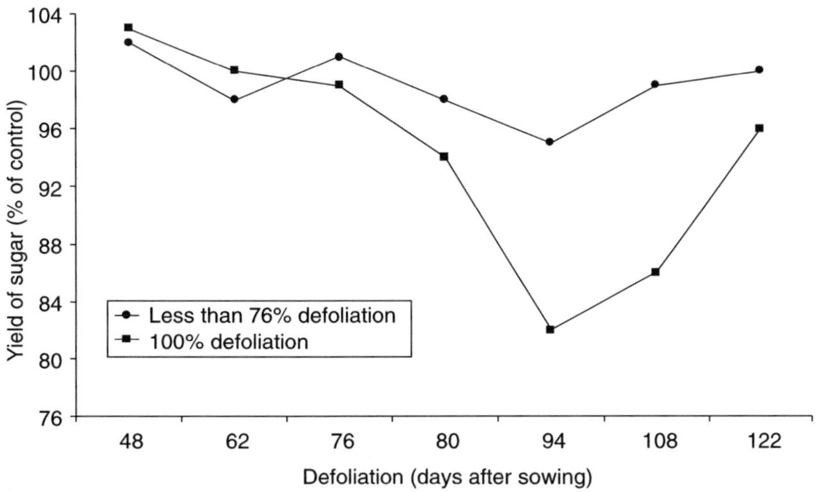

Fig. 12.4. Effect of defoliation on yield of sugar (after Afanasiev, 1964).

days after sowing. Even then, the depression was only 5%, indicating the immense recovery capacity of sugar beet. The largest yield reduction occurred when beets were completely defoliated between 90 and 108 days after sowing.

Soine (1967), working in the Red River Valley, found very similar results to those of Afanasiev. He also showed that there was no significant difference in purity for the 25, 50 and 75% defoliated treatment. However, there was a significant reduction in juice purity for the 100% defoliation at any time in the growing season.

More recently, Muro *et al.* (1998) showed similar results in Spain. The most critical growth state for the effect of defoliation on yield was between 1700 and 1800 degree-days, which usually occurred around the first of August. Both Afanasiev and Muro *et al.* suggested greater reductions in yield from traumatic versus non-traumatic defoliation (high pressure water or flailing with wooden sticks versus hand cutting with shears).

Carter *et al.* (1978), in Idaho, found that 75% defoliation in early August reduced yields by 17%. Increasing nitrogen supply increased leaf-area recovery rates, but they did not return to pre-defoliation levels. The authors found no indication that stored sugar was used for the regrowth of leaves.

Root damage

Most factors that decrease the efficiency of the root system also increase the need for fertilizer. Examples dealt with in previous chapters are soil compaction, which decreases their ability to explore soil fully, and ploughed-in straw, which during rotting releases toxins that damage roots. In both examples, more fertilizer was needed to maintain yield.

Nematodes also damage roots and decrease their ability to take up water and nutrients. The free-living nematodes *Longidorus* and *Trichodorus* feed on roots of seedlings and damage small plants. In the former genus, secondary roots may be pruned by the nematodes and, in the latter, the primary root may be destroyed, causing severe fanging of the taproots. The experiments of Dunning and Winder (1972) showed that the crop's requirement of nitrogen fertilizer was greatly increased to compensate. Cooke and Draycott (1971) found that, if the nematodes were killed with a partial soil sterilant before sowing sugar beet, the nitrogen requirement was normal but, without sterilant, much more nitrogen was needed (Fig. 12.5). Partial sterilization not only improved the health of roots but also increased the amount of available nitrogen in

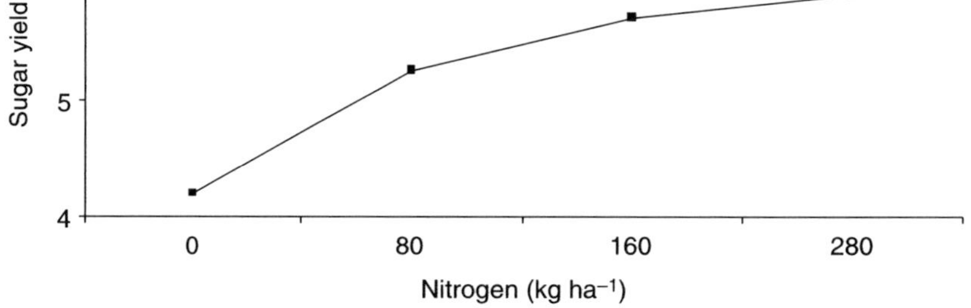

Fig. 12.5. Effect of nitrogen fertilizer without and with fumigation applied to control free-living nematodes. Average of 15 field experiments on sandy soils in the main sugar beet areas of the UK.

soil by speeding mineralization. It also decreased the amount lost by leaching by slowing nitrification of ammonium nitrogen (Draycott and Last, 1971).

Plant Population

Many experiments throughout the world have sought to determine the number of sugar beet plants per hectare needed to produce optimum sugar yield. The optimum number undoubtedly is influenced by site, season, variety and local cultural practices. The underlying principle is that the leaf canopy should expand rapidly to achieve complete cover as soon after sowing as possible. Leaf area should then stabilize at about LAI 3.

This ideal leaf canopy, first described as long ago as 1952 by Watson and frequently confirmed by later studies (Goodman, 1966; Scott and Jaggard, 1993), optimizes light interception. Plant number per unit area has a dominant effect, particularly early in the growing season. It is generally thought that 80,000–100,000 plants ha^{-1} produces the target LAI for maximum sugar yield.

Nitrogen supply also has a major effect on the development and stabilization of the ideal leaf canopy, as already described in Chapter 2. Not surprisingly, the way plant population and nitrogen (and other nutrients) interact has resulted in research aimed at improving sugar yield with a combination of plant populations and nitrogen fertilizer. In a few studies the second-order interaction with these and water supply has been determined, as described below.

UK

Harris (1969, 1970) described work at Reading University testing combinations of 64,000–160,000 plants ha^{-1} and 0–160 kg N ha^{-1}. Unfortunately, the crop did not respond to the fertilizer on his site but he noticed that the closer the plants, the lighter the colour of the leaves. He suggested that plants in the dense stands were competing for nitrogen, but giving more fertilizer produced no more sugar per hectare.

Later studies on a responsive site at Broom's Barn confirmed that the closer the beet were grown to each other, the smaller the concentration of nitrogen in the leaves (and roots). Populations of 22,000–133,000 plants ha^{-1} were tested in factorial combination with 0–225 kg N ha^{-1}. Both factors greatly influenced sugar yield but there was little evidence of an interaction, about 80,000 plants ha^{-1} and 125 kg N ha^{-1} being optimal at all levels of the other factor (Draycott and Webb, 1971).

In these same experiments, the effect of irrigation was superimposed over all plant populations and nitrogen treatments. Harris (1970) made a similar study of the second-order interaction, plant population × nitrogen × irrigation. All the work showed that the closer the beet were grown together, the greater the competition for nitrogen and water. The crop produced the best yields with 80,000 plants ha^{-1} and 125 kg N ha^{-1}, both without and with irrigation, and the second-order interaction was not significant.

It is concluded that competition for light in the dense stands was the main limiting factor to increased yield of sugar; giving more nitrogen and water failed to increase sugar yield and tests of other major nutrients showed that they were not limiting yield. Competition for light appeared to stimulate the growth of leaves and at harvest the proportion of dry matter partitioned to the roots was least with the largest plant density.

USA

Growers of sugar beet in the USA generally tend to aim for smaller plant populations than in Europe. Row widths are often greater in the USA, which may suit machinery and harvesting and aid flood irrigation. It is also thought that the larger roots produced from small populations may keep better when piled. In addition, nitrogen usage also tends to be greater than in the UK and some other EU countries. Experiments in many states have been made to investigate the ideal combination of plant population and nitrogen nutrition. Early work by Herron *et al.* (1964a) in Kansas tested 44,000–90,000 plants ha^{-1}, all grown in fairly wide rows (56 cm). All were given 0–175 kg N ha^{-1}. There was no interaction and the best result (11 t sugar ha^{-1}) was with 90,000 plants ha^{-1} given 80 kg N ha^{-1}, similar to the European experience.

In the arid climate of Arizona, Nelson (1969) irrigated sugar beet grown at a wide range of populations (37,000–148,000 plants ha^{-1}) and nitrogen applications (80–275 kg N ha^{-1}). Nelson expected that the narrow-spaced plants would need most nitrogen but they responded least; since the crop received adequate moisture, he suggested that a shortage of light was limiting yields of the closely spaced plants. In Utah, Haddock and Kelley (1948) recorded quite different results. They grew sugar beet in rows 25, 50 and 60 cm apart with a constant spacing of 30 cm (80,000, 65,000 and 47,000 plants ha^{-1}, respectively). Various fertilizers were tested, including 0, 90 and 180 kg N ha^{-1}. Nitrogen increased sugar yield significantly with all plant spacings and there was a large positive interaction between increased plant density and increased nitrogen application. The authors concluded that, to obtain maximum return from large dressings of nitrogen, it is necessary to have enough plants to utilize it fully.

Haddock and Kelley also investigated the effects of water and nitrogen applications for sugar beet grown at three plant densities. With adequate water, the closest spacing (90,000 plants ha^{-1}) produced the largest yield of sugar and needed the largest amount of nitrogen (175 kg N ha^{-1}) for maximum yield; giving less water decreased yield, but watering did not affect the amount of nitrogen needed. Haddock (1949) reported the results of soil-moisture measurements in the same experiments, which indicated that early watering was much more important than late watering with all plant stands.

Moraghan (1972) showed that increasing the number of plants from 23,400 to 70,700 ha^{-1} increased sugar yield by 20%, but increased water use by just 9.5%. He showed that the greatest effect of increasing population on water use was most evident in the 120–180 cm depth. He also showed that sugar beet utilized nitrate to at least 150 cm. In another paper from the same study, increasing plants per hectare did not increase the need for nitrogen fertilizer (Moraghan *et al.*, 1973). The optimum amount was 112 kg ha^{-1} across all populations studied.

Work in the Red River Valley shows some interesting results. In a 3-year study at two sites, Smith *et al.* (1990) showed that a population of 73,400 plants ha^{-1} was optimum for conditions in this non-irrigated sugar beet production area (Table 12.1). Yield and per cent harvestable beets (those removed with a conventional harvester) decreased with increasing population. There was little difference in yield or harvestability with uneven distribution of beets in the row compared with even distribution. In Nebraska, Yonts and Smith (1997) also showed that the amount of small beets produced increased markedly above 65,000 plants ha^{-1}, with an associated yield reduction.

Summary

In north-west Europe large yields of sugar beet are obtained without irrigation because rainfall during the growing season, plus soil

Table 12.1. Effect of plant population on root and sugar yield and harvestability of sugar beet (Smith *et al.*, 1990).

Population (1000 plants ha^{-1})	In-row spacing (cm)	Root yield (t ha^{-1})	Sugar yield (t ha^{-1})	Harvestable roots (%)
59	30	39.4	5.6	79
73	24	40.4	5.8	73
88	20	39.5	5.7	63
103	17	38.8	5.6	58
117	15	37.7	5.4	51

reserves of water, is nearly sufficient to satisfy the needs of the crop in most years. The deep and spreading root system of healthy sugar beet allows 150 mm or more of water to be taken from soil reserves and this is replenished by winter rainfall. Supplementary irrigation increases sugar yield in dry summers, particularly on sandy soils or where the root system is damaged by pests or diseases or where its development is impeded by poor soil physical conditions. Also, as yields continue to rise (Chapter 1), water shortage will become a major threat to further yield improvement, particularly if climate change decreases summer rainfall (Pidgeon, 2000).

Irrigation, even where it greatly increases yield, does not appear to affect the amount of fertilizer needed for maximum sugar yield. That the fertilizer requirement is unchanged is substantiated by experiments in drier climates, where changing the amount of water given to irrigated crops has little effect on the amount of fertilizer needed for maximum sugar yield. Although water shortage and nitrogen deficiency affect sugar beet growth in similar ways, the increases in sugar yield from water or nitrogen fertilizer are independent and additive.

As the number of plants per unit area is increased, the concentration of nitrogen in the dry matter is decreased; this effect can be observed in the field because leaves of plants in dense stands are light green in colour, compared with the dark green foliage of widely spaced plants. This indicates that plants in dense stands compete for nitrogen, and it has been assumed that more nitrogen fertilizer is needed if dense stands are to give maximum sugar yield. Experiments have shown that this is not so, for giving more nitrogen fertilizer increases the yield of tops without increasing the yield of roots or sugar. As with irrigation, changing the plant density appears to have little effect on the amount of nitrogen fertilizer needed for maximum sugar yield. It is concluded that competition for light in stands of more than 75,000 plants ha^{-1} is the main limiting factor to increased sugar yield. Giving more nitrogen fertilizer and/or water does not help; they simply stimulate the growth of tops without any commensurate increase in sugar yield.

13
Sugar Beet Quality

High-quality roots are critical in the extraction of sugar and directly affect the economics of both growing and processing the crop. To make pure crystalline (or liquid) sugar, some processors have introduced payment systems to encourage growers to produce roots with the required quality characteristics. There are many references in previous chapters showing that nutrition during crop growth plays a large part in final root quality. This chapter defines quality and identifies optimum nutrition to achieve it.

With most crops it is difficult to define quality because subjective criteria, such as texture, taste, shape and colour, are involved. Fortunately, with sugar beet, most of the important aspects of quality can be defined and measured. Physical characteristics important in harvesting and processing, such as root shape and ease of slicing (Rasmusson and Wiklund, 1960), are more difficult to define but are affected little by nutrition.

Root constituents are: (i) water; (ii) dry matter; (iii) total soluble solids; (iv) total insoluble solids; (v) sugar; (vi) non-sugar, soluble solids; (vii) soluble nitrogenous organic compounds; (viii) soluble, non-nitrogenous organic compounds; and (ix) soluble mineral matter (ash). Approximate relationships among these constituents and their relative concentration in the root are given in Fig. 13.1.

Thus only 20% of the weight of beet as harvested makes up the material in solution processed in the factory. Water and insoluble solids are easily removed but this soluble portion containing the sugar must be purified during processing, removing the non-sugar soluble solids. The quality of sugar beet is based on the compounds in this group of constituents. Since nutrition of the growing beet affects the quantity and nature of non-sugar soluble solids, it becomes a vital link in producing good-quality beets.

Sugar concentration and juice purity are the two most important components of beet as harvested (Carruthers and Oldfield, 1961). Sugar is expressed as a percentage of the fresh weight of the beet, commonly called the 'sugar percentage' or 'sugar content'. It is an important quality, affecting the amount of roots handled and transported and the factory throughput. Purity is the ratio of sugar to total soluble solids, expressed as a percentage. After soluble components are removed from beet, sugar is purified by a series of chemical and physical steps. Impurities decrease the extraction of sugar because some sugar is removed from solution with the impurities. The amount of sugar lost to molasses ranges from 1.5 to 1.8 parts for each part of impurity (non-sugar components). Harvey and Dutton (1993) have reviewed other aspects of root quality and processing.

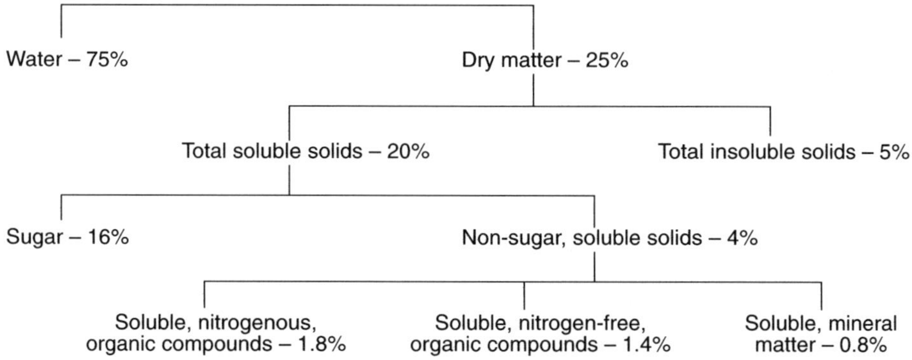

Fig. 13.1. Approximate composition of sugar beet roots by weight (from Alexander, 1971).

Determination of Sugar Percentage

The amount of sugar in roots is normally determined polarimetrically on an extract of fresh macerated root (brei) by the method used by Sachs and described by Le Docte (1927). Sugar percentage so measured is usually in the range 15–20%. Nutrients added as fertilizers may have considerable effects on the sugar percentage: some decrease it and some are beneficial. Where a nutrient decreases sugar percentage but increases root yield, it is important to know the 'break-even' point where the increase in root yield equals the decrease in sugar percentage, resulting in maximum sugar production.

Over the past 50 years, considerable strides have been made in increasing sugar percentage. Many of these improvements result from plant breeding efforts worldwide, mostly by decreasing water content (increasing dry matter). The sugar percentage of fresh roots is usually closely related to the amount of water in the roots. Consequently, climatic conditions before harvest may have a marked effect. Some nutrients also affect the water content of roots and thereby the sugar percentage. The concentration of sugar in roots expressed as a percentage of root dry matter removes the effects of changes in water content. In most of the literature this is not reported. A few reports have shown that some nutrients also slightly affect the amount of sugar in dry matter.

Determination of Juice Purity

Juice purity can only be determined directly in the laboratory by a time-consuming combination of refractometric and polarimetric measurements on juice expressed from fresh roots (Carruthers and Oldfield, 1961). Many laboratories have therefore adopted an alternative, indirect and quicker method of estimating juice purity (Carruthers *et al.*, 1962). Sodium, potassium and alpha-amino nitrogen concentrations in the extract (prepared for the determination of sugar percentage) are determined and the values used in a regression, e.g. juice purity = 97.0 – 0.0008(2.5K + 3.5Na + 10 alpha-amino N); potassium, sodium and nitrogen being expressed as mg 100 g^{-1} of sugar.

Carruthers *et al.* (1962) found a close relationship between the two methods of assessing juice purity (r^2 = 0.74). Harvey and Dutton (1993) provided a more detailed assessment of non-sugars in roots and methods of measurement.

Effect of Nitrogen Fertilizer

UK

The outstanding contributions of Carruthers and his team in the 1950s and 1960s greatly helped in understanding the criteria for sugar beet quality. His formulae have been tested, adapted and adopted in many countries. Often nitrogen fertilizer was mentioned by him because it decreased quality. This led to field trials to test the effects of fertilizers and how to optimize their use, not only for yield (as in the past) but for quality too (Adams, 1962; Draycott and Cooke, 1966; Collier, 1967; Boyd *et al.*, 1970). Tables 13.1 and 13.2 summarize this early work.

Generally a small amount of nitrogen fertilizer was found to have little, if any, deleterious effect on sugar percentage and in a few trials it was beneficial. Larger amounts (100 kg N ha^{-1} or more) always decreased sugar percentage (see Chapter 2). The mode of action was mainly through decreasing the dry-matter percentage of the roots.

Results with juice purity were different because even the smallest amount of nitrogen fertilizer (or organic manure) decreased it. Later work has shown that this was because all the nitrogen sources increased the concentration of nitrogen-containing components of roots at harvest.

Increasing nitrate concentration in the sugar beet plant would be expected to depress sugar percentage and juice purity. Last and Tinker (1968) examined this effect in detail. They varied nitrate concentration in sugar beet leaves and petioles using different quantities of nitrogen fertilizer. Sugar percentage and juice purity were measured in beet roots at harvest. Increases in nitrate nitrogen in petioles up to 700 p.p.m. (sampled in June) had little effect on sugar percentage. Above 700 p.p.m. an increase of 180 p.p.m. in petioles corresponded to a decrease of 1% in sugar percentage. Increasing plant nitrate concentration in June also had serious consequences for juice purity. The authors stressed that sugar percentage and juice purity depend on many factors and, although nitrate concentrations were related to sugar percentage and juice purity in single experiments, they found no dependable relationships between them for all experiments.

Table 13.1. Effect of nitrogen fertilizer and farmyard manure on sugar percentage and juice purity.

	kg N ha^{-1}		
	75	150	225
Without farmyard manure			
Sugar percentage	16.6	−0.4	−0.8
Juice purity	88.8	−0.5	−1.3
With farmyard manure			
Sugar percentage	16.4	−0.4	−0.8
Juice purity	88.2	−0.7	−1.4

Table 13.2. Effect of nitrogen fertilizer on sugar percentage and juice purity (from Boyd *et al.*, 1970).

		kg N ha^{-1}			
	No. fields	0	75	150	225
Sugar percentage	110	17.3	17.3	16.8	15.8
Juice purity	73	94.7	94.5	93.9	93.2

Nitrogen has a surprisingly similar effect on widely differing soils (Fig. 13.2). Roots of sugar beet grown on peat soils had an average sugar percentage of only 14.7 and a juice purity of 90.8%, whereas averages on sandy soil were 17.0 and 96.0%, respectively. A wide range of amounts of nitrogen fertilizer had parallel effects on all soil types for sugar percentage and juice purity.

Armstrong and Milford (1985) reported on a study of nitrogen uptake, sugar yield and amino N concentration. Their work supported the view (Last *et al.*, 1983; Burcky, 1991) that about 200 kg N ha^{-1} total uptake and no more is needed, even for very high-yielding crops, e.g. 15 t sugar ha^{-1}. Optimum amino N concentration in roots at harvest was more difficult to define.

Armstrong and Milford pointed out that, in an analysis of a large number of crops on a wide range of soils, amino N rose rapidly when total uptake exceeded 220 kg N ha^{-1}. For disease- and drought-free crops, an uptake of 200 kg N ha^{-1} coincided with an amino N concentration of less than 150 mg 100 g^{-1} sugar. In droughty years the equivalent amino N value was 200 mg 100 g^{-1} sugar, and with virus yellows infection 300 mg 100 g^{-1} sugar. On peaty soils, even in the absence of drought or disease, total uptakes were excessive (300 plus kg N ha^{-1}) and so were amino N concentrations (up to 450 mg 100 g^{-1} sugar). They thought that this effect on peaty soils was an inevitable result of the uptake of much nitrogen late into the

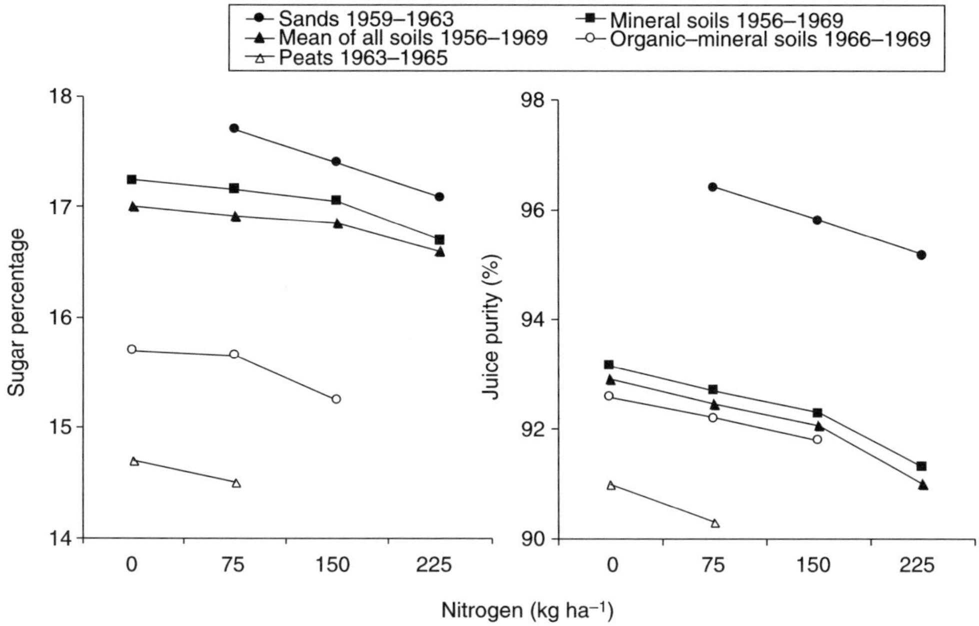

Fig. 13.2. Effect of nitrogen fertilizer and soil type on sugar percentage and juice purity (from Draycott *et al.*, 1971b).

autumn, which accumulated in roots, greatly exceeding that needed for growth.

In a later investigation by the same group (Pocock *et al.*, 1990; Allison *et al.*, 1996a), nitrogen (0–180 kg N ha^{-1}) increased amino N exponentially from 57 to 130 mg 100 g^{-1} sugar at the extremes of application. The increase was accompanied by a small increase in the other two impurities, sodium and potassium. In the report by Pocock *et al.* (1990), results from Belgium were also examined to extend the range of values of nitrogen uptake and amino N. Respectively, they ranged from 65 to 383 kg N ha^{-1} and 44 to 410 mg amino N 100 g^{-1} sugar. It appeared that, in both countries, optimum uptake was about 200 kg N ha^{-1} to produce best yield and quality. Unfortunately, on some very fertile soils, uptakes were 200–400 kg N ha^{-1} without any nitrogen fertilizer.

Europe

Results of 28 experiments in Ireland (McDonnell *et al.*, 1966) showed that 45 kg N ha^{-1} had no effect on sugar percentage. Further additions of fertilizer decreased sugar percentage by 0.1% for each 23 kg N ha^{-1}. Juice purity was reduced in a near-linear fashion from 96.4 to 95.3 over the range of 0 to 125 kg N ha^{-1}. Carolan (1960), also working in Ireland, found that nitrogen fertilizers caused large increases in the glutamine concentration of sugar beet and this partly accounted for the decrease in juice purity. Climatic conditions in Ireland are such that sugar beet is often still growing at harvest.

In France, Dubourg *et al.* (1957) examined the effect of a wide range of amounts (0–250 kg N ha^{-1}) on sugar percentage and concentration of amino acids in the soluble fraction of sugar beet roots. All amounts of nitrogen used decreased sugar percentage and increased amino acid concentration.

Von Lüdecke and Nitzsche (1967) investigated the effects of 'normal' and excessive amounts of fertilizer nitrogen on sugar percentage and juice purity in Germany. Normal fertilizing (125, 125 and 175 kg N, P_2O_5 and K_2O ha^{-1}, respectively) had little effect on quality, but overfertilizing (400, 400 and 570 kg N, P_2O_5 and K_2O ha^{-1}, respectively) decreased it drastically, partly by retarding certain metabolic processes in autumn. Winner (1966) also in Germany and Asselbergs *et al.* (1960) in The Netherlands described similar results.

In pot experiments von Müller *et al.* (1962) tested the influence of the N : K_2O ratio on the quality of sugar beet. When giving a large supply of nitrogen it was important to have a wide N : K_2O ratio (1 : 3) for maximum quality. When the ratio was narrow (1 : 1) maturity was retarded and quality decreased due to an increase in the roots. Heistermann (1968) confirmed these results, finding that a wide N : K_2O ratio increased the dry matter content of the roots, and hence the sugar percentage.

Denmark was one of the countries that led the way in measuring amino N on roots when delivered to factories. This was started in 1976 at one factory and adopted by all Danish factories in 1979. Amino N values were reported to individual growers (Marcussen, 1985).

As a result of investigations in their countries it was suggested that the optimum amino N concentration was between 110 and 130 mg 100 g^{-1} of sugar, this equating to a total crop uptake of nitrogen of 220–260 kg ha^{-1}. Growers responded by decreasing nitrogen fertilizer applications from 190 kg N ha^{-1} in the early 1970s to 140 kg N ha^{-1} in the mid-1980s.

Marcussen and Smed (1996) measured the effect of four amounts of potassium and four amounts of sodium in full factorial combination on the yield and quality of sugar beet in Denmark. Their soils contained little sodium because winter rain washed most of it away. Yield was maximal with about 120 kg K_2O ha^{-1} plus 100 kg Na ha^{-1}. The same combination increased sugar percentage by 0.17% and decreased amino N concentration significantly.

In Germany, Burba (1996) investigated the effect of 0–200 kg N ha^{-1} on the concentration of amino N and of invert sugars in roots at harvest. Amino N was always greatly increased by all amounts of fertilizer, 200 kg N ha^{-1} doubling its concentration compared with 0 kg N ha^{-1}. Burba listed the concentrations and proportions of the five main monosaccharides in roots as shown in Table 13.3. Interestingly, in his field experiments with 0–200 kg N ha^{-1}, concentrations of invert sugars were unaffected. The paper contains details of new chemical methods for determination of the parameters studied.

USA

Ryser (1966) made a study of the effect of nitrogen and other agronomic practices in sugar beet grown on farms in Oregon. In this survey of samples taken from grower deliveries, increasing nitrogen decreased sugar percentage (Table 13.4). Delaying harvest increased the sugar percentage and decreased the effect of nitrogen. In Minnesota (Ogden *et al.*, 1958) even a small amount of nitrogen caused a large decrease in sugar percentage and juice purity (Table 13.4). Both decreased linearly throughout the range tested and the effect on sugar percentage was so great that, despite an increase in root yield, sugar yield was not increased. However, the sugar beet followed a legume, so the residual nitrogen was presumably sufficient to satisfy the requirement of nitrogen by the sugar beet without further addition as fertilizer.

In California, Stockinger *et al.* (1963) confirmed that different cropping systems influenced both yield and quality by their effect on the availability of soil nitrogen. Cropping systems that added organic matter or left residual nitrogen from large fertilizer applications increased yield, but decreased sugar percentage and juice purity. With 470 kg N ha^{-1} there was no difference between cropping systems. The nitrogen from organic sources had no advantage over inorganic fertilizer nitrogen.

Table 13.3. Amounts and proportions of monosaccharides in roots (from Burba, 1996).

	mg kg^{-1}	%
Glucose	623	61
Fructose	227	22
Galactose	77	8
Ribose	77	8
Arabinose	17	1

Table 13.4. Effect of nitrogen fertilizer on sugar percentage and juice purity in the USA.

	kg N ha^{-1}						
	0	10–45	46–90	91–135	136–180	> 180	Reference
Sugar	17.0	–	−0.5	–	–	–	Haddock *et al.*, 1956
Purity	91.0	–	−1.3	–	–	–	
Sugar	16.8	−0.6	−0.9	–	–	–	Ogden *et al.*, 1958
Purity	83.2	–	−1.4	–	–	–	
Sugar	16.8	–	–	–	−0.2	−1.3	Haddock *et al.*, 1959
Purity	90.2	–	–	–	−0.7	−2.8	
Sugar	–	16.4	−0.1	−0.2	−0.5	–	Ryser, 1966
Sugar	16.4	+0.2	−0.2	−0.4	−0.8	−1.0	Giles, 1994
Sugar	17.7	−0.1	−0.1	−0.3	−0.5	−0.5	Christenson *et al.*, 1985
Purity	95.3	−0.2	−0.4	−0.5	−0.6	−1.2	
Sugar	18.3	−0.1	+0.1	−0.2	−0.5	−0.5	Christenson and Butt, 2000

Haddock *et al.* (1956) in Utah investigated the effects of nitrogen fertilizer on sugar beet quality in more detail. They measured a range of nitrogen containing constituents in the sugar beet root, showing that decreased quality was associated with large concentrations of glutamine and ammonium nitrogen. Glutamine in particular was closely related to quality; where nitrogen fertilizer or other cultural factors increased glutamine concentration, quality always decreased. Even with modest additions of nitrogen, there was a marked decrease in both sugar percentage and juice purity (Table 13.4).

In a later report, Haddock *et al.* (1959) established that nitrogen concentration in sugar beet tissue was inversely related to sugar percentage and juice purity. Where sugar percentage and juice purity were small, top : root ratios were large. Large applications of nitrogen (> 135 kg N ha^{-1}) depressed purity more than sugar percentage, both by greater than 1% when more than 180 kg N ha^{-1} was added (Table 13.4).

A summary of 86 studies in the Red River Valley is shown in Table 13.4 (Giles, 1994, index to reports). These results confirm all the other reports above, showing a near-linear decline in sugar percentage with increasing nitrogen applied (Table 13.4). Purity was depressed by over 1% when application exceeded 180 kg N ha^{-1}. Similar results were found in Michigan (Christenson *et al.*, 1981; Christenson and Butt, 2000). Excessive nitrogen applied to maize prior to sugar beet depressed both sugar percentage and juice purity (Christenson *et al.*, 1981). Sugar percentage was depressed 0.6% and juice purity 0.9% by excessive nitrogen applied to maize the year prior to sugar beet.

Eckhoff (1999) applied 75, 100 and 125% of the recommended amount of nitrogen to irrigated fields in eastern Montana, USA; the recommended amount being based on residual nitrogen to 120 cm, estimated nitrogen mineralized from organic matter and expected yield goal. Eckhoff reported greater sugar content and extraction with 75% of the recommended rate than with greater amounts of fertilizer nitrogen.

The mean effects of nitrogen on sugar and purity are given in Table 13.5. Adequate amounts (100–120 kg N ha^{-1}) generally have small effects on sugar percentage and juice purity. However, there is a striking effect at the largest application, particularly for juice purity.

Phosphorus

Adams (1962) reported that phosphorus fertilizer had no effect on the sugar percentage or juice purity of sugar beet in 41 experiments on commercial farms. In experiments on peaty soils, Tinker (1970) reported that 40 kg P_2O_5 ha^{-1} increased sugar percentage by 0.1% but had no effect on juice purity. McDonnell *et al.* (1966) described the effect of superphosphate (18% P_2O_5) on the sugar per-

Table 13.5. Mean effect of nitrogen fertilizer on sugar percentage and juice purity in the USA.

kg N ha^{-1}	Sugar (%)	Juice purity (%)
0	17.0	93.1
75	−0.2	−0.5
150	−0.3	−0.9
225	−0.5	−1.5

centage and juice purity of sugar beet in Ireland. The smallest amount increased sugar percentage, but no additional increase was shown with larger amounts (Table 13.6). Juice purity was increased significantly in the first year, but had no effect in the second year. Gericke (1966) found surprisingly large responses to phosphorus fertilizer in respect of the sugar percentage of sugar beet in Germany. It also improved the feeding value of the sugar beet leaves, for protein concentration was increased considerably and oxalic acid concentration decreased by a third.

There are conflicting reports on the effect of phosphorus fertilizer on sugar beet quality from the USA. In an experiment in Minnesota, Ogden *et al.* (1958) reported a significant decrease in sugar percentage from large amounts of phosphorus, but juice purity was affected little (Table 13.6). However, in Kansas, phosphorus fertilizer had no effect on sugar percentage when tested at 0 and 130 kg P_2O_5 ha^{-1} (Herron *et al.*, 1964b).

In a study with increasing quantities of phosphorus fertilizer, Christenson *et al.* (1975) found improvement of both sugar percentage and juice purity on a soil with little available phosphorus (Table 13.6). The study was continued and different residual concentrations of phosphorus established prior to growing sugar beet again. Both purity and sugar percentage were suppressed with increasing residual phosphorus when 55 kg P_2O_5 ha^{-1} was used at sowing (Table 13.6).

Thus phosphorus fertilizer usually has a positive effect on sugar percentage and, on soils with small concentrations of available phosphorus, can be of the order of +0.3% or more. However, on soils with larger concentrations of available phosphorus, it is unlikely that sugar percentage will be improved. All the evidence suggests that phosphorus fertilizers have little, if any, effect on juice purity.

Potassium and Sodium

Potassium and sodium are two serious impurities that decrease the extraction of white sugar in the processing of roots. It is expected that giving these elements in fertilizers will

Table 13.6. Effect of phosphorus fertilizer and extractable soil phosphorus on sugar percentage and juice purity.

	kg P_2O_5 ha^{-1}					
	0	10–100	101–200	201–300	> 300	Reference
Sugar	16.6	–	–	−0.3	−0.5	Ogden *et al.*, 1958
Purity	81.8	–	–	+0.1	+0.0	
Sugar	15.1	–	+0.0	–	–	Herron *et al.*, 1964b
Purity	89.8	–	−0.7	–	–	
Sugar	16.5	+0.2	+0.2	+0.2	–	McDonnell *et al.*, 1966
Purity	95.8	+0.1	+0.1	+0.2	–	
Sugar	16.1	−0.1	+0.2	+0.2	–	Christenson *et al.*, 1975
Purity	94.2	0.01	+0.4	+0.2	–	Low extractable phosphorus
	mg P kg^{-1} soil					
	12	18	22	40	110	
Sugar	17.4	−0.1	+0.1	−0.2	−0.3	Christenson *et al.*, 1981
Purity	96.2	−0.2	−0.1	−0.4	−0.6	55 kg P_2O_5 ha^{-1} applied to all

increase their concentration in the root. However, most workers have found mostly positive effects from potassium application on sugar percentage (Table 13.7). McDonnell *et al.* (1966) reported experiments in Ireland showing that potassium fertilizer increased sugar percentage. Likewise Winner (1966) in Germany, Gascho *et al.* (1969) in Michigan and Moraghan (1979) in the Red River Valley all reported increases in sugar percentage from additions of potassium fertilizer. The experiment in Michigan was on a soil with more than 100 mg kg^{-1} exchangeable potassium. Moraghan's work was on a site showing leaf symptoms of potassium deficiency. In these studies, it appeared that even excessive potassium supply did not have a deleterious effect on sugar percentage. However, one exception is that Simon *et al.* (1966) found that an excess of potassium in Belgian soils reduced sugar percentage by 0.4%, a dubious result, which has not been repeated.

Increasing potassium supply did not have a large effect on juice purity (Table 13.7). McDonnell *et al.* (1966) showed a 0.1% decrease in purity with greater than 300 kg K_2O ha^{-1}. Gascho *et al.* (1969) found no effect, but Moraghan (1979), on a potassium-deficient site, found that additional potassium increased purity.

Carruthers *et al.* (1956) reported that sodium decreased juice purity. In Great Britain, Draycott *et al.* (1970a) determined the sodium, potassium and alpha-amino nitrogen (another impurity in sugar beet juice) in crops that had been treated with a wide range of sodium (0–950 kg NaCl ha^{-1}) and potassium (0–800 kg KCl ha^{-1}) fertilizer. As expected, each fertilizer increased the concentration of that element in roots by significant but, especially in the case of large applications, surprisingly small amounts. What is also of interest is that each decreased the concentration of alpha-amino nitrogen, which probably accounts for the small effect of potassium and sodium fertilizers on juice purity.

Huijbregts *et al.* (1996) reported work in The Netherlands to an IIRB session devoted to factors affecting sugar beet quality. They showed that, over the previous 20 years, quality had greatly improved, mainly due to breeding for a lower concentration of amino N, sodium and potassium, coupled with improved use of fertilizer. In field trials they found that nitrogen (0–250 kg N ha^{-1}) had little effect on sodium and potassium but doubled amino-nitrogen concentrations. Sodium fertilizer (0–300 kg Na_2O ha^{-1}) had no effect on potassium and amino nitrogen but increased sodium concentration by half. Potassium fertilizer (0–450 kg K_2O ha^{-1}) slightly decreased sodium, had no effect on amino nitrogen and, interestingly, only slightly increased potassium concentration.

In parallel work in the UK with sodium and potassium fertilizers, Jarvis and Bee (1996) found that sodium (0–225 kg Na ha^{-1}) increased sodium concentration on average from 64 to 75 mg Na 100 g^{-1} sugar and potassium (0–300 kg K ha^{-1}) from 1080 to 1140 mg K 100 g^{-1} sugar. Thus, in both The Netherlands and the UK, potassium fertilizer has a relatively small effect on quality, somewhat less than sodium fertilizer.

Table 13.7. Effect of potassium fertilizer on sugar percentage and juice purity.

	kg K_2O ha^{-1}					
	0	10–100	101–200	201–300	> 300	Reference
Sugar	16.4	+0.3	–	+0.3	+0.5	McDonnell *et al.*, 1966
Purity	95.9	+0.0	–	+0.0	−0.1	
Sugar	–	17.1	–	–	−0.4	Simon *et al.*, 1966
Sugar	18.1	+0.3	+0.3	+0.6	–	Winner, 1966
Sugar	14.6	–	–	+0.1	–	Gascho *et al.*, 1969
Purity	95.0	–	–	+0.0	–	
Sugar	12.7	+0.2	–	+0.7	–	Moraghan, 1979
Purity	93.6	+0.2	–	+0.4	–	

Magnesium

Jorritsma (1956) reported that magnesium fertilizer improved the juice purity of sugar beet grown in The Netherlands. Tinker (1967b) tested magnesium sulphate on sugar beet grown on fields where deficiency of magnesium was likely. On average in 17 fields, the sugar percentage was increased by about 0.2% and juice purity also improved slightly. In 19 similar experiments (Draycott and Durrant, 1969a), the average increase from 630 kg kieserite ha^{-1} was 0.1 in sugar percentage with no effect on juice purity. However, on fields where magnesium deficiency was severe (80% of plants with symptoms on leaves), sugar percentage was increased greatly (+0.5%) with an accompanying increase in juice purity (+0.2%).

Lachowski (1966) grew a sugar beet seed crop in soil with low exchangeable magnesium (34 mg kg^{-1} soil) with and without 100 kg $MgSO_4$ ha^{-1}. Magnesium improved the sugar percentage in the roots of the progeny, presumably due to an improvement in magnesium concentration in the seed.

Effects of Fertilizers on Keeping Quality of Sugar Beet

In most sugar beet-growing countries, harvested roots are stored for varying periods before processing. For example, in the UK, roots may be kept in storage heaps and clamps for a few days at the start of harvest through to several months when temperature has cooled in January/February and for short periods at the sugar factory (Jaggard *et al.*, 1997b). In the northern USA, most of the crop is stored in very large piles, either close to the factory or in satellite storage facilities. The length of storage time may be over 120 days.

In the Red River Valley, to extend the processing season, a portion of the crop is completely frozen in late December by drawing air through the piles when ambient air temperature reaches −15°C. Due to the mass of beets in the frozen pile, warming in late March and April does not cause significant thawing of roots. Swift (1980) reported sugar loss of 5.8% from deep-frozen piles and 7.6% for all stored beets at one site over a 2-year period. The storage piles contained over 100,000 t of beet. The length of storage averaged 182 days.

We have shown considerable effects from nutrients on sugar beet quality. It would seem to follow that nutrition and cultural practices should affect the storage properties of the roots. However, surprisingly few investigations have been made.

Martens and Oldfield (1970) reviewed the storage of sugar beet in 16 European countries. In the USA, Larmer (1937) measured the effect of phosphorus on the keeping quality of sugar beet. He grew the crop under conditions of adequate and inadequate phosphorus supply on fields known to be deficient in the element. Roots from plots receiving phosphorus or complete fertilizer showed less decay than roots from unfertilized plots. In these tests the roots were kept in coarse-meshed sacks and exposed to conditions of commercial storage heaps. Besides the reduction of loss by decay, the phosphorus fertilizer decreased the loss of sugar due to transpiration.

In another experiment, Larmer found indications that nitrogen fertilizer and farmyard manure improved keeping quality. Also in Utah, Stout and Smith (1950) grew sugar beet with and without a commercial application of fertilizer (exact details of the fertilizer are not given). It had little effect on the respiration rate of the stored roots but there was no response in yield to the fertilizer. Large roots did, however, respire more slowly than an equal weight of smaller roots.

In Colorado, Gaskill (1950) tested 0, 55 and 170 kg N ha^{-1}, 0 and 55 kg K_2O ha^{-1} and 0, 110 and 220 kg P_2O_5 ha^{-1}. Roots from the field experiment were stored at 7 and 18°C for 3 and 4 months, and those with nitrogen kept better at 7°C than those without. At 18°C there were no significant differences attributable to nitrogen. The effects of potassium and phosphorus were small, which the author considered not surprising, as there was a negligible yield response to potassium and relatively little to phosphorus.

Dexter *et al.* (1966) reported on the effect of fertilizer nitrogen on sugar percentage and juice purity in sugar beet stored for 77 days at 35°C. Loss of sugar and decline in juice purity were greater when excessive nitrogen (260 kg ha^{-1}) was given than with lesser amounts (34 and 100 kg N ha^{-1}).

It seems from this sparse experimental evidence that a commercial amount of nitrogen is likely to improve the keeping quality of roots. Boron applications are also important where there is a shortage of the element in soil, because roots of deficient plants deteriorate rapidly in storage. Where soils have little available phosphorus, fertilizer is also likely to improve keeping quality but, where yield responses are small, quality will change little. The effects of other major and minor nutrient elements remain to be determined, as does the effect of excessive amounts of nutrients.

Value of Improvements in Quality

Figure 13.3 shows the average effect of nitrogen fertilizer on sugar percentage and juice purity over many fields. Figure 13.4 shows the average amount of fertilizer used in the UK in 1955–1995 in British Sugar surveys. The remarkable improvement in appearance of sugar beet led to overuse in the 1960s but, since then (due to better knowledge of crop requirements, described in Chapter 2), use has fallen by about 60 kg N ha^{-1} over the past 30 years (Fig. 13.4). From Fig. 13.3 this would result in an improvement in sugar percentage of about 0.6% and in juice purity of 0.5%. It is difficult to quantify the many advantages of the rise in sugar percentage but the effect on purity would improve UK sugar extraction by about 25,000 t with greatly improved sugar colour (Draycott, 1999).

Summary

The mean effect of nitrogen fertilizer on sugar percentage and juice purity over many experiments shows a steady decline in both from increasing amounts of nitrogen. The first 75 kg N ha^{-1} decreases sugar percentage by only 0.2% but more has a progressively larger effect, and 225 kg N ha^{-1} normally decreases it by about 1.0%. If available soil nitrogen is very small, the first 75 kg N ha^{-1} may increase the sugar percentage slightly. Nitrogen decreases juice purity linearly, each 75 kg ha^{-1} by about 0.5% throughout the whole range.

Optimum nitrogen supply should stimulate leaf growth to an LAI of 3 as quickly as possible, keep the crop growing until the late

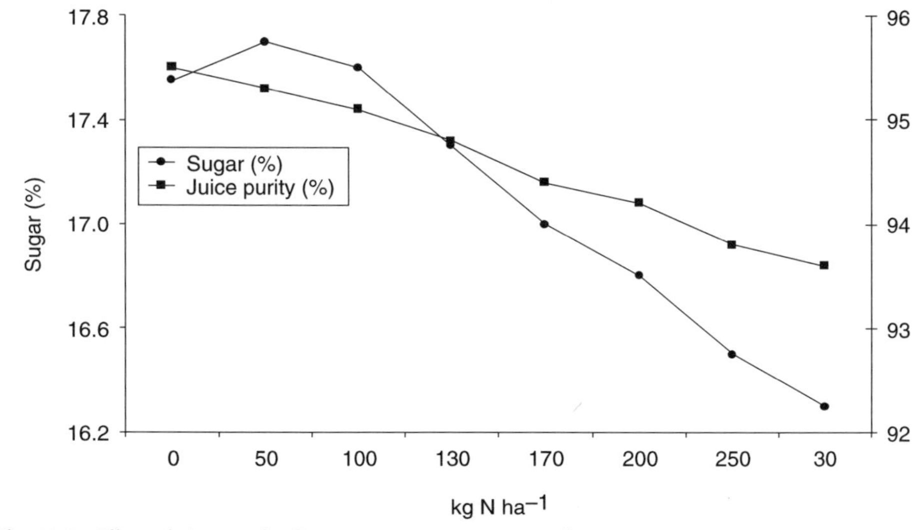

Fig. 13.3. Effect of nitrogen fertilizer on sugar percentage and juice purity.

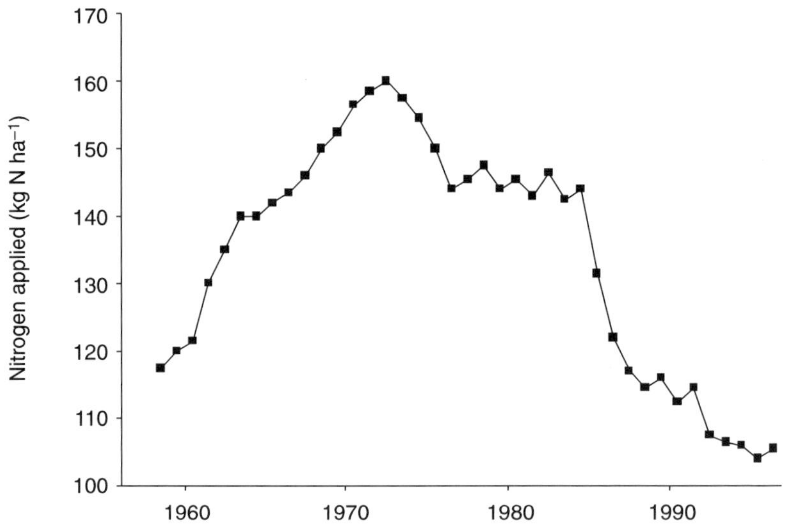

Fig. 13.4. Amount of nitrogen fertilizer applied to UK sugar beet, 1955–1995 (British Sugar data).

season and then decline so as to produce slight nitrogen deficiency symptoms prior to harvest. Generally, this is in the range of 100 to 120 kg N ha^{-1} in the absence of large reserves of available soil nitrogen.

There are many reports of harmful effects on quality where uptake is excessive prior to harvest. Large reserves of nitrogen, e.g. from unused fertilizer, from organic manure or from fen, peat and muck soils, result in total nitrogen uptake exceeding what we believe to be the optimum of 200 kg N ha^{-1} (there are reports of uptakes of more than 400 kg N ha^{-1}). Inevitably amino-nitrogen concentration is increased and hence purity is greatly depressed.

Phosphorus fertilizer increases the sugar percentage slightly on severely deficient soils but on moderately fertile fields neither increases nor decreases it. Generally juice purity is not affected by increasing quantities of phosphorus fertilizer. There were a few cases where juice purity was increased with increasing supply of phosphorus fertilizer. However, just one study showed that both juice purity and sugar percentage were decreased with increasing residual phosphorus, as indicated by extractable-phosphorus concentrations.

In most situations potassium and sodium fertilizers increase sugar percentage, probably by decreasing the amount of water in the root. There are conflicting reports of their effects on juice purity. Some report that potassium has no effect on juice purity and some that sodium, too, has no effect. But others report that both nutrients depress juice purity slightly. Both elements are serious impurities in the root juice, so increasing their concentration in the plant would be expected to decrease purity. However, increasing the concentration of sodium or potassium may decrease the concentration of some other impurity, e.g. amino nitrogen. On average, these cations improve sugar percentage by 0.1–0.2% with little effect on purity. Magnesium additions improve both sugar percentage and juice purity, this being especially noticeable where there is severe magnesium deficiency.

In an early review of over 400 experiments from 1934 to 1949, the effects of nitrogen, phosphorus and potassium on sugar percentage were: nitrogen, −0.38; phosphorus, +0.02; and potassium, +0.24. Results collected since that time are not at great variance with these. The best guide for the optimum root quality of sugar beet is to supply adequate, but not excessive, amounts of nutrients. Soil analysis, fertilizer recommendations and plant analysis to monitor growth are the best methods for meeting this goal, all of which have been covered in previous chapters.

14
Nutrient Requirements of Fodder Beet and of the Beet Seed Crop

FODDER BEET

Where climate and soils are suitable for sugar beet production, fodder beet will thrive because the plants are similar botanically (both are subspecies of *Beta vulgaris*) (Winner, 1993) and in requirements of water and nutrients. On livestock farms with dairy cows, beef cattle, pigs or sheep, fodder beet is often the highest-yielding crop that can be grown. Table 14.1 shows the typical yield of several forage crops in a temperate climate. Fodder beet fresh root yields of 50–75 t ha^{-1} are common and, where grown well with adequate plant population, nutrients and water, yields exceed 100 t ha^{-1}. Root dry-matter percentage differs depending on fodder beet variety type (Elliott and Weston, 1993). Those with taproots that grow deep in the soil like sugar beet often contain about 20% dry matter. Those growing out of the ground like mangolds usually contain about 12% dry matter. Roots are all high in energy but low in protein, just like sugar beet. Their high sugar concentration makes them very palatable. This increases total dry-matter intake by stock of fodder beet roots. The crop also produces 10–20 t ha^{-1} of leaf material, which, in contrast to the roots, contains high crude protein (15%). The roots plus tops together give a good nutritional balance. Modern varieties of fodder beet all carry the monogerm characteristic, which

Table 14.1. Forage crop yields.

	Typical fresh	Typical dry matter	Potential dry matter
		t ha^{-1}	
Fodder beet			
Roots only	65	10	16
Roots plus tops	80	12	20
Kale	40	6	10
Maize	40	12	16
Grass silage			
1 cut per year	23	6	8
3 cuts per year	48	12	18
Whole-crop silage	30	9	15

allows the crop to be sown to a stand and grown without hand labour. Weeds, diseases and other pests can be controlled with materials developed for sugar beet. Similarly, machinery needs are fulfilled by those produced for sugar beet, e.g. drills and harvesters.

Soil Requirements

Like sugar beet, all fodder-beet types need deep soil with an adequate macro- and micronutrient concentration. Ideally, soil should allow fibrous roots to develop to 2 m (Durrant *et al.*, 1973) to provide some 500 mm water (soil water, plus rainfall during the growth period). Plant nutrients and concentrations for optimum production are similar or slightly greater than for sugar beet (Draycott and Hollies, 2001).

Nutrient Uptake and Offtake

A recent 3-year study in the UK (Hollies, 2001) provides new data on the uptake and offtake of nutrients for fodder beet. Table 14.2 shows the uptake of a crop yielding 70–90 t of roots ha^{-1} and total dry matter of roots plus tops of 16–20 t ha^{-1}. Potassium uptake was very large and in one 2-week period in the study the crop was taking up potassium at a rate of 18 kg K_2O ha^{-1} day^{-1}.

Nutrient offtake of phosphorus and potassium can be calculated by multiplying fresh root yield per hectare by 0.6 kg P_2O_5 t^{-1} and by 4.0 kg K_2O t^{-1}, respectively. If tops are also removed, the factors are 1.7 kg P_2O_5 ha^{-1} and 7.5 kg K_2O ha^{-1}.

Table 14.2. Nutrient uptake by fodder beet yielding 70–90 t roots ha^{-1} averaged over 3 years in the UK.

kg ha^{-1}						
N	P_2O_5	SO_3	K_2O	Na_2O	CaO	MgO
250	90	50	580	120	50	60

Nutrient Application

Nitrogen

With sugar beet, the aim is to utilize nitrogen from soil, fertilizer and manure to produce a crop with maximum extractable sugar per unit area. Chapters 2 and 13 defined how this can be achieved and the importance of limiting fertilizer usage to obtain maximum sugar production. In the case of fodder beet, the objective is somewhat different, because the target is usually maximum dry-matter production per unit area, not sugar. For sugar beet, the total uptake of nitrogen is ideally about 200 kg N ha^{-1}. With fodder beet, Table 14.2 shows an uptake of 250 kg N ha^{-1}. Thus, for sandy soils with small reserves of nitrogen (e.g. 50 kg N ha^{-1} in the top metre of soil), optimum fertilizer application is in the region of 150–200 kg N ha^{-1}.

There is a paucity of evidence to suggest optimum timing, but the technique of applying a portion (40 kg N ha^{-1}) immediately after sowing and the balance at full emergence avoids damage to germinating seeds and leaching losses, as described in Chapter 2.

On farms growing fodder beet, it is likely that some form of organic manure will be used and also that soil reserves of nitrogen will provide a significant portion of the crop need. Details of how to account for these two sources to modify fertilizer application are in Chapters 2 and 8 and MAFF (2000).

Phosphorus

Phosphorus requirements of fodder beet are similar to those of sugar beet, and the optimum fertilizer applications in Tables 3.4 and 3.5 can be used in relation to soil-available phosphorus. Uptake of the element is relatively small but a sufficiency is needed during establishment to ensure rapid seedling growth and early leaf cover. For soils with 0–15 mg P kg^{-1}, fertilizer should be applied during seedbed preparation. In all other situations, application can be at any time, aiming to keep the soil with reserves of 16–25 mg P kg^{-1}.

Potassium and sodium

Fodder beet has a large requirement for the two cations potassium and sodium, as shown in Table 14.2, potassium being taken up in greater quantities than any other nutrient. The two elements are partially interchangeable (see Chapter 4). Table 14.3 shows optimum requirements of potassium and sodium based on the amount in the exchangeable form in soil. As with nitrogen, the contribution made from recent applications of organic manures should be deducted from these amounts, as outlined in Chapter 8.

Soil pH, and other plant nutrients

In common with sugar beet, fodder beet must be grown in neutral or slightly alkaline soil, and correction of acidity, described in Chapter 5, is crucial. Other plant macronutrients, such as magnesium and sulphur, and micronutrients, such as boron and manganese, are all needed in similar quantities and situations as for sugar beet.

NUTRIENT REQUIREMENTS OF THE SEED CROP

Sugar beet is a biennial plant, producing seed only after vernalization and when grown with a suitable day length. The root crop described in previous chapters is grown in a single season but plants grown for seed production need a second year to put up flowering stems and set seed. In a comprehensive review of worldwide seed production, Bornscheuer *et al.* (1993) rightly pointed out that quality seed is one of the most important aspects of a successful sugar beet root crop.

The last time the nutrient requirement of the seed crop was reviewed (Draycott, 1972) the use of monogerm seed for the root crop was only just becoming widely adopted. Thus much of the experimental work with fertilizers had been done with multigerm seed crops. Now that monogerm seed is almost universally used for the root crop, with a low seed rate and no subsequent thinning, seed quality is even more crucial. Another big change over the past 30 years has been the movement of the European seed crop from northern countries to southern France and northern Italy, in which two areas it is now concentrated. In the 1950s and 1960s, much of the seed for the EU was grown in Denmark, The Netherlands, Germany and the UK (Scott, 1968). Today these countries produce very little seed because cool and wet autumns were shown to result in poor quality (see Bornscheuer *et al.*, 1993). Current seed production in the USA is consolidated in the Willamette Valley of western Oregon. There is a small amount produced in Utah and at various other locations as a part of variety development programmes.

The seed crop is produced either by sowing the mother seed in the summer of the first year and transplanting the 'stecklings' the following spring or by direct (*in situ*) sowing without transplanting. The former system is the most common method of multiplication for European countries. The direct-sown crop is sometimes grown under a cover crop, which protects the sugar beet plants from diseases and other pests.

Production in the Willamette Valley is by direct sowing supplemented with steckling transplanting as necessary. Often direct sowing includes two pollinators. When the results of variety trials are evaluated, the

Table 14.3. Optimum amounts of potassium and sodium fertilizer for fodder beet as determined by soil analysis.

Soil potassium[a] (mg K kg^{-1})	Optimum fertilizer (kg K_2O ha^{-1})
0–60	300
61–120	250
120–240	150
> 240	100
Soil sodium[a] (mg Na kg^{-1})	**Optimum fertilizer (kg Na_2O ha^{-1})**
0–25	200
26–50	100
> 50	0

[a]Extracted in a molar solution of ammonium nitrate.

undesired pollinator is removed, either with herbicides or by mechanical means. In other cases, stecklings are transplanted into the seed-production field.

Nitrogen

Transplanted crops

Ellerton (1947) did much of the early work with multigerm crops in the UK and worked out some of the principles of producing high yields of good quality seeds from stecklings. He found that nitrogen fertilizer was needed in March to stimulate early growth, yields being increased by up to 25%. Nitrogen given at flowering in late June was less effective, increasing yield by 8%. Ellerton found that nitrogen fertilizer increased seed germination percentage. Mann and Barnes (1945) found that early nitrogen was important for yield but not for the germination percentage of the resultant seed. Sneddon (1963) warned that nitrogen fertilizer could delay the ripening of lower clusters on the seed stem and decrease germination, especially in wet years.

Direct-sown crops

Sneddon (1963) compared two times of application of top dressing of 125 kg N ha^{-1} on direct-sown crops. Nitrogen was given either before bolting commenced or after the plants had produced maximum extension and had developed flower buds. The nitrogen greatly increased yield in all 3 years of the experiment. Applying the nitrogen before bolting was slightly more effective than when given before flowering. Nitrogen only affected the seed size in one experiment, when the proportion of clusters over 4 mm was 40% compared with 35% for plots given no nitrogen top dressing. Nitrogen, however, consistently decreased germination from about 75% to 67%.

Scott (1969) described eight experiments in four areas testing nitrogen fertilizer applied in late February or in March/April of the harvest year on direct-sown crops. Two experiments were on deep, organic, silty clay loam soil in Lincolnshire, two on a clay loam in Northamptonshire, three on shallow Cotswold limestone brash soil and one on Bunter sandstone soil in Nottinghamshire. Although seed yields were similar without nitrogen on all fields, the response to nitrogen was much greater on the deep silty clay loam than on the shallow limestone soil. Not only did nitrogen nearly double yield on the silt but more nitrogen fertilizer was needed for maximum yield. The crop on the clay soil gave an intermediate response. Scott concluded that the seed crop on deep organic silt responded to more nitrogen than on the shallow limestone soils, because water limited yield on the latter. Nitrogen fertilizer never decreased yield significantly.

Increasing the fertilizer supply increased yield because plants produced more clusters, not larger or heavier ones. Only in one experiment, where the crop was affected by adverse weather conditions, did nitrogen decrease the germination percentage of the seed produced from 82 to 75%. Where the weather was good and the plots harvested at the normal time (August/September), nitrogen did not depress germination.

Scott (1967) described an experiment to test the effect of 0–200 kg N ha^{-1} for an *in situ* crop. The nitrogen increased seed yield (Table 14.4) up to the largest amount tested. However, ripening was delayed and the largest dressing decreased germination percentage.

Table 14.4. The effect of nitrogen fertilizer given in spring to an *in situ* seed crop (after Scott, 1967).

	N applied (kg ha^{-1})				
	0	50	100	150	200
Seed yield (t ha^{-1})	2.75	2.81	2.89	2.89	3.19
Germination percentage	81.5	82.3	81.3	81.3	74.5

Longden and Johnson (1977) experimented with monogerm (and multigerm) seed crops. They examined the effect of single or split applications of nitrogen in February, March, April and May. Yield was not affected much by any timing of nitrogen but late application decreased the germination percentage of the seed.

Phosphorus, Potassium and Sodium

There is little evidence that the seed crop responds to applications of phosphorus fertilizer (Ellerton, 1947; Scott, 1969) in respect of either yield or quality, probably because, in most areas where the seed crop is grown, soils have been farmed well with a sufficiency of fertilizer and manure. In the case of potassium and sodium, both elements often improve performance (Longden, 1970). Thus there are close parallels with the sugar beet root crop, which is also very responsive to these two cations.

Nutrient Concentrations in Seed

Longden (1970) reported the effects of fertilizer on the chemical composition of seeds produced. Nitrogen at 190, 250 and 310 kg ha^{-1} gave seed with 2.04, 2.13 and 2.26% N, respectively, in the dry matter. Phosphorus fertilizer did not affect the concentration of the element in the dry matter. When samples of the seeds were sown in the field, there were no effects on emergence or seedling weight. He found that, on average, seeds should contain 2.0% N, 0.45% P and 2.0% K in dry matter, leaves 2.7% N, 0.39% P and 2.9% K and roots 1.7% N, 0.36% P and 1.7% K. For a 2.5 t ha^{-1} crop of seed with 5 t ha^{-1} straw and 1 t ha^{-1} roots, the crop at harvest contains 200 kg N ha^{-1}, 75 kg P_2O_5 ha^{-1} and 260 kg K_2O ha^{-1}.

Seed Crop in North America

Early work established the importance of adequate nutrition in several states growing the seed crop for the home market and later for export. Pultz (1937) made experiments with direct-sown seed crops in Utah and New Mexico to determine the response to nitrogen given either in September of the sowing year or March of the harvest year or on both occasions. Nitrogen increased yield greatly but there were no consistent differences in the quality of the crop.

Chemical analysis showed that the concentrations of carbohydrates and nitrogenous compounds in the roots were maximal during winter dormancy. After seed stalk development began in spring, sugars and nitrogen were rapidly withdrawn from the roots. Loss of nitrogen from roots was only slightly affected by fertilizer treatment and continued until most of the nitrogen had been withdrawn.

However, the amount of sugar removed from the root depended on the supply of nitrogen from soil. When nitrogen was continuously available during the fruiting period, the sugar percentage in the root decreased steadily until the seed matured and this resulted in a long period of flower production and a large seed yield. Where nitrogen became a limiting factor during fruiting, loss of sugar from roots ceased, flowering stopped prematurely and seed yield from such plants was small.

Campbell (1968) reviewed sugar beet seed production in Oregon on fertile, deep loam and silt soils. The crop was direct-sown and irrigated in the first year. Heavy winter rainfall leached nitrogen and large dressings were needed in the spring, either as soon as ground conditions permitted or from aircraft.

Snyder (1959) compared the vigour and percentage germination of seeds from crops grown with various amounts of nitrogen. The amount of nitrogen supplied to parent plants had no pronounced effect and it was concluded that the best amount of nitrogen was the one which gave the largest yield of seed. Pendleton (1954) also tested nitrogen, phosphorus and potassium on seed yield and germination. Seed yield without fertilizer was small and germination was poor. Nitrogen fertilizer more than doubled yield. Nitrogen and phosphorus increased germination percentage, together by 8%.

Other early parallel studies confirmed the need for a continuous supply of available nitrogen to ensure maximum yield. Some reports show improvements in seed quality. A deep, fertile, water-retentive soil for seed production is specified and additions of other major (phosphorus, sulphur, potassium) and trace elements (particularly boron) where soil supply was small (Overpeck and Elcock, 1937; Tolman, 1943; Pendleton *et al.*, 1950; Pendleton, 1954; Lehnhardt and Bonk, 1991).

Commercial Use of Fertilizer on the Seed Crop

The last survey of sugar beet seed production in the EU and USA was that of Scott (1968). Table 14.5 shows the amounts of fertilizer being used at that stage. In the first year of steckling production (EU) or at establishment in autumn (USA), similar amounts of nitrogen and phosphorus were given, plus sown potassium. In the second year, when the crop was producing seed, nearly double the amounts of nitrogen were applied in the USA as in the EU, following differences in experimental findings in the two continents. Potassium applications were somewhat similar. Scott noted that in many areas amounts of phosphorus and potassium varied widely to suit local conditions and was often similar to root crop applications in the seed crop production year.

Bornscheuer *et al.* (1993) reviewed seed production methods worldwide and included some information on nutrition. For the indirect (EU) method, they defined an optimum steckling size of 100–150 g with a top diameter of 4–5 cm, these producing the best seed crops. Recent information on nutrition of steckling beds from France (L. Dyer, UK, 2002, personal communication) shows the optimum at 100 kg N ha^{-1}, 175 kg P_2O_5 ha^{-1} and 85 kg K_2O ha^{-1}.

For the seed production year of the steckling system, both Bornscheuer *et al.* and Dyer stress the importance of deep, fertile and water retentive soils, clayey and chalky ones being preferred. Excessive nitrogen, coupled with excess precipitation, always caused problems. In dry areas, however, Lejealle (1986) reported that 200–300 kg N ha^{-1} gave best results in southern France, without affecting seed quality.

Amounts of phosphorus and potassium are increasingly decided from soil analysis, as with the root crop. Table 14.6 shows the average use in three important seed producing countries. Both elements are always applied after ploughing and worked into the soil as the field is being prepared for transplanting the stecklings. About one-third of the nitrogen is also applied at this stage to ensure rapid early growth. The final nitrogen applications are then made before the beginning of bolting, to avoid any risk of delayed maturity and reduced seed quality.

Seed producing fields in the USA generally receive 90 kg N, 65 kg P_2O_5, 45 kg K_2O and 45 kg S ha^{-1} in the autumn prior to the sowing of the seed crop. In late March or early April, an additional 90 kg N and 30 kg S are applied. A second spring application of a similar amount of nitrogen and sulphur goes on about 1 month after the first application. Adjustments may be made, depending on crop history, crop residues and soil analysis.

Table 14.5. Fertilizer used for seed crops by the indirect method in the EU (means of 13 countries) and for direct-sown crops in the USA (after Scott, 1968).

	EU		USA	
($kg\ ha^{-1}$)	1st year	2nd year	1st year	2nd year
N	76	131	73	258
P_2O_5	96	140	98	300
K_2O	132	212	45	152

Table 14.6. Amounts of phosphorus and potassium fertilizer applied in the three principal producing areas of Europe.

	kg ha^{-1}	
	P_2O_5	K_2O
France	100	240
Italy	150	0
Turkey	120	60

Both Longden (1970) and Bornscheuer *et al.* (1993) have similarly reviewed the use of major nutrients on the direct-sown seed crop. A summary is shown in Table 14.7. The same principles apply to good seed production as with the steckling system: adequate nutrition and water, but not excessive fertilizer, in both the first and the second years.

Table 14.7. Optimum fertilizer applications for direct-sown crops.

	kg ha^{-1}			
Growing method and timing of fertilizer application	N	P_2O_5	K_2O	NaCl
After Longden (1970)				
Under-sown crops				
First year (as top dressing in autumn)	88	112	175	–
Second year (as top dressing in spring)	176	25	40	–
Open-sown crops				
First year (in seedbed)	75	112	140	375
Second year (as top dressing in spring)	163	90	90	–
After Bornscheuer *et al.* (1993)				
Cover crop				
Before sowing	–	130–180	As needed,	
With cover crop sowing	40		varying	
At cover crop harvest	50		between	
After winter (February/March)	60		0 and 350	
Total	150			
Open-sowing method				
Before sowing	50	130–180	As needed,	
After winter (February/March)	50		varying	
At bolting (April)	50		between	
Total	150		0 and 350	

References

Adams, F. and Evans, C.E. (1962) A rapid method for measuring lime requirement of Red–Yellow Podzolic soils. *Soil Science Society of America Proceedings* 26, 355–357.

Adams, S.N. (1960) The value of calcium nitrate and urea for sugar beet and the effect of late nitrogenous top dressings. *Journal of Agricultural Science, Cambridge* 54, 395–398.

Adams, S.N. (1961a) The manuring of sugar beet. *Chemistry and Industry*, 564–566.

Adams, S.N. (1961b) The effect of time of application of phosphate and potash on sugar beet. *Journal of Agricultural Science, Cambridge* 56, 127–130.

Adams, S.N. (1961c) The effect of sodium and potassium fertilizer on the mineral composition of the sugar beet. *Journal of Agricultural Science, Cambridge* 56, 383–388.

Adams, S.N. (1962) The response of sugar beet to fertilizer and the effect of farmyard manure. *Journal of Agricultural Science, Cambridge* 58, 219–226.

Adepoju, A.Y., Pratt, P.F. and Mattigod, S.V. (1982) Availability and extractability of phosphorus from soil having high residual phosphorus. *Soil Science Society of America Journal* 46, 583–588.

Adriano, D.C. (2001) *Trace Elements in Terrestrial Environments – Biogeochemistry, Bioavailability, and Risks of Metals*. Springer-Verlag, New York, 867 pp.

Afanasiev, M.M. (1964) The effect of simulated hail injuries on yield and sugar content of beets. *Journal of American Society of Sugar Beet Technologists* 13, 225–237.

Agerberg, L.S. (1969) Results of three different cropping systems at different plant nutrient levels. *Lantbr Högsk Meddn Ser A* 117, 1–42.

Albasal, N., Dor, Z., Carmeli, R. and Kafkafi, U. (1970) Growth of sugar beet (var. *polyrave*) in relation to petiole nitrate content. *Experimental Agriculture* 6, 151–155.

Alexander, J.T. (1971) Factors affecting quality. In: Johnson, R.T., Alexander, J.T., Rush, G.E. and Hawkes, G.R. (eds) *Advances in Sugarbeet Production: Principles and Practices*. Iowa State University Press, Ames, Iowa, pp. 371–399.

Allen, S.E. (1984) Slow-release fertilizers. In: Hauck, R.D. (ed.) *Nitrogen in Crop Production*. American Society of Agronomy, Madison, Wisconsin, pp. 195–206.

Allison, M.F. (1989) Sugar beet, straw incorporation and nitrogen. *British Sugar Beet Review* 57(2), 37–39.

Allison, M.F. (1992) Early nitrogen applications. In: *Rothamsted Experimental Station Report for 1991*, Lawes Trust, Harpenden, pp. 73–74.

Allison, M.F. (1996) Micronutrient needs of sugar beet. *British Sugar Beet Review* 64(4), 26–29.

Allison, M.F. and Armstrong, M.J. (1991) The nitrate leaching problem – are catch crops the solution? *British Sugar Beet Review* 59(3), 8–11.

Allison, M.F. and Armstrong, M.J. (1995) Can the fertilizer nitrogen requirement of sugar beet be predicted? *British Sugar Beet Review* 63(4), 6–9.

Allison, M.F. and Chapman, J. (1995) Role of phosphate in beet crop nutrition. *British Sugar Beet Review* 63(4), 12–14.

Allison, M.F. and Hetschkun, H.M. (1995) Five years of straw incorporation and its effect on growth, yield and nitrogen nutrition of sugar beet (*Beta vulgaris*). *Journal of Agricultural Science, Cambridge* 125, 61–68.

Allison, M.F., Jaggard, K.W. and Last, P.J. (1992) Effects of straw incorporation on the yield, nitrogen fertilizer and insecticide requirements of sugar beet (*Beta vulgaris*). *Journal of Agricultural Science, Cambridge* 118, 199–206.

Allison, M.F., Brown, S. and Pilbrow, J. (1993) Nitrogen catch crops and sugar beet. *British Sugar Beet Review* 61(2), 16–19.

Allison, M.F., Jaggard, K.W. and Armstrong, M.J. (1994) Time of application and chemical form of potassium, phosphorus, magnesium and sodium fertilizers and effects on the growth, yield and quality of sugar beet. *Journal of Agricultural Science, Cambridge* 123, 61–70.

Allison, M.F., Armstrong, M.J., Jaggard, K.W., Milford, G.F.J. and Todd, A.D. (1996a) An analysis of the agronomic, economic and environmental effects of applying N fertilizer to sugar beet. *Journal of Agricultural Science, Cambridge* 127, 475–486.

Allison, M.F., Last, P.J. and Bean, K.M.R. (1996b) Response by sugar beet to foliar sprays of copper. *Journal of the Science of Food and Agriculture* 72, 219–225.

Allison, M.F., Armstrong, M.J., Jaggard, K.W. and Todd, A.D. (1998a) Integration of nitrate cover crops into sugar beet rotations. I. Management and effectiveness of nitrate cover crops. *Journal of Agricultural Science, Cambridge* 130, 53–60.

Allison, M.F., Armstrong, M.J., Jaggard, K.W. and Todd, A.D. (1998b) Integration of nitrate cover crops into sugar beet rotations. II. Effect of cover crops on growth yield and N requirement of sugar beet. *Journal of Agricultural Science, Cambridge* 130, 61–67.

Allmaras, R.R., Kraft, J.M. Jr and Smucker, A.J.M. (1988) Soil compaction and crop residue management effects on root diseases of annual food legumes. In: Summerfield, R.J. (ed.) *World Crops: Cool Season Food Legumes*. Kluwer Academic Publishers, London, pp. 627–647.

Alloway, B.J. and Tills, A.R. (1984) Copper deficiency in world crops. *Outlook on Agriculture* 13, 32–34.

Amaducci, M.T., Rosso, F., Venturi, G., Barbanti, L. and Meriggi, P. (1998) Effect of different cover crops on the dynamics of nitrogen and on yield and quality of succeeding sugar beet. *Institut International de Recherches Betteravières Proceedings*, 87–102.

Analogides, D.A. (1983) Estimating sugar beet responses to fertilizer nitrogen on the basis of soil nitrogen indices. *Institut International de Recherches Betteravières Proceedings*, 423–431.

Analogides, D.A. (1987a) Seasonal uptake of P, K and Na by irrigated sugar beet as related to growth and soil nutrient supply. *Institut International de Recherches Betteravières Proceedings*, 305–324.

Analogides, D.A. (1987b) Estimating response by irrigated sugar beet to P and K fertilization in relation to soil fertility variables. *Institut International de Recherches Betteravières Proceedings*, 325–340.

Anderson, F.N. and Peterson, G.A. (1978) Optimum starter fertilizer placement for sugarbeet seedlings as determined by uptake of radioactive ^{32}P isotope. *Journal of American Society of Sugar Beet Technologists* 20, 19–24.

Anderson, W.B. (1982) Diagnosis and correction of iron deficiency in field crop – an overview. *Journal of Plant Nutrition* 5, 785–795.

Andkjaer, B., Thomsen, J.N., Marcussen, C. and Jorgensen, A.M. (1994) N recommendations in sugar beet through Danish 'square grid forecasts' and amino N values. *Institut International de Recherches Betteravières Proceedings*, 259–270.

Anon. (2001) *Glossary of Soil Science Terms*. Soil Science Society of America, Madison, Wisconsin, 135 pp.

Archer, F.C. and Hodgson, I.H. (1987) Total and extractable trace element contents of soils in England and Wales. *Journal of Soil Science* 38, 421–431.

Archer, J.R. (1980) *Trace Element Deficiencies in Field Crops*. MAFF Booklet 2197, MAFF, London, 12 pp.

Armstrong, M.J. (1985) Sulphur nutrition. In: *Rothamsted Experimental Station Report for 1984*, Lawes Trust, Harpenden, p. 47.

Armstrong, M.J. and Draycott, A.P. (1983) Prospects for improving use of nitrogen fertilizer for the grower. *British Sugar Beet Review* 51, 9–12.

Armstrong, M.J. and Jaggard, K. (1990) Timing the application of P, K, Na and Mg fertilizers. *British Sugar Beet Review* 58(2), 26–28.

Armstrong, M.J. and Milford, G.F.J. (1985) The nitrogen nutrition of sugar beet. *British Sugar Beet Review* 53(4), 42–44.

Armstrong, M.J. and Milford, G.F.J. (1987) Correct timing of N fertilizer applications for the beet crop. *British Sugar Beet Review* 55(1), 41–43.

Armstrong, M.J. and Stillingfleet, N. (1993) Nitrogen fertilizer recommendations for 1994. *British Sugar Beet Review* 61(4), 4–6.
Armstrong, M.J., Milford, G.F.J., Biscoe, P.V. and Last, P.J. (1983) Influences of nitrogen on physiological aspects of sugar beet productivity. *Institut International de Recherches Betteravières Proceedings*, 53–61.
Armstrong, M.J., Milford, G.F.J., Pocock, T.O., Last, P.J. and Day, W. (1986) The dynamics of nitrogen uptake and its remobilization during the growth of sugar beet. *Journal of Agricultural Science, Cambridge* 107, 145–154.
Armstrong, M.J., Jarvis, P.J., Milford, G.F.J., Bellett-Travers, D.M. and Leigh, R.A. (1998) Potassium offtakes in sugar beet: their relation to yield and beet quality. *Aspects of Applied Biology* 52, 53–56.
Armstrong, M.J., Milford, G.F.J. and Hollies, J. (1999) Potassium requirement of sugar beet. *British Sugar Beet Review* 67(3), 4–6.
Armstrong, M.J., Milford, G.F.J., Johnston, A.E. and Hollies, J. (2000) Revised fertilizer recommendations for the sugar beet crop. *British Sugar Beet Review* 68(3), 2–6.
Arnon, D.E. and Whatley, F.R. (1949) Is chloride a coenzyme of photosynthesis? *Science* 110, 554–556.
Arvidsson, J. (1998) Soil compaction caused by heavy sugar beet harvesters. In: *Advances in Sugar Beet Research*. Monograph Vol. 1, Institut International de Recherches Betteravières, Brussels, pp. 35–42.
Asselbergs, C.J., van der Poel, P.W., Verhaart, M.L. and de Visser, N.H. (1960) Rohsaftgewinnung im laboratorium zum Studium des Techinschen wertes der Zuckerrübe. In: *Proceedings XIth Session Committee International Technical Sucre*, pp. 78–91.
Aubertot, J.N., Dürr, C., Kiêu, K. and Richard, G. (1999) Characterization of sugar beet seedbed structure. *Soil Science Society of America Journal* 63, 1377–1384.
Baird, B.L., Bonnemann, J.J. and Richards, A.W. (1954) The use of chemical additives to control soil crusting and increase emergence of sugar beet seedlings. *Proceedings of American Society of Sugar Beet Technologists* 8, 136–142.
Bakhsh Kelarestaghi, K., Khalili, A. and Ahmadi, G. (2002) Effect of micro nutrients on sugar beet in Djovain area of Iran. *Institut International de Recherches Betteravières Proceedings*, 389–395.
Bakken, L.R., Børresen, T. and Njøs, A. (1987) Effect of soil compaction by tractor traffic on soil structure, denitrification and yield of wheat (*Triticum aestivum* L.) *Journal of Soil Science* 38, 541–552.
Balls, W.L. and Holton, F.S. (1915) Causes of variation in yield. *Philosophical Transactions of the Royal Society, London, B* 208, 103–180, 403–480.
Barber, S.A. (1959) The influence of alfalfa, bromegrass and corn on soil aggregation and crop yield. *Soil Science Society of America Proceedings* 23, 258–259.
Barber, S.A. (1995) *Soil Nutrient Bioavailability: a Mechanistic Approach*. John Wiley & Sons, New York, 414 pp.
Barrow, V.L. (1955) Use of activated sludge as a fertilizer. *World Crops* 7, 435–437.
Bartens, A.P. (2001) *Sugar and Sweetener Economy*. Bartens, Berlin, 416 pp.
Baver, L.D. (1949) Practical values from physical analyses of soils. *Soil Science* 68, 1–14.
Baver, L.D. and Farnsworth, R.B. (1940) Soil structure effects in the growth of sugar beets. *Soil Science Society of America Proceedings* 5, 45–48.
Bear, F.E. and Prince, A.L. (1947) *Agricultural Value of Sewage Sludge*. Bulletin 733, New Jersey Agricultural Experiment Station, Rutgers University, New Brunswick, New Jersey.
Beckers, R. (1999) Sulphur and sodium nutrition in sugar beet. *Institut International de Recherches Betteravières Info* 4, 6–7.
Berger, K.C. (1950) Sugar beet fertilization in Wisconsin. *Proceedings of American Society of Sugar Beet Technologists* 6, 440–444.
Berger, K.C. and Truog, E.C. (1939) Boron determination in soils and plants. *Industrial and Engineering Chemistry, Analytical Edition* 11, 540–545.
Beringer, H. (1987) Recent data about P, K and Na fertilizer application to sugar beet in Central Europe. *Institut International de Recherches Betteravières Proceedings*, 103–133.
Berrow, M.L. and Burridge, J.C. (1980) Trace element levels in soils: effects of sewage sludge. In: *Inorganic Pollution and Agriculture*. MAFF Reference Book 326, HMSO, London, pp. 159–183.
Berry, W.L. and Ulrich, A. (1968) Cation absorption from culture solution by sugar beets. *Soil Science* 106, 303–308.
Berry, W.L. and Ulrich, A. (1970) Calcium nutrition of sugar beets as affected by potassium. *Soil Science* 110, 389–394.
Bertrand, A.R. and Kohnke, H. (1957) Subsoil conditions and their effects on oxygen supply and growth of corn roots. *Soil Science Society of America Proceedings* 21, 135–140.

Bihery, M.A. and Lachmar, T.E. (1994) Groundwater quality degradation as a result of over-pumping the delta Wadi El-Arish Sinai Peninsula. *Egypt Environmental Geology* 24, 293–305.

Binford, G.G., Hergert, G.W. and Blumenthal, J.M. (2000) Sugar beets. In: Ferguson, R.B. (ed.) *Nutrient Management for Agronomic Crops in Nebraska*. Extension Circular EC 155, University of Nebraska, Lincoln, Nebraska, pp. 127–130.

Birch, J.A. Devine, J.R. and Holmes, M.R.J. (1966) Field experiments on the magnesium requirement of cereals, potatoes and sugar beet in relation to nitrogen and potassium application. *Journal of Science of Food and Agriculture* 17, 76–81.

Björling, K. (1954) Yellowing in beets caused by magnesium deficiency. *Socker* 8, 147–156.

Blake, G.R., Ogden, D.B., Adams, E.P. and Boelter, D.H. (1960) Effect of soil compaction on development and yield of sugar beets. *Journal of American Society of Sugar Beet Technologists* 11, 236–242.

Blake, G.R., Nelson, W.W. and Allmaras, R.R. (1976) Persistence of subsoil compaction in a mollisol. *Soil Science Society of America Journal* 40, 943–947.

Blake, L. and Johnston, A.E. (1999) The retention and release of phosphorus from soil. *IMPHOS Phosphate Newsletter* 8–9, 6.

Blaylock, A.D., Belden, K. and Hough, H.W. (1996) *Guide to Wyoming Fertilizer Recommendations*. Cooperative Extension Service Bulletin B-1045, University of Wyoming, Laramie, Wyoming.

Blumenthal, J.M. (2001) Fertilizing sugarbeet. In: Wilson, R.G., Smith, J.A. and Miller, S.D. (eds) *Sugarbeet Production Guide*. Cooperative Extension EC01–156, University of Nebraska, Lincoln, Nebraska, pp. 75–80.

Boawn, L.C. and Viets, F.G., Jr (1956) Zinc fertilizer tests on sugar beets in Washington. *Journal of American Society of Sugar Beet Technologists* 9, 212–216.

Boawn, L.C., Viets, F.G., Jr, Crawford, C.L. and Nelson, J.L. (1960a) Effect of nitrogen carrier, nitrogen rate, zinc rate and induced soil changes on zinc uptake by sorghum, potatoes and sugar beets. *Soil Science* 90, 329–337.

Boawn, L.C., Nelson, C.E., Viets, F.G., Jr and Crawford, C.L. (1960b) *Nitrogen Carrier and Nitrogen Rate Influence on Soil Properties and Nutrient Uptake by Crops*. Bulletin 614, Washington State Agricultural Experiment Station, Pullman, Washington.

Boawn, L.C., Viets, F.G., Jr, Nelson, C.E. and Crawford, C.L. (1961) Yield and zinc content of sugar beets as affected by nitrogen source, rate of nitrogen and zinc application. *Journal of American Society of Sugar Beet Technologists* 11, 279–286.

Bolton, E.F. and Aylesworth, J.W. (1968) Effect of soil physical condition and fertility on yield of sugar beets on a Brookston clay soil. *Journal of American Society of Sugar Beet Technologists* 14, 664–670.

Bolton, J. and Penny, A. (1968) The effects of potassium and magnesium fertilizers on yield and composition of successive crops of ryegrass, clover, sugar beet, potatoes, kale and barley on sandy soil at Woburn. *Journal of Agricultural Science, Cambridge* 70, 303–311.

Boon, R. and Vanstallen, R. (1983) Nitrogen fertilization advice for sugar beet based on soil analysis. *Institut International de Recherches Betteravières Proceedings*, 433–445.

Bornscheuer, E. Meyerholz, K. and Wunderlich, K.H. (1993) Seed production and quality. In: Cooke, D.A. and Scott, R.K. (eds) *The Sugar Beet Crop*. Chapman & Hall, London, pp. 120–155.

Boyd, D.A. (1959) The effect of farmyard manure on fertilizer responses. *Journal of Agricultural Science, Cambridge* 52, 384–391.

Boyd, D.A. (1961) Current fertilizer practice in relation to manurial requirements. *Proceedings of Fertiliser Society* 65, pp. 1–35.

Boyd, D.A. (1968) Experiments with ley and arable farming systems. In: *Rothamsted Experimental Station Report for 1967*, Lawes Trust, Harpenden, pp. 316–331.

Boyd, D.A., Garner, H.V. and Haines, W.B. (1957) The fertilizer requirements of sugar beet. *Journal of Agricultural Science, Cambridge* 48, 464–476.

Boyd, D.A., Tinker, P.B.H., Draycott, A.P. and Last, P.J. (1970) Nitrogen requirement of sugar beet grown on mineral soils. *Journal of Agricultural Science, Cambridge* 74, 37–46.

Brady, N.C. and Weil, A.R. (1999) *The Nature and Properties of Soils*, 12th edn. Prentice Hall, Englewood Cliffs, New Jersey, 881 pp.

Brandenburg, E. (1931) Die Herz- und Trockenfäule der Rüben als Bormangelerscheinung. *Phytopathologische Zeitschrift* 3, 499–517.

Brandenburg, E. (1939) Ueber die Grundlagen der Boranwendung in der Landwirtschaft. *Phytopathologische Zeitschrift* 12, 1–112.

Bravo, S., Lee, G.S. and Schmehl, W.R. (1989) The effect of planting date, nitrogen fertilizer rate and harvest date on seasonal concentration and total content of six macronutrients in sugar beet. *Journal of American Society of Sugar Beet Technologists* 26, 34–49.

Bravo, S., Lee, G.S. and Schmehl, W.R. (1992) Effect of planting date, nitrogen fertilizer and harvest date on seasonal concentrations and total content of five micronutrients in sugar beet. *Journal of American Society of Sugar Beet Technologists* 29, 45–57.

Bray, R.H. and Kurtz, L.T. (1945) Determination of total, organic, and available forms of phosphorus in soils. *Soil Science* 59, 39–45.

Britton, J. (1994) Commercial use of the EUF soil nitrogen test in Ireland on sugar beet and other crops. *Institut International de Recherches Betteravières Proceedings*, 353–361.

Bronner, H. (1983) Estimating the optimal nitrogen fertilizer requirements by means of nitrogen extraction from soil with hot water. *Institut International de Recherches Betteravières Proceedings*, 447–454.

Bronner, H. and Bachler, W. (1980) Evaluating the nitrogen requirement of sugarbeet from hydrolysable soil nitrogen. *Soil Science* 130, 303–306.

Brown, A.L., Hills, F.J. and Krantz, B.A. (1968) Lime, P, K and Mn interactions in sugar beets and sweet corn. *Agronomy Journal* 60, 427–429.

Brown, K.F. and Biscoe, P.V. (1985) Fibrous root growth and water use of sugar beet. *Journal of Agricultural Science, Cambridge* 105, 679–691.

Brown, K.F. and Dunham, R.J. (1989) Recent progress on the fibrous root system of sugar beet. In: Licht, F.O. (ed.) *World Sugar and Sweetener Handbook*. GmbH, Ratzburg, pp. F5–F13.

Brown, K.F., Messem, A.B., Dunham, R.J. and Biscoe, P.V. (1987) Effect of drought on growth and water use of sugar beet. *Journal of Agricultural Science, Cambridge* 109, 421–435.

Brown, R.J. (1943) Sampling sugar beet petioles for measurement of soil fertility. *Soil Science* 56, 213–222.

Broyer, T.C., Carlton, A.B., Johnson, C.M. and Stout, P.R. (1954) Chlorine – a micronutrient element for higher plants. *Plant Physiology* 29, 526–532.

Brummer, V. (1966) Effect of autumn and spring applications of fertilizer on sugar beets. *Maatalous ja Koetoiminta* 20, 91–100.

Brutlag, A., Smith, G.H., Giles, J.F. and Cattanach, A.W. (1989) Modified ridge till systems for sustainable sugarbeet production. In: *1988 Sugarbeet Research and Extension Reports* 20. North Dakota State University, Fargo, North Dakota, pp. 165–170.

Bunting, A.H. (1963) Experiments on organic manures, 1942–1949. *Journal of Agricultural Science, Cambridge* 60, 121–140.

Burba, M. (1996) Invert sugar and harmful nitrogen as quality parameters of sugar beet. *Institut International de Recherches Betteravières Proceedings*, 369–383.

Burcky, K. (1991) Dynamics of nitrogen in soil, its uptake and utilization by beet. *Institut International de Recherches Betteravières Proceedings*, 287–296.

Bush, D.S. (1995) Calcium regulation in plant cells and its role in signaling. *Annual Review of Plant Physiology* 46, 95–122.

Buzás, I. and Johnston, A.E. (1999) Balanced plant nutrition – I. Sugar beet cropping systems for high yield and quality. In: *Report of a Workshop at the Beta Research Institute*. Budapest, Hungary, pp. 1–40.

Campbell, C.A., Ellert, B.H. and Jame, Y.W. (1993) Nitrogen mineralization potential of soils. In: Carter, M.R. (ed.) *Soil Sampling and Methods of Analysis*. Lewis Publishers, Boca Raton, Florida, pp. 341–349.

Campbell, S.C. (1968) Sugar beet seed production in Oregon, USA. *Journal of the International Institute of Sugar Beet Research* 3, 165–174.

Carolan, R.J. (1960) Nonsugars in factory juices with special reference to effective alkalinity. In: *Proceedings XIth Session Committee International Technical Sucre*, pp. 203–220.

Carruthers, A. and Oldfield, J.F.T. (1961) Methods for the assessment of beet quality. *International Sugar Journal* 63, 72–74, 103–105, 137–139.

Carruthers, A., Oldfield, J.F.T. and Teague, H.J. (1956) A comparison of the effects on juice quality of nitrate of soda, sulphate of ammonia and salt. In: *Proceedings of the International Institute of Sugar Beet Research XIX Congress*. Brussels.

Carruthers, A., Oldfield, J.F.T. and Teague, H.J. (1962) Assessment of beet quality. Paper presented to the XVth Annual Technical Conference of the British Sugar Corporation, pp. 1–28.

Carter, J.N. (1982) Effect of nitrogen and irrigation levels, location and year on sucrose concentration of sugarbeets in Southern Idaho. *Journal of American Society of Sugar Beet Technologists* 21, 286–306.

Carter, J.N. and Traveler, D.J. (1981) Effect of time and amount of nitrogen uptake on sugarbeet growth and yield. *Agronomy Journal* 73, 665–671.

Carter, J.N., Jensen, M.E. and Bosma, S.M. (1971) Interpreting the rate of change in nitrate nitrogen in sugarbeet petioles. *Agronomy Journal* 63, 669–674.

Carter, J.N., Jensen, M.E., Ruffing, B.J., Bosma, S.M. and Richards, A.W. (1972) Effect of nitrogen and irrigation on sugarbeets production in Southern Idaho. *Journal of American Society of Sugar Beet Technologists* 17, 5–14.

Carter, J.N., Jensen, M.E. and Bosma, S.M. (1974) Determining nitrogen fertilizer needs for sugarbeets from residual soil nitrate and mineralizable nitrogen. *Agronomy Journal* 66, 319–323.

Carter, J.N., Traveller, D.J. and Bosma, S.M. (1978) Sugarbeet yield and seasonal growth characteristics as affected by hail damage and nitrogen level. *Journal of American Society of Sugar Beet Technologists* 20, 73–83.

Carter, J.N., Jensen, M.E. and Traveller, D.J. (1980a) Effect of mid to late season water stress on sugarbeet growth and yield. *Agronomy Journal* 72, 806–815.

Carter, J.N., Traveller, D.J. and Rosenau, R.C. (1980b) Root and sucrose yields of sugarbeets as affected by mid to late season water stress. *Journal of American Society of Sugar Beet Technologists* 20, 583–596.

Cassel Sharmasarkar, F., Sharmasarkar, S., Held, L.J., Miller, S.D., Vance, G.F. and Zhang, R. (2001a) Agroeconomic analyses of drip irrigation for sugarbeet production. *Agronomy Journal* 93, 517–523.

Cassel Sharmasarkar, F., Sharmasarkar, S., Miller, S.D., Vance, G.F. and Zhang, R. (2001b) Assessment of drip and flood irrigation on water and fertilizer used efficiencies for sugar beets. *Agricultural Water Management* 46, 241–251.

Cathcart, J.B. (1980) World phosphate reserves and resources. In: Khasawneh, F.E., Sample, E.C. and Kamprath, E.J. (eds) *The Role of Phosphorus in Agriculture*. American Society of Agronomy, Madison, Wisconsin, pp. 1–18.

Cattanach, A. (1991) *Boron Fertilization of Sugarbeets in the Red River Valley*. 1990 Sugarbeet Research and Extension Reports 21, North Dakota State University, Fargo, North Dakota, 118 pp.

Chabannes, J. (1959) Long-term effects of repeated additions of borates. *Belgian Industrial Chemistry* Suppl. No. 2, 697–698.

Chambers, B.J. and Smith, K. (1999) Utilising organic manures with minimum risk to sugar beet quality. *British Sugar Beet Review* 67(2), 25–28.

Chambers, B.J., Lord, E.I., Nicholson, F.A. and Smith, K.A. (1999) Predicting nitrogen availability and losses from application of organic manures to arable land: MANNER. *Soil Use and Management* 15, 137–143.

Cherney, J.H. and Robinson, D.L. (1982) A comparison of plant digestion methods for identifying contamination of plant tissue by Ti analysis. *Agronomy Journal* 75, 145–147.

Christensen, N.W., Taylor, R.G., Jackson, T.L. and Mitchell, B.L. (1981) Chloride effects on water potentials and yield of winter wheat infected with take-all root rot. *Agronomy Journal* 73, 1053–1058.

Christenson, D.R. (1986) Evaluation of urea based fertilizers including urea phosphates. *Report to Tennessee Valley Authority for Work Conducted from 1984 to 1986*. Crop and Soil Sciences Department, Michigan State University, East Lansing, Michigan, 55 pp.

Christenson, D.R. (1997) Soil organic matter in sugar beet and dry bean cropping systems in Michigan. In: Paul, E.A., Paustain, K., Elliott, E.T. and Cole, C.V. (eds) *Soil Organic Matter in Temperate Agroecosystems – Long-Term Experiments in North America*. CRC Press, New York, pp. 151–159.

Christenson, D.R. (2000) *Agrotain*. Departmental Fact Sheet, Crop and Soil Sciences Department, Michigan State University, East Lansing, Michigan.

Christenson, D.R. and Butt, M.B. (1997) Mineralization potential as affected by cropping system. *Communications in Soil Science and Plant Analysis* 28, 1047–1058.

Christenson, D.R. and Butt, M.B. (1998) Nitrogen mineralization of soils from Michigan's Saginaw Valley and Thumb region. *Communications in Soil Science and Plant Analysis* 29, 2355–2363.

Christenson, D.R. and Butt, M.B. (2000) Response of sugarbeet to applied nitrogen following field bean (*Phaseolus vulgaris* L.) and corn (*Zea mays* L.). *Journal of Sugar Beet Research* 37, 1–16.

Christenson, D.R. and Zinati, G.M. (1991) Nitrogen uptake by sugar beets from various depths in the soil. In: *Abstracts of the 26th Biennial Meeting, American Society of Sugar Beet Technologists*. Denver, Colorado, p. 6.

Christenson, D.R., Lucas, R.E. and Doll, E.C. (1972) *Fertilizer Recommendations for Michigan Vegetable and Field Crops*. Cooperative Extension Bulletin E-550, Michigan State University, East Lansing, Michigan.

Christenson, D.R., Bricker, C., Reisen, J., Voth, R., Mertz, M., Mezoghi, M., Almustafa, W., Oaks, J. and Kestner, R. (1975) Soil fertility and management for the production of sugar beets, navy beans and corn. In: *Saginaw Valley Bean and Beet Research Farm Report*. Michigan State University, East Lansing, Michigan, pp. 3–28.

Christenson, D.R., Bricker, C.E. and Reisen, J. (1979) Soil fertility and management for the production of sugar beets, navy beans, corn, wheat and soybeans. In: *Saginaw Valley Bean and Beet Research Farm Report*. Michigan State University, East Lansing, Michigan, pp. 4–26.

Christenson, D.R., Bricker, C. and Reisen, J. (1981) Soil fertility and management for the production of sugar beets, navy beans, corn and soybeans. In: *Saginaw Valley Bean and Beet Research Farm Report*. Michigan State University, East Lansing, Michigan, pp. 5–47.

Christenson, D.R., Bricker, C. and Johnson, R. (1985) Soil fertility and management for the production of sugar beets, navy beans and corn. In: *Saginaw Valley Bean and Beet Research Farm Report*. Michigan State University, East Lansing, Michigan, pp. 5–35.

Christenson, D.R., Warncke, D.D. and Leep, R. (1988) *Lime for Michigan Soils*. Cooperative Extension Service Extension Bulletin E-471, Michigan State University, East Lansing, Michigan.

Christenson, D.R., Bricker, C.E. and Schmenk, R.E. (1990) Effect of sodium nitrate on yield and quality of sugar beets. In: *Saginaw Valley Bean and Beet Research Farm Report*. Michigan State University, East Lansing, Michigan, pp. 24–26.

Christenson, D.R., Bricker, C.E. and Hubbell, L. (1991a) Yield and quality of sugar beets as affected by applied boron. In: *Agricultural Experiment Station Research Report* 51. Michigan State University, East Lansing, Michigan.

Christenson, D.R., Bricker, C.E. and Gallagher, R.S. (1991b) Crop yields as affected by cropping systems and rotation. In: *Agricultural Experiment Station Research Report* 51, Michigan State University, East Lansing, Michigan.

Christenson, D.R., Warncke, D.D., Vitosh, M.L., Jacobs, L.W. and Dahl, J.G. (1992) *Fertilizer Rcommendations for Field Crops in Michigan*. Cooperative Extension Service Extension Bulletin E-550A, Michigan State University, East Lansing, Michigan, 36 pp.

Christenson, D.R., Bricker, C.E., Siler, L., Zinati, G.M. and Butt, M. (1993) *Soil and Management for Sugar Beet Production. Report to Michigan Sugar Company and Monitor Sugar Company for Work Done from 1989 to 1993*. Crop and Soil Sciences Department, Michigan State University, East Lansing, Michigan.

Christenson, D.R., Brimhall, P.B., Hubbell, L. and Bricker, C.E. (2000) Yield of sugar beet, soybean, corn, field bean and wheat as affected by lime application on alkaline soils. *Communications in Soil Science and Plant Analysis* 31, 1145–1154.

Christmann, J. (1963) Results of trials on the application of nitrogenous fertilizers to sugar beet conducted by the Institut Technique de la Betterave from 1954 to 1962. In: *Proceedings of Institute Sugar Beet Research XXVI Congress*. Brussels, pp. 73–89.

Christmann, J. (1978) Recent research in France on dynamics of mineral nitrogen in soil. *Institut International de Recherches Betteravières Proceedings*, 73–87.

Clarkson, D.T., Sanderson, J. and Russell, R.S. (1968) Ion uptake and root age. *Nature* 220, 805–806.

Clarkson, D.T., Robards, A.W. and Sanderson, J. (1971) The tertiary endodermis in barley roots: fine structure in relation to radial transport of ions and water. *Planta* 96, 292–305.

Clement, C.R. and Williams, T.E. (1962) An incubation technique for assessing the nitrogen status of soils newly ploughed from leys. *Journal of Soil Science* 13, 82–91.

Collier, P.A. (1967) *ICI fertilizer Trials on Sugar Beet 1962–1965*. ICI Farming Service, London.

Connors, G. (2000) Kieserite – the superior magnesium and sulphur fertilizer. In: *Kali und Salz*. Brochure for Potash Limited, Hertford, UK, pp. 1–14.

Cook, R.L. (1948) Symptoms of nutritional disorders in sugarbeets. *Proceedings of American Society of Sugar Beet Technologists* 5, 316–328.

Cook, R.L. and Millar, C.E. (1953) *Plant Nutrient Deficiencies Diagnosed by Plant Symptoms, Tissue Tests, Soil Tests*. Agricultural Experiment Station Special Bulletin 353, Michigan State College, East Lansing, Michigan, 84 pp.

Cook, R.L., Millar, C.E. and Robertson, L.S. (1945) A crop rotation layout with an illustration of the statistics involved in combining several years data. *Soil Science Society of America Proceedings* 10, 213–218.

Cook, R.L., Davis, J.F., Frakes, M.G., Nichol, G.E. and Reeve, P.A. (1957) Fertilizers for sugar beets. *Michigan State University Quarterly Bulletin* 39, 524–535.

Cook, R.L., Davis, J.F. and Frakes, M.G. (1959) An analysis of production practices of sugar beet farmers in Michigan – 1958. *Michigan State University Agricultural Experiment Station Quarterly Bulletin* 42, 401–420.

Cooke, D.A. and Draycott, A.P. (1971) The effects of soil fumigation and nitrogen fertilizers on nematodes and sugar beet in sandy soils. *Annals of Applied Biology* 69, 253–264.

Cooke, D.A., Jaggard, K.W., Draycott, A.P., Scott, R.K., Webb, D.J. and Golding, M.J. (1982) Setting up and managing an experimental farm: the first 20 years' experience at Broom's Barn. *Experimental Agriculture* 18, 105–123.

Cooke, G.W. (1949) Placement of fertilizers for row crops. *Journal of Agricultural Science, Cambridge* 39, 359–373.

Cooke, G.W. (1951) Placement of fertilizers for sugar beet. *Journal of Agricultural Science, Cambridge* 41, 174–178.

Cooke, G.W. (1954) Recent developments in the use of fertilizers. *Agricultural Progress* 29, 110.

Cooke, G.W. (1967) *The Control of Soil Fertility*. Crosby Lockwood and Son, London, 526 pp.

Couvreur, L., Guiot, J. and Hermann, O. (1998) Set-aside and nitrogen availability for sugar beet. *Institut International de Recherches Betteravières Proceedings*. 117–129.

Crohain, A. and Rixhon, L. (1967) Practical fertilizing value of sugar beet leaves and crowns. *Bulletin Recherche Agronomique Gembloux* 2, 397–428.

Croll, B. (1990) Nitrate and water supplies in the United Kingdom. *British Sugar Beet Review* 58(4), 33–37.

Crookston, R.K. and Kurle, J.E. (1989) Corn residue effect on the yield of corn and soybean grown in rotation. *Agronomy Journal* 81, 229–232.

Crowther, E.M. and Yates, F. (1941) Fertilizer policy in war-time: the fertilizer requirements of arable crops. *Empire Journal of Experimental Agricultural* 9, 77–97.

Davis, J.F., Sundquist, W.B. and Frakes, M.G. (1959) The effect of fertilizers on sugar beets including an economic optima study of response. *Journal of American Society of Sugar Beet Technologists* 10, 424–434.

Davis, J.F., Nichol, G. and Thurlow, D. (1961) The effect of phosphorus fertilization and time of application of chemical composition and on yield, sucrose content and percent purity of sugar beet roots. *Journal of American Society of Sugar Beet Technologists* 11, 406–412.

Davis, J.F., Nichol, G. and Thurlow, D. (1962) The interaction of rates of phosphate application with fertilizer placement and fertilizer applied at planting time on the chemical composition of sugar beet tissue, yield, percent sucrose and apparent purity of sugar beet roots. *Journal of American Society of Sugar Beet Technologists* 12, 259–267.

de Boodt, M., De Leenheer, L. and Kirkham, D. (1961) Soil aggregate stability indexes and crop yields. *Soil Science* 91, 138–146.

DEFRA (2001a) *Guidelines for Farmers in Nitrate Vulnerable Zones*. Bulletin 5505, UK Department for Environment, Food and Rural Affairs, London, 32 pp.

DEFRA (2001b) *Arable Cropping and the Environment – a Guide*. UK Department for Environment, Food and Rural Affairs, London, 1–36.

Deibert, E.J., Giles, J.F., Lizotte, D. and Cattanach, N. (1984) Reduced tillage for sugarbeet production. In: *1983 Sugarbeet Research and Extension Reports* 14, North Dakota State University, Fargo, North Dakota, pp. 120–123.

Department of Environment (1989) Code of practice for agricultural use of sewage sludge. *Bulletin C.20.9/89. HMSO*, London, pp. 1–12.

Desprez, V. (1963) Influence of the spraying of urea on the yield and content of sugar beet. In: *Proceedings of the International Institute of Sugar Beet Research XXVI Congress*. Brussels.

Devine, J.R. (1962) The comparative agronomic value of fertilizers in solid and liquid form. In: *External Bulletin Document* 33. International Superphosphate Manufacturers Association, Levington, 17–28.

Devine, J.R. and Holmes, M.R.J. (1963) Field experiments on the value of urea as a fertilizer for barley, sugar beet, potatoes, winter wheat and grassland in Great Britain. *Journal of Agricultural Science, Cambridge* 61, 391–396.

Dexter, S.T., Frakes, M.G. and Nichol, G. (1966) The effect of low, medium and high nitrogen rates on the storage of sugar beet at high and low temperatures. *Journal of American Society of Sugar Beet Technologists* 14, 145–159.

Dibb, D.W. and Thompson, W.R., Jr (1985) Interaction of potassium with other nutrients. In: Munson, R.D. (ed.) *Potassium in Agriculture*. American Society of Agronomy, Madison, Wisconsin, pp. 515–533.

Doll, E.C. and Lucas, R.E. (1973) Testing soils for potassium, calcium and magnesium. In: Walsh, L.M. and Beaton, J.D. (eds) *Soil and Plant Analysis*. American Society of Agronomy, Madison, Wisconsin, pp. 133–151.

Dragović, S., Maksimović, L. and Karagić, Dj. (1996) Effect of stand density on formation of leaves and leaf area of sugar beet under irrigation. *Journal of Sugar Beet Research* 33, 45–54.

Draycott, A.P. (1969) The effect of farmyard manure on the fertilizer requirements of sugar beet. *Journal of Agricultural Science, Cambridge* 73, 119–124.

Draycott, A.P. (1971) Anhydrous ammonia compared with solid fertilizer for sugar beet. In: *Anhydrous Ammonia Symposium 1970*. IPC Business Press, Silsoe, pp. 69–71.

Draycott, A.P. (1972) *Sugar Beet Nutrition*. Applied Science Publishers, London, 250 pp.

Draycott, A.P. (1983) Trends in use of fertilizers for sugar beet. *British Sugar Beet Review* 51, 7–9.

Draycott, A.P. (1993) Nutrition. In: Cooke, D.A. and Scott, R.K. (eds) *The Sugar Beet Crop – Science into Practice*. Chapman & Hall, London, pp. 239–278.

Draycott, A.P. (1995) *Potash for Sugar Beet*. Leaflet 12, 5/95, Potash Development Association, 11 pp.

Draycott, A.P. (1996) *Fertilizing for High Yield and Quality Sugar Beet*. Bulletin 15, International Potash Institute, Basle, 52 pp.

Draycott, A.P. (1999) Factors affecting performance of sugar beet, particularly nutrition, soil and water. DSc thesis, University of Leeds.

Draycott, A.P. and Allison, M.F. (1998) Magnesium fertilizers in soil and plants: comparisons and usage. *Proceedings of Fertilizer Society, London* 412, 1–28.

Draycott, A.P. and Bugg, S.M. (1982) Response by sugar beet to various amounts and times of application of sodium chloride fertilizer in relation to soil type. *Journal of Agricultural Science, Cambridge* 98, 579–592.

Draycott, A.P. and Cooke, G.W. (1966) The effects of potassium fertilizers on quality of sugar beet. In: *Potassium Symposium 1966*. International Potash Institute, Bern, pp. 131–135.

Draycott, A.P. and Durrant, M.J. (1969a) The effects of magnesium fertilizers on the yield and chemical composition of sugar beet. *Journal of Agricultural Science, Cambridge* 72, 319–324.

Draycott, A.P. and Durrant, M.J. (1969b) Magnesium fertilizers for sugar beet (Part I). *British Sugar Beet Review* 37, 175–179.

Draycott, A.P. and Durrant, M.J. (1970a) Magnesium fertilizers for sugar beet (Part II). *British Sugar Beet Review* 38, 175–180.

Draycott, A.P. and Durrant, M.J. (1970b) The relationship between exchangeable soil magnesium and response by sugar beet to magnesium sulphate. *Journal of Agricultural Science, Cambridge* 75, 137–143.

Draycott, A.P. and Durrant, M.J. (1971a) Effects of nitrogen fertilizer, plant population and irrigation on sugar beet. Part II. Nutrient concentration and uptake. *Journal of Agricultural Science, Cambridge* 76, 269–275.

Draycott, A.P. and Durrant, M.J. (1971b) Effects of nitrogen fertilizer, plant population and irrigation on sugar beet. Part III. Water consumption. *Journal of Agricultural Science, Cambridge* 76, 277–282.

Draycott, A.P. and Durrant, M.J. (1972a) Comparisons of kieserite and calcined magnesite for sugar beet grown on sandy soils. *Journal of Agricultural Science, Cambridge* 79, 455–461.

Draycott, A.P. and Durrant, M.J. (1972b) The immediate and long-term value of some magnesium fertilizers for sugar beet. *Journal of Agricultural Science, Cambridge* 79, 463–471.

Draycott, A.P. and Durrant, M.J. (1976a) Response by sugar beet to superphosphate, particularly in relation to soils containing little available phosphorus. *Journal of Agricultural Science, Cambridge* 86, 181–187.

Draycott, A.P. and Durrant, M.J. (1976b) Response by sugar beet to potassium and sodium fertilizers, particularly in relation to soils containing little exchangeable potassium. *Journal of Agricultural Science, Cambridge* 87, 105–112.

Draycott, A.P. and Farley, R.F. (1971) The effect of sodium and magnesium fertilizers, and irrigation on growth, composition and yield of sugar beet. *Journal of the Science of Food and Agriculture* 22, 559–563.

Draycott, A.P. and Farley, R.F. (1973) Response by sugar beet to soil dressings and foliar sprays of manganese. *Journal of the Science of Food and Agriculture* 24, 675–685.

Draycott, A.P. and Holliday, R. (1970) Comparisons of liquid and solid fertilizers and anhydrous ammonia for sugar beet. *Journal of Agricultural Science, Cambridge* 74, 139–145.

Draycott, A.P. and Hollies, J. (2001) *Fodder Beet Fertilizer Requirements*. Leaflet 16, Potash Development Association, Laugharne, 1–12.

Draycott, A.P. and Last, P.J. (1970) Effect of previous cropping and manuring on the nitrogen fertilizer needed by sugar beet. *Journal of Agricultural Science, Cambridge* 74, 147–152.

Draycott, A.P. and Last, P.J. (1971) Some effects of partial sterilization on mineral nitrogen in a light soil. *Journal of Soil Science* 22, 152–157.

Draycott, A.P. and Last, P.J. (1978) Changes in soil mineral nitrogen in relation to time and nitrogen uptake by sugar beet. *Institut International de Recherches Betteravières Proceedings*, 35–43.

Draycott, A.P. and Martindale, W. (2000a) Effective use of nitrogen fertilizer. *British Sugar Beet Review* 68(2), 18–21.

Draycott, A.P. and Martindale, W. (2000b) Phosphate – its origin and role for sugar beet. *British Sugar Beet Review* 68(4), 5–10.

Draycott, A.P. and Messem, A.B. (1977) Response by sugar beet to irrigation, 1965–1975. *Journal of Agricultural Science, Cambridge* 89, 481–493.

Draycott, A.P. and Messem, A.B. (1979) Soil acidity: the need for a systematic approach to liming. *British Sugar Beet Review* 47(2), 21–23.

Draycott, A.P. and Webb, D.J. (1971) Effects of nitrogen fertilizer, plant population and irrigation on sugar beet. Part I. Yields. *Journal of Agricultural Science, Cambridge* 76, 261–267.

Draycott, A.P., Hodgson, D.R. and Holliday, R. (1967) Recent research on the value of fertilizers in solution. *Agricultural Progress* 42, 68–81.

Draycott, A.P., Marsh, J.A.P. and Tinker, P.B.H. (1970a) Sodium and potassium relationships in sugar beet. *Journal of Agricultural Science, Cambridge* 74, 568–573.

Draycott, A.P., Hull, R., Messem, A.B. and Webb, D.J. (1970b) Effects of soil compaction on yield and fertilizer requirement of sugar beet. *Journal of Agricultural Science, Cambridge* 75, 533–537.

Draycott, A.P., Durrant, M.J. and Boyd, D.A. (1971a) The relationship between soil phosphorus and response by sugar beet to phosphate fertilizer on mineral soils. *Journal of Agricultural Science, Cambridge* 77, 117–121.

Draycott, A.P., Durrant, M.J. and Last, P.J. (1971b) Effects of cultural practices and fertilizers on sugar beet quality. *Journal of the International Institute of Sugar Beet Research* 5, 169–185.

Draycott, A.P., Durrant, M.J., Hull, R. and Webb, D.J. (1972a) Yields of sugar beet and barley in contracting crop rotations at Broom's Barn, 1965–70. In: *Rothamsted Experimental Station Report for 1971*, Part 2, Lawes Trust, Harpenden, pp. 149–154.

Draycott, A.P., Durrant, M.J. and Webb, D.J. (1972b) Long term effects of fertilizers at Broom's Barn 1965–70. In: *Rothamsted Experimental Station Report for 1971*, Part 2, Lawes Trust, Harpenden, pp. 155–164.

Draycott, A.P., Durrant, M.J., Davies, D.B. and Vaidyanathan, L.V. (1976) Sodium and potassium fertilizers in relation to soil physical properties and sugar beet yield. *Journal of Agricultural Science, Cambridge* 87, 633–642.

Draycott, A.P., Durrant, M.J., Hull, R. and Messem, A.B. (1977) Changes in Broom's Barn Farm soils, 1960–75. In: *Rothamsted Experimental Station Report for 1976*, Part 2, Lawes Trust, Harpenden, pp. 33–52.

Draycott, A.P., Durrant, M.J., Hull, R. and Webb, D.J. (1978a) Yields of sugar beet and barley in contrasting crop rotations at Broom's Barn 1971–76. In: *Rothamsted Experimental Station Report for 1977*, Part 2, Lawes Trust, Harpenden, pp. 5–13.

Draycott, A.P., Durrant, M.J. and Webb, D.J. (1978b) Long-term effects of fertilizer at Broom's Barn, 1971–76. In: *Rothamsted Experimental Station Report for 1977*, Part 2. Lawes Trust, Harpenden, pp. 15–30.

Draycott, A.P., Allison, M.F. and Armstrong, M.J. (1997) Changes in fertilizer use in sugar beet production. *Institut International de Recherches Betteravières Proceedings*, 39–54.

Dubetz, S. and Oosterveld, M. (1976) Effects of weather variables on the yields of sugar beets grown in an irrigated rotation for fifty years. *Journal of American Society of Sugar Beet Technologists* 19, 143–149.

Dubourg, J., Saunier, R. and Devillers, P. (1957) Influence des engrais azotés sur la teneur des Betteraves en constituants azotés et particulièrement en acid glutamique. *Industrial Agriculture* 74, 883–888.

Dunham, R.J. (1991a) Effect of starter fertilizers close to seeds. *Institut International de Recherches Betteravières Proceedings*, 369–384.

Dunham, R.J. (1991b) Fertilizer placement. *British Sugar Beet Review* 59(2), 21–25.

Dunham, R.J. (1993) Water use and irrigation. In: Cooke, D.A. and Scott, R.K. (eds) *The Sugar Beet Crop – Science into Practice*. Chapman & Hall, London, pp. 279–309.

Dunham, R.J., Draycott, A.P. and Messem, A.B. (1993) Irrigation of sugar beet in the United Kingdom. *Institut International de Recherches Betteravières Proceedings*, 25–40.

Dunning, R.A. and Winder, G.H. (1972) Some effects, especially on yield, of artificially defoliating sugar beet. *Annals of Applied Biology* 70, 89–98.

Dürr, C. and Aubertot, J.N. (2000) Emergence of seedlings of sugar beet (*Beta vulgaris* L.). *Plant and Soil* 219, 211–220.

Dürr, C. and Boiffin, J. (1995) Sugar beet seedling growth from germination to first leaf stage. *Journal of Agricultural Science, Cambridge* 124, 427–435.

Dürr, C. and Mary, B. (1998) Effects of nutrient supply on pre-emergence growth and nutrient absorption in wheat and sugar beet. *Annals of Botany* 81, 665–672.

Dürr, C., Boiffin, J., Fleury, A. and Coulomb, I. (1992) Analysis of the variability of sugar beet (*Beta vulgaris* L.) growth during the early stages. II. Factors influencing seedling size in field conditions. *Agronomie* 12, 527–535.

Durrant, M.J. and Draycott, A.P. (1971) Uptake of magnesium and other fertilizer elements by sugar beet grown on sandy soils. *Journal of Agricultural Science, Cambridge* 77, 61–68.

Durrant, M.J., Love, B.J.G., Messem, A.B. and Draycott, A.P. (1973) Growth of crop roots in relation to soil moisture extraction. *Annals of Applied Biology* 74, 387–394.

Durrant, M.J., Draycott, A.P. and Boyd, D.A. (1974a) The response of sugar beet to potassium and sodium fertilizers. *Journal of Agricultural Science, Cambridge* 83, 427–434.

Durrant, M.J., Draycott, A.P. and Payne, P.A. (1974b) Some effects of sodium chloride on germination and seedling growth of sugar beet. *Annals of Botany* 38, 1045–1051.

Durrant, M.J., Draycott, A.P. and Milford, G.F.J. (1978) Effect of sodium fertilizer on water status and yield of sugar beet. *Annals of Applied Biology* 88, 321–328.

Duval, R. (2000) Long term effects of catch crop management on nitrate leaching in chalk soil. *Institut International de Recherches Betteravières Info* 7, 14–16.

Dyke, G.V. (1965) Green manuring for sugar beet. *British Sugar Beet Review* 34, 94–98.

Eaton, F.M. (1944) Deficiency, toxicity and accumulation of boron in plants. *Journal of Agricultural Research* 69, 237–277.

Eck, H.V., Winter, S.R. and Smith, S.J. (1990) Sugar beet yield and quality in relation to residual beef feedlot waste. *Agronomy Journal* 82, 250–254.

Eckhoff, J.L.A. (1999) Sugarbeet response to nitrogen at four harvest dates. *Journal of Sugar Beet Research* 36, 33–45.

Edinburgh and East of Scotland College of Agriculture (1957) *Experiments on Level and Time of Application of Nitrogenous Fertilizer for Sugar Beet 1954–1956*. Rural Advisory Leaflet 40, Edinburgh, 3 pp.

Ellerton, S. (1947) An experiment to show the effect of nitrogenous fertilizer and of 'topping' the stem on the yield and quality of sugar beet seed. In: *Proceedings of International Institute of Sugar Beet Research X Congress*, IIRB, Brussels.

Elliott, M.C. and Weston, G.D. (1993) Biology and physiology of the sugar beet plant. In: Cooke, D.A. and Scott, R.K. (eds) *The Sugar Beet Crop*. Chapman & Hall, London, pp. 37–66.

Englestad, O.P. and Terman, G.L. (1980) Agronomic effectiveness of phosphate fertilizers. In: Khasawneh, F.E., Sample, E.C. and Kamprath, E.J. (eds) *The Role of Phosphorus in Agriculture*. American Society of Agronomy, Madison, Wisconsin, pp. 311–332.

Epstein, E. (1972) *Mineral Nutrition of Plants: Principles and Perspectives*. John Wiley & Sons, New York, 412 pp.

Erickson, A.E. and Ellis, B.G. (1971) *The Nutrient Content of Drainage Water from Agricultural Land*. Agricultural Experiment Station Research Report 31, Michigan State University, East Lansing, Michigan.

Eslami M, S., Lee, G.S. and Schmehl, W.R. (1988) Nutrient concentrations in sugarbeet senescing leaves during the season and in six plant parts at harvest. *Journal of Sugar Beet Research* 25, 11–27.

Etchevers, J.D. and Moraghan, J.T. (1983) Response of sugarbeets grown under dryland conditions to phosphorus fertilizer. *Journal of American Society of Sugar Beet Technologists* 22, 17–28.

Ezekari, M., Boukhal, A. and Halim, M. (1993) Effect of irrigation frequency, nitrogen dose rate and its split application on sugar beet yield and quality in the Tadla Valley. In: *Institut International de Recherches Betteravières 56 Congress Proceedings*. Brussels, pp. 69–83.

Farley, R.F. (1980) Manganous oxide as a seed pellet additive for controlling manganese deficiency in sugar beet seedlings. *Plant and Soil* 54, 451–459.

Farley, R.F. and Draycott, A.P. (1973) Manganese deficiency of sugar beet in organic soils. *Plant and Soil* 38, 235–244.

Farley, R.F. and Draycott, A.P. (1974) Growth and yield of sugar beet in relation to potassium and sodium supply. *Journal of the Science of Food and Agriculture* 26, 385–392.

Farley, R.F. and Draycott, A.P. (1976) Diagnosis of manganese deficiency in sugar beet and response to manganese applications. *Journal of the Science of Food and Agriculture* 27, 991–998.

Farley, R.F. and Draycott, A.P. (1978) Manganese deficiency in sugar beet and the incorporation of manganese in the coating of pelleted seed. *Plant and Soil* 49, 71–83.

Ferguson, R.B. (2001) Sensing soil properties for precision agriculture. In: *Society of Chemical Industry Proceedings*, London, pp. 7–8.

Findenegg, G.R., Nelemans, J.A. and Arnozis, P.A. (1989) Effect of external pH and Cl^- on the accumulation of NH_4^+ ions in the leaves of sugar beet. *Journal of Plant Nutrition* 12, 593–601.

Finkner, R.E., Ogden, D.B. Hanzas, P.C. and Olson, R.F. (1958) The effect of fertilizer treatment on the calcium, sodium, potassium, raffinose, galactinol, nine amino acids, and total amino acid content of three varieties of sugar beets grown in the Red River Valley of Minnesota. *Journal of American Society of Sugar Beet Technologists* 10, 272–280.

Fiskell, J.G.A. and Mourkedes, G.A. (1955) A comparison of manganese sources using tomato plants grown on marl, peat and sand soils. *Plant and Soil* 6, 313–331.

Fitts, J.B., Gammon, N., Jr and Forbes, R.B. (1967) Relative availability of manganese from several sources. *Proceedings of Soil and Crop Science Society of Florida* 27, 243–251.

Fixen, P.E. (1993) Crop responses to chloride. *Advances in Agronomy* 50, 107–150.

Fixen, P.E., Gelderman, R.H., Gerwing, J.R. and Farber, B.G. (1987) Calibration and implementation of a soil Cl test. *Journal of Fertilizer Issues* 4, 91–97.

Flocker, W.J. and Nielsen, D.R. (1962) The absorption of nutrient elements by tomatoes associated with levels of bulk density. *Soil Science Society of America Proceedings* 26, 183–186.

Flocker, W.J., Lingle, J.C. and Vomocil, J.A.(1959) Influence of soil compaction on phosphorus absorption by tomato plants from an applied phosphate fertilizer. *Soil Science* 88, 247–250.

Foth, H.D. and Ellis, B.G. (1997) *Soil Fertility*, 2nd edn. CRC Press, Boca Raton, Florida, 290 pp.

Foy, C.D. (1984) Effects of aluminium on plant growth. In: Carson, E.W. (ed.) *The Plant Root and its Environment*. University Press of Virginia, Charlottesville, Virginia, pp. 601–642.

Francis, S. (2002) Two hundred years of sugar beet. *British Sugar Beet Review* 70(3), 32–35.

Frank, K., Beegle, D. and Denning, J. (1998) Phosphorus. In: Brown, J.R. (ed.) *Recommended Chemical Soil Test Procedures for the North Central Region*. North Central Regional Research Publication No. 221, University of Missouri, Columbia, Missouri, pp. 21–30.

Frankinet, M., Raimond, Y., Destain, J.P. and Rixhon, L. (1987) Influence of organic manures and incorporation of harvest residues on mineral PK fertilization in large scale cropping. *Institut International de Recherches Betteravières Proceedings*, 163–176.

Franzen, D.W. (1999a) *What is Site-specific Farming?* Extension Service SF-1176 (1), North Dakota State University, Fargo, North Dakota.

Franzen, D.W. (1999b) *Soil Sampling and Variable-rate Fertilizer Application*. Extension Service SF-1176 (2), North Dakota State University, Fargo, North Dakota.

Franzen, D.W. (1999c) *Yield Mapping*. Extension Service SF-1176 (3), North Dakota State University, Fargo, North Dakota.

Franzen, D.W. (1999d) *Site-specific Farming and the Environment*. Extension Service SF-1176 (4), North Dakota State University, Fargo, North Dakota.

Franzen, D.W. and Cihacek, L.J. (1996) *North Dakota Fertilizer Recommendation: Tables and Equations*. Extension Service SF-882, North Dakota State University, Fargo, North Dakota.

Fried, M. and Broeshart, H. (1967) *The Soil Plant System in Relation to Inorganic Nutrition*. Academic Press, New York, 358 pp.

Friessleben, G., Lori, K. and Friessleben, H. (1988) Herbstadammformung für die Zuckerrübenaussaat. *Agrartechnik, Berlin* 38, 10–11.

Fuehring, H.D. and Finkner, R.E. (1973) Interrelationships of applied nitrogen, applied zinc, plant population, and frequency of irrigation on yield and quality of sugarbeets. *Journal of American Society of Sugar Beet Technologists* 17, 358–374.

Fuehring, H.D., Hashimi, M.A., Haddad, K.S., Hussieni, K.K. and Makhdoom, M.U. (1969) Nutrient interacting effects on sucrose yield of sugar beets (*Beta vulgaris*) on a calcareous soil. *Proceedings of Soil Science Society of America* 33, 718–721.

Furrer, O.J. (1973) Results of tillage trials with sugar beets in Switzerland. *Institut International de Recherches Betteravières* 6(2), 94–108.

Fürstenfeld, F. and Bürcky, K. (2000) Current results concerning boron supplies for sugar beet in Southern Germany. *Institut International de Recherches Betteravières Proceedings*, 415–417.

Gallagher, P.A. (1967) *Fertilizer Experiments on Sugar Beet in Eire*. Information Leaflet 98, Nitrate Corporation of Chile, London.

Gallagher, R.N. (1975) The occurrence of calcium in plant tissue as crystals of calcium oxalate. *Communications in Soil Science and Plant Analysis* 6, 315–330.

Gardner, R. and Robertson, D.W. (1942) The nitrogen requirement of sugar beets. *Colorado Agricultural Experiment Station Technical Bulletin* 28, Fort Collins, Colorado.

Garner, H.V. (1950) Sugar beet irrigation. *British Sugar Beet Review* 18, 145–150.

Garner, H.V. (1966) Experiments on the direct, cumulative and residual effects of town refuse manures and sewage sludge at Rothamsted and other centres 1940–1947. *Journal of Agricultural Science, Cambridge* 67, 223–234.

Gascho, G.J., Davis, J.F., Fogg, R.A. and Frakes, M.G. (1969) The effects of potassium carriers and levels of potassium and nitrogen fertilization on the yield and quality of sugar beets. *Journal of American Society of Sugar Beet Technologists* 15, 298–305.

Gaskell, M., Mitchell, J., Smith, R., Koike, S. and Fouche, C. (2000) *Soil Fertility Management for Organic Crops*. Publication 7249, University of California, Davis, California.

Gaskill, J.O. (1950) Progress report on the effects of nutrition, bruising and washing upon rotting of stored sugar beets. *Proceedings of American Society of Sugar Beet Technologists* 6, 680–685.

Gasser, J.K. (1964) Some factors affecting losses of ammonia from urea and ammonium sulfate applied to soil. *Journal of Soil Science* 15, 258–272.

Geering, H.R., Hodgson, J.F. and Sdano, C. (1969) Micronutrient cation complexes in soil solution: IV. The chemical state of manganese in soil solution. *Soil Science Society of America Proceedings* 33, 81–85.

Gelata, S., Sabbagh, G.J., Stone, J.F., Elliott, R.L., Mapp, H.P., Bernardo, D.J. and Watkins, K.B. (1994) Importance of soil and cropping systems in the development of regional water quality policies. *Journal of Environmental Quality* 23, 36–42.

Gembarzewski, H. and Korzeniowska, J. (1990) Simultaneous extraction of B, Cu, Fe, Mn, Mo and Zn from mineral soils and an estimation of results. In: *Agrobiological Research Report* 43 (2). Institute of Soil Science and Plant Cultivation, Breslau, Poland, pp. 115–127.

Gemtos, T.A. and Lellis, T. (1997) Effects of soil compaction, water and organic matter contents on emergence and initial plant growth of cotton and sugar beet. *Journal of Agricultural Engineering Research* 66, 121–134.

Gericke, S. (1966) Phosphate fertilising and quality of the sugar beet crop. *Zucker* 24, 663–667.

Germida, J.J., Karamanos, R.E. and Stewart, J.W.B. (1985) A simple microbial bioassay for plant available manganese. *Soil Science Society of America Journal* 49, 1411–1415.

Ghariani, S.A. (1981) Impact of variable irrigation water supply on yield-determining parameters and seasonal water-use efficiency of sugar beets. PhD thesis, University of California, Davis, 153 pp.

Giardini, L., Pimpini, F., Borin, M. and Gianquinto, G. (1992) Effects of poultry manure and mineral fertilizers on yield of crops. *Journal of Agricultural Science, Cambridge* 118, 207–213.

Gilbert, W.A., Ludwick, A.E. and Westfall, D.G. (1981) Predicting in-season N requirements of sugarbeets based on soil and petiole nitrate. *Agronomy Journal* 73, 1018–1022.

Giles, J.F. (1994) Subject matter index, 1970–1992. In: *1993 Sugarbeet and Extension Reports* 24, North Dakota State University, Fargo, North Dakota, pp. 1–67.

Giles, J.F., Ludwick, A.E. and Reuss, J.O. (1977) Prediction of late season nitrate-nitrogen content of sugarbeet petioles. *Agronomy Journal* 69, 85–88.

Giles, J.F., Cattanach, A.W. and Cattanach, N.R. (1991) Effect of boron and Triggrr™ on sugarbeet yield and quality. In: *1990 Sugarbeet Research and Extension Reports* 21. North Dakota State University, Fargo, North Dakota, pp. 119–121.

Giles, J.F., Cattanach, N.R. and Cattanach, A.W. (1993) Modified ridge-till system for sugar beet production. In: *1992 Sugarbeet Research and Extension Reports* 23. North Dakota State University, Fargo, North Dakota, pp. 223–227.

Gill, W.R. and Miller, R.D. (1956) A method for study of the influence of mechanical impedance and aeration on the growth of seedling roots. *Soil Science Society of America* 20, 154–157.

Gilmour, J.T., Clark, M.D. and Daniel, S.M. (1996) Predicting long-term decomposition of biosolids with a seven-day test. *Journal of Environmental Quality* 25, 766–770.

Gilmour, J.T., Cogger, C.G., Jacobs, L.W., Wilson, S.A., Evanylo, G.K. and Sullivan, D.M. (2000) *Estimating Plant-available Nitrogen in Biosolids*. Water Environment Research Foundation, Alexandria, Virginia.

Glenn, D.M. and Dotzenko, A.D. (1978) Minimum versus conventional tillage in commercial sugar beet production. *Agronomy Journal* 70, 341–344.

Goodman, P.J. (1966) Effect of varying plant populations on growth and yield of sugar beet. *Agricultural Progress* 41, 89–107.

Goos, R.J., Johnson, B.E., Peterson, R.A. and Kobes, N. (2001) Effect of sugarbeet by-products on the solubility and availability of ferrous sulfate in soil. *Journal of Sugar Beet Research* 38, 153–172.

Graham, R.D. and Webb, M.J. (1991) Micronutrients and disease resistance and tolerance in plants. In: Mortvedt, J.J., Cox, F.R., Shuman, L.M. and Welch, R.M. (eds) *Micronutrients in Agriculture*, 2nd edn. Soil Science Society of America, Madison, Wisconsin, pp. 329–370.

Greenwood, D.J., Cleaver, T.J., Turner, M.K., Hunt, J., Niendorf, K.B. and Loquens, S.M.H. (1980) Comparison of the effects of nitrogen fertilizer on the yield, nitrogen content and quality of 21 different vegetable and agricultural crops. *Journal of Agricultural Science, Cambridge* 95, 471–485.

Gregg, C.M. and Harrison, C.M. (1950) A study of the effects of some different sods and fertilizers on sugar beet yields. *Proceedings of American Society of Sugar Beet Technologists* 6, 306–310.

Gregory, P.J. (1990) Soil physics and irrigation: tapping the potential for drip. *Agricultural Water Management* 17, 159–169.

Grimes, D.W. (1959) Effect of crop rotation, manure and commercial fertilizers upon yield, percent sugar and gross sugar production of sugar beets in southwestern Kansas. *Journal of American Society of Sugar Beet Technologists* 10, 364–370.

Groves, S., Chambers, B.J. and Williams, J. (1999) Potash fertilizer value of farm manures for sugar beet. *British Sugar Beet Review* 67(4), 12–15.

Guenzi, W.D. and McCalla, T.M. (1966) Phenolic acids in oats, wheat, sorghum and corn residues and their phytotoxicity. *Agronomy Journal* 58, 303–304.

Guenzi, W.D., McCalla, T.M. and Norstadt, F.A. (1967) Presence and persistence of phytotoxic substances in wheat, oat, corn and sorghum residues. *Agronomy Journal* 59, 163–165.

Guérif, M., Duke, C. and Dürr, C. (1995) Assessment of sugar beet growth and yield. *Institut International de Recherches Betteravières Proceedings*, 433–441.

Gummerson, R.J. (1989) Seed-bed cultivations and sugar beet seedling emergence. *Journal of Agricultural Science, Cambridge* 112, 159–169.

Günter, I. (1978) Seasonal changes of mineral nitrogen in beet soils. *Institut International de Recherches Betteravières Proceedings*, 23–34.

Gupta, U.C. (1979) Boron nutrition of crops. *Advances in Agronomy* 31, 273–307.

Gupta, U.C. and Cutcliffe, J.A. (1978) Effects of method of boron application on leaf tissue concentrations and control of brown-heart in rutabaga. *Canadian Journal of Plant Science* 58, 63–68.

Gurevich, S.M. and Boronina, I.I. (1964) Uptake and removal of nutrients by sugar beet in relation to the level of nutrition. *Agrokhimiya* 10, 73–81.

Guttay, J.R., Cook, R.L. and Robertson, L.S. (1958) Sugar beet production in Michigan as affected by cropping sequence and fertility level. *Journal of American Society of Sugar Beet Technologists* 10, 66–75.

Haddock, J.L. (1949) The influence of plant population, soil moisture, and nitrogen fertilisation on the sugar content and yield of sugar beets. *Agronomy Journal* 41, 79–84.

Haddock, J.L. (1952) The influence of soil moisture condition on the uptake of phosphorus from calcareous soils by sugar beets. *Soil Science Society of America Proceedings* 16, 235–238.

Haddock, J.L. (1959) Yield, quality and nutrient content of sugar beets as affected by irrigation regime and fertilizers. *Journal of American Society of Sugar Beet Technologists* 10, 344–355.

Haddock, J.L. and Kelley, O.J. (1948) Inter-relations of moisture, spacing and fertility to sugar beet production. *Proceedings of American Society of Sugar Beet Technologists* 5, 378–396.

Haddock, J.L., Linton, D.C. and Hurst, R.C. (1956) Nitrogen constituents associated with reduction of sucrose percentage and purity of sugar beets. *Journal of American Society of Sugar Beet Technologists* 9, 110–117.

Haddock, J.L., Smith, P.B., Downie, A.R., Alexander, J.T., Easton, B.E. and Jensen, V. (1959) The influence of cultural practices on the quality of sugar beets. *Journal of American Society of Sugar Beet Technologists* 10, 290–301.

Haise, H.R., Jensen, L.R. and Alessi, J. (1955) The effect of synthetic soil conditioners on soil structure and production of sugar beets. *Soil Science Society of America Proceedings* 19, 17–19.

Håkansson, I. (1990) A method for characterizing the state of compactness of the plough layer. *Soil and Tillage Research* 16, 105–120.

Håkansson, I., Voorhees, W.B. and Riley, H. (1988) Vehicle and wheel factors influencing soil compaction and crop response in different traffic regimes. *Soil and Tillage Research* 11, 239–282.

Hale, J.B., Watson, M.A. and Hull, R. (1946) Some causes of chlorosis and necrosis of sugar beet foliage. *Annals of Applied Biology* 33, 13–28.

Hale, V.Q. and Miller, R.J. (1966) Relationships between NO_3^--N in petioles during the growing season and yield components of sugar beets (*Beta vulgaris*). *Agronomy Journal* 58, 567–569.

Haley, S., Lockley-Jolly, F. and Suarez, N. (2001) *Sugar and Sweetener Situation and Outlook Report (May)*. Economic Research Service, United States Department of Agriculture, Washington, DC.

Hall, A.D. (1902) The continuous growth of mangolds for twenty-seven years on the same land, Barnfield, Rothamsted. *Journal of the Royal Agricultural Society* 63, 27–59.

Halvorson, A.D. and Hartman, G.P. (1975) Long-term nitrogen rates and sources influence sugar beet yield and quality. *Agronomy Journal* 67, 389–393.
Halvorson, A.D. and Hartman, G.P. (1984) Reduced seedbed tillage effects on irrigated sugar beet yield and quality. *Agronomy Journal* 76, 603–606.
Hamence, J.H. and Oram, P.A. (1964) Effects of soil and foliar applications of sodium borate to sugar beet. *Journal of the Science of Food and Agriculture* 15, 565–579.
Hammes, J.K. and Berger, K.C. (1960) Manganese deficiency in oats and correlation of plant manganese with various soil tests. *Soil Science* 90, 239–244.
Hanley, F. and Ridgman, W.J. (1963) An investigation into long-term effects of combine drilling on the yield of arable crops. *Experimental Husbandry* 9, 19–27.
Hanson, B., Grattan, S.R. and Fulton, A. (1999) *Agricultural Salinity and Drainage*. Division of Agriculture and Natural Resources Publication 3375, University of California, Davis, California.
Hao, X., Chang, C., Conner, R.L. and Bergen, P. (2001) Effect of minimum tillage and crop sequence on crop yield and quality under irrigation in a southern Alberta clay loam soil. *Soil and Tillage Research* 59, 45–55.
Harris, P.M. (1969) A study of the interaction between method of establishing a method of harvesting the sugar beet crop. *Journal of the International Institute of Sugar Beet Research* 4, 84–103.
Harris, P.M. (1970) The interaction between plant density and irrigation in the sugar beet crop. In: *Proceedings of the International Institute of Sugar Beet Research XXXIII Congress*. Brussels.
Harvey, C.W. and Dutton, J.V. (1993) Root quality and processing. In: Cooke, D.A. and Scott, R.K. (eds) *The Sugar Beet Crop*. Chapman & Hall, London, pp. 571–617.
Harvey, P.N. (1959) The disposal of cereal straw. *Journal of Royal Agricultural Society* 120, 55–63.
Havlin, J.L., Westfall, D.G. and Golus, H.M. (1984) Six years of phosphorus and potassium fertilization of irrigated alfalfa on calcareous soils. *Soil Science Society of America Journal* 48, 331–336.
Headden, W.P. (1912) *Deterioration in the Quality of Sugar Beets due to Nitrates Formed in the Soil*. Bulletin 183, Colorado Agricultural Experiment Station, Fort Collins, Colorado.
Heathcote, G.D. (1970) Effect of plant spacing and time of sowing of sugar beet on aphid infestation and spread of virus yellows. *Plant Pathology* 19, 32–39.
Heathcote, G.D. (1972) Influence of cultural factors on incidence of aphids and yellows in beet. *Journal of International Institute of Sugar Beet Research* 6, 6–14.
Heathcote, G.D. (1974) The effect of plant spacing, nitrogen fertilizer and irrigation on the appearance of symptoms and spread of virus yellows in sugar beet crops. *Journal of Agricultural Science, Cambridge* 82, 53–60.
Hebblethwaite, P.D. and McGowan, M. (1980) The effects of soil compaction on the emergence, growth and yield of sugar beet and peas. *Journal of the Science of Food and Agriculture* 31, 1131–1142.
Hébert, J. (1987) Value of soil tests for assessing PKNaMg fertilization of sugar beet. *Institut International de Recherches Betteravières Proceedings*, 51–70.
Heistermann, P. (1968) Yield and quality of sugar beet and potatoes as affected by fertilizer N : K ratio. *Zeszyty Problemowe Posteppow Nauk Rolnicsych* 84, 273–288.
Henderson, D.W., Hill, F.J., Loomis, R.S. and Nourse, E.F. (1968) Soil moisture conditions, nutrient uptake and growth of sugar beet as related to method of irrigation of an organic soil. *Journal of American Society of Sugar Beet Technologists* 15, 35–48.
Henkens, C.H. and Jongman, E. (1965) The movement of manganese in the plant and the practical consequences. *Netherlands Journal of Agricultural Science* 13, 392–407.
Henkens, C.H. and Smilde, K.W. (1966) Evaluation of glassy frits as micronutrient fertilizers. I. Copper and molybdenum frits. *Netherlands Journal of Agricultural Science* 14, 165–177.
Henkens, C.H. and Smilde, K.W. (1967) Evaluation of glassy frits as micronutrient fertilizers. II. Manganese frits. *Netherlands Journal of Agricultural Science* 15, 21–30.
Henriksson, L. and Håkansson, I. (1993) Soil management and crop establishment. In: Cooke, D.A. and Scott, R.K. (eds) *The Sugar Beet Crop – Science into Practice*. Chapman & Hall, London, pp. 157–177.
Hera, C., Davidescu, D. and Vines, I. (1961) The effect of ammonia liquor and anhydrous ammonia on yields of sugar beet, silage maize and oats. *Lucrat. Stiintifice Institutul Agronomic 'Nicolae Balcescu' Series A Agronomie* 5, 163–169.
Herlihy, M. (1986) Crop response and soil test calibration of P and K for sugar beet. In: *Annual Report on Sugar Beet Research Programme*, An Foras Taluntais, Ireland, pp. 1–9.
Herron, G.M., Grimes, D.W. and Finkner, R.E. (1964a) Effect of plant spacing and fertilizer on yield, purity, chemical constituents and evapotranspiration of sugar beets in Kansas, Parts I and II. *Journal of American Society of Sugar Beet Technologists* 12, 699–714.

Herron, G.M., Grimes, D.W. and Finkner, R.E. (1964b) Effect of plant spacing and fertilizer on yield, purity, chemical constituents and evapotranspiration of sugar beets in Kansas: I. Yield of roots, purity, percent sucrose and evapotranspiration. *Journal of American Society of Sugar Beet Technology* 12, 686–698.

Hewitt, E.J. (1948) Relation of manganese and some other metals to the iron status of plants. *Nature London* 161, 489–490.

Hewitt, E.J. (1953) Metal inter-relations in plants. Part I. Effects of some metal toxicities on sugar beet, tomato, oat, potato and marrowstem kale grown in sand culture. *Journal of Experimental Botany* 4, 59–64.

Hilde, D.J., Levos, R.W. and Ellingson, R.L. (1987) *Progress in Sugarbeet Quality Improvement Red River Valley 1980–1986*. American Crystal Sugar Company, Moorhead, Minnesota.

Hill, J. (1999) A review of the role of boron in the nutrition of oil seed rape and sugar beet. In: *Boron in Agriculture, Writtle College Report*. Writtle College, Chelmsford, UK, pp. 1–66.

Hill, K.W. (1946) Yield, percentage of sucrose, and coefficients of apparent purity of sugar beets as affected by rotational, manurial, and fertilizer practices at the Dominion Experimental Station, Lethbridge, Alberta. *Proceedings of American Society of Sugar Beet Technologists* 4, 63–72.

Hills, F.J. and Axtell, J.D. (1950) The effect of several nitrogen sources on beet sugar yields in Kern County, California. *Proceedings of American Society of Sugar Beet Technologists* 6, 356–361.

Hills, F.J. and Ulrich A. (1971) Nitrogen nutrition. In: Johnson, R.T., Alexander, J.T., Rush, G.E. and Hawkes, G.R. (eds) *Advances in Sugarbeet Production: Principles and Practices*. Iowa State University Press, Ames, Iowa, pp. 111–136.

Hills, F.J., Ferry, G.V., Ulrich, A. and Loomis, R.S. (1963) Marginal nitrogen deficiency of sugar beets and the problems of diagnosis. *Journal of American Society of Sugar Beet Technologists* 12, 476–484.

Hills, F.J., Sailsbery, R.L., Ulrich, A and Sipitanos, K.M. (1970) Effect of phosphorus on nitrate in sugar beet (*Beta vulgaris*). *Agronomy Journal* 62, 91–92.

Hills, F.J., Abshahi, A., Broadbent, F.E. and Peterson G.A. (1981) Effect of nitrapyrin on uptake of nitrogen by sugarbeet from labeled ammonium sulfate. *Journal American Society of Sugar Beet Technologists* 21, 150–158.

Hills, F.J., Sailsbery, R. and Ulrich, A. (1982) *Sugarbeet Fertilization*. Cooperative Extension Bulletin 1891, University of California, Davis, California.

Hocking, T. (1995) The role of sulphur in sugar beet. *British Sugar Beet Review* 63(1), 10–13.

Hodgson, J.F., Gerring, H.R. and Norvell W. A. (1965) Micronutrient cation complexing in soil solution: I. Partitioning between complexed forms by solvent extraction. *Soil Science Society of America Proceedings* 29, 665–669.

Hodgson, J.F., Lindsay, W.L. and Trierweiler, J.F. (1966) Micronutrient cation complexing in soil solution: II. Complexing zinc and copper in displaced solution from calcareous soils. *Soil Science Society of America Proceedings* 30, 723–726.

Hoeft, R.G. (1984) Current status of nitrification inhibitor use in US agriculture. In: Hauck, R.D. (ed.) *Nitrogen in Crop Production*. American Society of Agronomy, Madison, Wisconsin, pp. 561–570.

Hoffman, C. (1998) Influence of continuous reduced tillage on the growth of sugar beet. *Zuckerindustrie* 123, 225–226.

Hoitink, H.A., VanDoren, D.M. Jr and Schmitthenner, A.F. (1977) Suppression of *Phytophthora cinnamomi* in a composted hardwood bark potting medium. *Phytopathology* 67, 560–565.

Hoitink, H.A., Inbar, Y. and Boehm, M.J. (1991) Status of compost-amended potting mixes naturally suppressive to soil-borne diseases of floricultural crops. *Plant Disease* 75, 869–873.

Hollies, J. (1997) *Phosphate and Potash Removal by Crops*. Leaflet 10, Potash Development Association, Laugharne, 4 pp.

Hollies, J. (2000) *Phosphate and Potash Removal by Crops*. Revision Leaflet 10/00, Potash Development Association, Laugharne, 4 pp.

Hollies, J. (2001) *Fodder Beet – P and K Offtake*. Leaflet 27, Potash Development Association, Laugharne, 8 pp.

Hollies, J., Draycott, A.P. and Chambers, B.J. (2001) Beet nutrition. *The Agronomist*, 1, 5–6.

Holmes, J.C., Gill, W.D., Rodger, J.B.A., White, G.R. and Lawley, D.N. (1961) Experiments with salt and potash on sugar beet in south-east Scotland. *Experimental Husbandry* 6, 1–7.

Holmes, J.C., Lang, R.W. and Hunter, E.A. (1973) The effect of method and rate of application of common salt and muriate of potash on sugar beet. *Journal of Agricultural Science, Cambridge* 80, 239–244.

Holmes, M.R.J. and Whitear, J.D. (1976) Nitrogen requirement of sugar beet in relation to irrigation. *Journal of Agricultural Science, Cambridge* 87, 559–566.

Horn, D. (1994) Nitrogen dynamics of sugar beet soils in Eastern Germany. *Institut International de Recherches Betteravières Proceedings*, 271–279.

Horn, R. (1998) Effect of aggregation and tillage systems on soil deformation by mechanical stress and strain. In: *Advances in Sugar Beet Research*. Monograph Vol. 1, Institut International de Recherches Betteravières, Brussels, pp. 7–20.

Hossner, L.R. and Richards, G.E. (1968) The effect of phosphorus source on the movement and uptake of band applied manganese. *Soil Science Society of America Proceedings* 32, 83–85.

Houba, V.J.G., Novozamsky, I., Huijbregts, A.W.M. and van der Lee, J.J. (1994) Advantages and disadvantages of Nmin/EUF/$CaCl_2$ analytical methods to recommend N fertilizer for sugar beet. *Institut International de Recherches Betteravières Proceedings*, 295–303.

Hoyt, P.B. (1968) The effect of soil conditioners on growth of sugar beet in a sandy loam soil. *Experimental Husbandry* 16, 70–72.

Huijbregts, A.W.M., Glattkowski, H., Houghton, B.J. and Hadjiantoniou, D. (1996) Effect of agronomic factors on parameters used in formulas to estimate extractable sugar in sugar beets. *Institut International de Recherches Betteravières Proceedings*, 353–368.

Hull, R. (1960) *Sugar Beet Diseases*, 2nd edn. Technical Bulletin of the Ministry of Agriculture, Fisheries and Food 142, HMSO, London.

Hull, R. (1965) Control of sugar beet yellows: symposium on some approaches towards integrated control of British insect pests. *Annals of Applied Biology* 56, 345–347.

Hull, R. and Watson, M. (1947) Factors affecting the loss of yield of sugar beet caused by beet yellows virus. II. Nutrition and variety. *Journal of Agricultural Science, Cambridge* 37, 301–310.

Hull, R. and Webb, D.J. (1967) The effect of subsoiling and different levels of manuring on yields of cereals, lucerne and sugar beet. *Journal of Agricultural Science, Cambridge* 69, 183–187.

Hull, R. and Wilson, A.R. (1946) Distribution of violet root rot (*Helicobasidium purpureum* Pat.) of sugar beet and preliminary experiments on factors affecting the disease. *Annals of Applied Biology* 33, 420–433.

IIRB (1998) Soil compaction and compression in relation to sugar beet production. In: *Advances in Sugar Beet Research*. Monograph Vol. 1, Institut International de Recherches Betteravières, Brussels, pp. 1–103.

Jacob, A. (1958) *Magnesium, the Fifth Major Plant Nutrient*. Staples Press, London.

Jacobs, L.W. (1995a) *Utilization of Animal Manure for Crop Production Part I. Management of Manure Nutrients and Water Quality*. Extension Bulletin MM-1, Michigan State University, East Lansing, Michigan.

Jacobs, L.W. (1995b) *Utilization of Animal Manure for Crop Production Part II. Manure Application to Cropland*. Extension Bulletin MM-2, Michigan State University, East Lansing, Michigan.

Jacobs, L.W. (1995c) *Utilization of Animal Manure for Crop Production Part III. Worksheet to Calculate Manure Application Rates*. Extension Bulletin MM-3, Michigan State University, East Lansing, Michigan.

Jaggard, K.W. (1977) Effects of soil density on yield and fertilizer requirement of sugar beet. *Annals of Applied Biology* 86, 301–312.

Jaggard, K.W. (1984) Predrilling land work and yield loss. *British Sugar Beet Review* 52(2), 9–10.

Jaggard, K.W., Allison, M.F. and Todd, A.D. (1997a) Nutrient availability and yield variation in beet rotations: implications for precision farming. *Institut International de Recherches Betteravières Proceedings*, 303–306.

Jaggard, K.W., Clark, C.J.A., May, M.J., McCullagh, S. and Draycott, A.P. (1997b) Changes in the weight and quality of sugar beet roots in storage clamps on farms in England. *Journal of Agricultural Science, Cambridge* 129, 287–310.

James, D.W., Leggett, G.E. and Dow, A.I. (1967) *Phosphorus Fertility Relationships of Central Washington Irrigated Soils*. Agricultural Experiment Station Bulletin 688, Washington State University, Pullman, Washington.

James, D.W., Kidman, D.C., Weaver, W.H. and Reeder, R.L. (1968) Potassium fertilization of sugar beets in central Washington. *Journal of American Society of Sugar Beet Technologists* 14, 682–694.

Jameson, H.R. (1959) Liquid nitrogenous fertilizers. *Journal of Agricultural Science, Cambridge* 53, 333–338.

Jansson, S.L. (1987) Optimum supply of P, K and Na to sugar beet in North West Europe. *Institut International de Recherches Betteravières Proceedings*, 85–101.

Jarvis, P. (1997) Poultry manure and how to use it. *British Sugar Beet Review* 66(2), 23–25.

Jarvis, P. and Bee, P. (1996) Influence of potassium and sodium fertilizer on beet yields and quality. *British Sugar Beet Review* 64(3), 25–27.

Jensen, M.E. and Erie, L.J. (1971) Irrigation and water management. In: John, R.T., Alexander, J.T., Rush, G.E. and Hawkes, G.R. (eds) *Advances in Sugarbeet Production: Principles and Practices.* Iowa State University Press, Ames, Iowa, pp. 189–222.

Johnson, B.S. and Erickson, A.E. (1991) Sugar beet response to sub-soiling and wheel traffic. *Agronomy Journal* 83, 386–390.

Johnson, B.S., Erickson, A.E. and Voorhees, W.B. (1989) Physical conditions of a lake plain soil as affected by deep tillage and wheel traffic. *Soil Science Society of America Journal* 53, 1545–1551.

Johnston, A.E. (1986) Soil organic matter, effects on soils and crops. *Soil Use and Management* 2, 97–105.

Johnston, A.E. (1989) Potable waters and the nitrate problem. *British Sugar Beet Review* 57(4), 22–23.

Johnston, A.E. (1997) The value of long term field experiments in agricultural, ecological and environmental research. *Advances in Agronomy* 59, 291–333.

Johnston, A.E. (2001) Principles of crop nutrition for sustainable food production. *International Fertilizer Society Proceedings* 459, 1–40.

Johnston, A.E. and Goulding, K. (1992) K concentration in surface and ground-water and the loss of K in relation to land use. In: *Potassium in Ecosystems: Biogeochemical Fluxes of Cations in Agro and Forest Systems.* International Potash Institute, Basle, pp. 135–158.

Johnston, A.E. and Milford, G.F.J. (2001) The effect of nutrients in the subsoil on crop responses to applied potassium and phosphorus. *British Sugar Beet Review* 69(2), 24–29.

Johnston, A.E., Warren, R.G. and Penny, A. (1970) The value of residues from long-period manuring at Rothamsted and Woburn. V. The value to arable crops of residues accumulated from potassium fertilizers. In: *Rothamsted Experimental Station Report for 1969*, Part 2. Lawes Trust, Harpenden, pp. 69–90.

Johnston, A.E., Mattingly, G.E.G. and Poulton, P.R. (1976) Effect of phosphate residues on soil P values and crop yields. In: *Rothamsted Experimental Station Report for 1975*, Part 2. Lawes Trust, Harpenden, pp. 5–35.

Johnston, A.E., Lane, P.W., Mattingly, G.E.G., Poulton, P.R. and Hewitt, M.V. (1986) Effects of soil and fertilizer P on yields of potatoes, sugar beet, barley and winter wheat on a sandy clay loam soil at Saxmundham, Suffolk. *Journal of Agricultural Science, Cambridge* 106, 155–167.

Johnston, A.E., Goulding, K.W.T., Poulton, P.R. and Chalmers, A.G. (2001) Reducing fertilizer inputs: endangering arable soil fertility? *International Fertilizer Society Proceedings* 487, 1–44.

Jønsson, L. (1969) Effect of phosphate fertilising on the efficiency of nitrogen for sugar beet. *LantbrHögsk Meddn* 120A, 18.

Jorritsma, J. (1956) The manuring of sugar beet. I. N-K-Mg trial fields 1946–51 inclusive. *Mededeling Instituut Rationele Suikerproductive* 26, 227.

Jourdan, O., Bourrié, B. and Etourneaud, F. (1992) Elaboration of curves of absorption of mineral elements. In: *La Betterave.* Ministère de l'Agriculture et de la Pêche, Aspach-le-Bas, France, pp. 1–8.

Judy, W., Melton, J., Lessman, G., Ellis, B. and Davis, J. (1964) *Zinc Fertilization of Pea Beans, Corn and Sugar Beets in 1964.* Agricultural Experiment Station Research Report 33, Michigan State University, East Lansing, Michigan.

Kaffka, S.R., Corwin, D.L. and Lesch, S.M. (2002) Sugar beet yield and quality variation at the field scale in response to soil salinity and correlated properties in California. *Institut International de Recherches Betteravières Proceedings*, 159–170.

Kamprath, E.J. (1970) Exchangeable aluminium as a criterion for liming leached mineral soils. *Soil Science Society of America Proceedings* 34, 252–254.

Kamprath, E.J. (1984) Crop response to lime on soils in the tropics. In: Adams, F. (ed.) *Soil Acidity and Liming*, 2nd edn. Agronomy 12, American Society of Agronomy, Madison, Wisconsin, pp. 349–368.

Kauss, H. (1987) Some aspects of calcium dependent regulation in plant metabolism. *Annual Review of Plant Physiology* 38, 47–72.

Kay, B.D., Grant, C.D. and Groenevelt, P.H. (1985) Significance of ground freezing on soil bulk density under zero tillage. *Soil Science Society of America Journal* 49, 973–978.

Keener, H.M., Dick, W.A. and Hoitink, H.A. (2000) Composting and beneficial utilization of composted by-product materials. In: Dick, W.A. (ed.) *Land Application of Agricultural, Industrial and Municipal By-products.* Soil Science Society of America, Madison, Wisconsin.

Keeney, D.R. (1982) Nitrogen availability indexes. In: Page, A.L., Miller, R.H. and Keeney, D.R. (eds) *Methods of Soil Analysis*, Part 2. Agronomy Monograph 9, American Society of Agronomy, Madison, Wisconsin, pp. 711–733.

Kirkby, E.A., Armstrong, M.J. and Milford, G.F.J. (1987) Absorption and physiological roles of P and K in the sugar beet plant with reference to the functions of Na and Mg. *Institut International de Recherches Betteravières Proceedings*, 1–23.

Kluge, R. (1990) Uptake of boron by sugar beets during vegetative growth on loess soils with a high supply of boron. *Bodenkultur* 41, 195–203.

Koch, H.J. (1998) Rotational set-aside preceding sugar beet. *Institut International de Recherches Betteravières Proceedings*, 103–115.

Kolenbrander, G.J. (1978) The nitrogen cycle and fertilizer requirements. *Institut International de Recherches Betteravières Proceedings*, 1–11.

Kozera, H. and Lachowski, J. (1959) Some effects of preharvest foliage sprays on sugar beet plants. *Biuletyn Isntytut Hodowli i Aklimatyzacji Roslin*, 73–84.

Krügel, C., Dreyspring, C. and Lotthammer, R. (1938) Leaching experiments with borates. *Superphosphate* 8 and 9, 141–150, 161–166.

Kuipers, H. (1955) A regional investigation into the relation between soil structure, yield of sugar beets and nitrogen supply. *Netherlands Journal of Agricultural Science* 3, 170–181.

Kuter, G.A., Hoitink, H.A. and Chen, W. (1988) Effects of municipal sludge compost curing time on suppression of *Pythium* and *Rhizoctonia* diseases of ornamental plants. *Plant Disease* 72, 751–756.

Lachowski, J. (1959) Comparison of pulverised granulated superphosphate in fertilising sugar beets. *Roczniki Nauk Rolniczych* 80, 134–147.

Lachowski, J. (1961) The effect of boron, manganese and copper on the production value of sugar beet in Poland. *Roczniki Nauk Rolniczych* 84A, 63–88.

Lachowski, J. (1966) The effect of magnesium upon the growth and crop of root plants of sugar beet, and the progeny obtained after them. *Hodowla Roslin Aklimatyzacja i Nasiennictwo* 10, 37–59.

Lachowski, J. and Wesolowski, F. (1964) The influence of iron sulphate on yield of sugar beets in Poland. *Roczniki Nauk Rolniczych* 89A, 547–564.

Lamb, J.A. (1989) Sulfur for sugarbeet in Minnesota and North Dakota. In: *Sugarbeet Research and Extension Report* 19. North Dakota State University, Fargo, North Dakota, pp. 98–99.

Lamb, J.A. and Cattanach, A.W. (1990) Zinc application on sugarbeet. In: *1989 Sugarbeet Research and Extension Reports* 20. North Dakota State University, Fargo, North Dakota, pp. 84–86.

Lamb, J.A. and Rehm, G.W. (1999) Agronomic and economic evaluation of grid cell sized need for nitrogen recommendations for sugar beet in southern Minnesota. In: *1998 Sugarbeet Research and Extension Reports*. North Dakota State University, Fargo, North Dakota, pp. 106–112.

Lamb, J.A., Schmitt, M.A., Bredehoeft, M., Roehl, S. and Fisher, J. (2001) Management of turkey and swine manure derived nitrogen in a sugar beet cropping system. In: *2000 Sugarbeet Research and Extension Reports* 32. North Dakota State University, Fargo, North Dakota, pp. 125–134.

Lamers, J.G., Perdok, U.D., Lumkes, L.M. and Klooster, J.J. (1986) Controlled traffic farming systems in The Netherlands. *Soil and Tillage Research* 8, 65–76.

Larmer, F.G. (1937) Keeping quality of sugar beets as influenced by growth and nutritional factors. *Journal of Agricultural Research* 54, 185–198.

Larson, W.E. (1954a) Response of sugar beets to potassium fertilization in relation to soil physical and moisture conditions. *Soil Science Society of American Proceedings* 18, 313–317.

Larson, W.E. (1954b) Effect of method of application of double superphosphate on yield and phosphorus uptake by sugar beets. *Proceedings of American Society of Sugar Beet Technologists* 8(1), 25–31.

Last, P.J. and Bean, K. (1990) Manganese deficiency and the adjuvant connection. *British Sugar Beet Review* 58(3), 15–16.

Last, P.J. and Bean, K.M.R. (1991) Controlling manganese deficiency in sugar beet with foliar sprays. *Journal of Agricultural Science, Cambridge* 116, 351–358.

Last, P.J. and Draycott, A.P. (1971) Predicting the amount of nitrogen fertilizer needed for sugar beet by soil analysis. *Journal of the Science of Food and Agriculture* 22, 215–220.

Last, P.J. and Draycott, A.P. (1972) Top-dressing of nitrogen for sugar beet. *Experimental Husbandry* 22, 82–88.

Last, P.J. and Draycott, A.P. (1975a) Growth and yield of sugar beet on contrasting soils in relation to nitrogen supply. I. Soil nitrogen analyses and yield. *Journal of Agricultural Science, Cambridge* 85, 19–26.

Last, P.J. and Draycott, A.P. (1975b) Growth and yield of sugar beet on contrasting soils in relation to nitrogen supply. II. Growth, uptake and leaching of nitrogen. *Journal of Agricultural Science, Cambridge* 85, 27–37.

Last, P.J. and Tinker, P.B.H. (1968) Nitrate nitrogen in leaves and petioles of sugar beet in relation to yield of sugar and juice purity. *Journal of Agricultural Science, Cambridge* 71, 383–392.

Last, P.J., Draycott, A.P. and Webb, D.J. (1981) Effect of green manures on yield and nitrogen requirement of sugar beet. *Journal of Agricultural Science, Cambridge* 97, 159–170.

Last, P.J., Draycott, A.P., Messem, A.B. and Webb, D.J. (1983) Effects of nitrogen fertilizer and irrigation on sugar beet at Broom's Barn 1973–1978. *Journal of Agricultural Science, Cambridge* 101, 185–205.

Last, P.J., Webb, D.J., Bugg, R.B., Bean, K.M.R., Durrant, M.J. and Jaggard, K.W. (1985) Long-term effects of fertilizers at Broom's Barn, 1965–1982. In: *Rothamsted Experimental Station Report for 1984*, Part 2, Lawes Trust, Harpenden, pp. 231–249.

Last, P.J., Draycott, A.P. and Allison, M.F. (1994) Effects of soil mineral nitrogen, soil type and organic manure use on the nitrogen fertilizer requirement and quality of sugar beet. In: *Proceedings of 57th Winter Congress of Institut International de Recherches Betteravières*. Brussels, pp. 343–351.

Lawton, K. (1945) The influence of soil aeration on the growth and absorption of nutrient by corn plants. *Soil Science Society of America Proceedings* 10, 263–268.

Lawton, K., Erickson, A.E. and Robertson, L.S. (1954) Utilization of phosphorus by sugar beets as affected by fertilizer placement. *Agronomy Journal* 46, 262–264.

Leach, L.D. and Davey, A.E. (1942) Reducing southern sclerotium rot of sugar beets with nitrogenous fertilizers. *Journal of Agricultural Research* 64, 1–18.

Leamer, R.W. (1963) Residual effects of phosphorus fertilizer in an irrigated rotation in the southwest. *Soil Science Society of America Proceedings* 27, 65–68.

Le Docte, A. (1927) Commercial determination of sugar in the beet root using the Sachs–Le Docte process. *International Sugar Journal* 29, 483–487, 488–492.

Lehnhardt, L. and Bonk, M. (1991) Effect of boron fertilization on flowering sugar beet and the development and properties of their seeds. *Annual Report of Agricultural Institute, Potsdam, Germany* 35(1), 65–71.

Leifert, C. and Golden, M.H. (2000) *A Re-evaluation of the Beneficial and Other Effects of Dietary Nitrate.* International Fertiliser Society, London, 22 pp.

Lejealle, F. (1986) Effect of method of production and nitrogen fertilizer on seed quality. *Institut International de Recherches Betteravières Proceedings*, 101–120.

Lemon, E.R. and Erickson, A.E. (1952) The measurement of oxygen diffusion in the soil with a platinum electrode. *Soil Science Society of America Proceedings* 16, 160–163.

Lewis, A.H. (1941) The placement of fertilizers. I. Root crops. *Journal of Agricultural Science, Cambridge* 31, 295–307.

Lexander, K. (1993) Present understanding of the physiology of sugar beet seed germination. *Institut International de Recherches Betteravières Proceedings*, 387–394.

Licht, F.O. (2000) *World Sugar and Sweetener Yearbook.* Licht, Ratzeburg, 500 pp.

Lichthardt, J.J. and Jacobsen, J.S. (1992) *Fertilizer Guidelines for Montana.* Extension Service EB 104, Montana State University, Bozeman, Montana.

Liebig, M.A., Jones, A.J., Mielke, L.N. and Doran, J.W. (1993) Controlled wheel traffic effects on soil properties in ridge tillage. *Soil Science Society of America Journal* 57, 1061–1066.

Lill, J.G., Salter, R.M., Cumings, G.A. and Hurst, L.A. (1938) Fertilizer placement test with sugar beets at Holgate, Ohio. *Proceedings of Annual Meeting National Joint Committee on Fertilizer Application* 14, 132–133.

Lill, J.G., Salter, R.M., Cumings, G.A. and Hurst, L.A. (1940) Fertilizer placement test with sugar beets at Holgate Ohio. *Proceedings of Annual Meeting National Joint Committee on Fertilizer Application* 16, 117–120.

Limb, R. and McAughtrie, J. (2000) Home grown organic sugar beet. *British Sugar Beet Review* 68(3), 7–10.

Lindén, B. and Nouno, S. (1983) Nitrogen fertilization prognoses for sugar beet based on determination of the mineral N content in soil profiles. *Institut International de Recherches Betteravières Proceedings*, 473–492.

Lindsay, W.L. (1972) Inorganic phase equilibria of micronutrients in soils. In: Mortvedt, J.J., Giordano, P.M. and Lindsay, W.L. (eds) *Micronutrients in Agriculture.* Soil Science Society of America, Madison, Wisconsin, pp. 41–57.

Lindsay, W.L. (1974) Role of chelation in micronutrient availability. In: Carson, E.W. (ed.) *The Plant Root and Its Environment.* University Press of Virginia, Charlottesville, Virginia, pp. 506–524.

Longden, P.C. (1970) Manuring the beet seed crop growth in England. *NAAS Quarterly Review* 87, 112–118.

Longden, P.C. and Johnson, M.G. (1977) Effects of single or split application of nitrogen fertilizer to *in situ* sugar beet seed plants in the spring of their second year's growth. *Journal of Agricultural Science, Cambridge* 89, 609–620.

Loomis, R.S. and Worker, G.F. Jr (1963) Responses of the sugar beet to low soil moisture at two levels of nitrogen nutrition. *Agronomy Journal* 55, 509–515.

Loomis, R.S., Brickey, J.H., Broadbent, F.E. and Worker, G.F. Jr (1960) Comparisons of nitrogen source materials for midseason fertilization of sugar beet. *Agronomy Journal* 52, 97–101.

López-Bellido, L., Castillo, J.E. and Fuentes, M. (1994) Nitrogen uptake by autumn sown sugar beet. *Fertilizer Research* 38, 101–109.

Low, A.J. (1955) Improvements in the structural state of soils under leys. *Journal of Soil Science* 6, 179–199.

Lucas, R.E. and Knezek, B.D. (1972) Climatic and soil conditions promoting micronutrient deficiencies in plants. In: Mortvedt, J.J., Giordano, P.M. and Lindsay, W.L. (eds) *Micronutrients in Agriculture*. Soil Science Society of America, Madison, Wisconsin, pp. 265–288.

Ludwick, A.E., Sharpee, K.W. and Attoe, O.J. (1968) Manganese–sulfur fusions as a source of manganese for crops. *Agronomy Journal* 60, 232–234.

Ludwick, A.E., Gilbert, W.A. and Westfall, D.G. (1980) Sugarbeet quality as related to KCl fertilization. *Agronomy Journal* 72, 453–456.

Lumsden, R.D., Lewis, J.A. and Millner, P.D. (1983) Effect of composted sewage sludge on several soil-borne pathogens and diseases. *Phytopathology* 73, 1543–1547.

McDonnell, P.M., Gallagher, P.A., Kearney, P. and Carroll, P. (1966) Fertilizer use and sugar beet quality in Ireland. In: *Potassium Symposium 1966*, International Potash Institute, Bern, pp. 107–126.

McEnroe, P. and Coulter, B. (1964) Effect of soil pH on sugar content and yield of sugar beet. *Ireland Journal of Agriculture Research* 3, 63–69.

McEwen, J. and Johnston, A.E. (1979) The effects of subsoiling and deep incorporation of P and K fertilizers on the yield and nutrient uptake of barley, potatoes, wheat and sugar beet grown in rotation. *Journal of Agricultural Science, Cambridge* 92, 695–702.

McGrath, S.P. (2000) Practical field test for sulphur deficiency. *Arable Farming*, 8 July, 6.

McGrath, S.P. and Loveland, P.J. (1992) *The Soil Geochemical Atlas of England and Wales*. Blackie, Glasgow.

McGrath, S.P., Zhao, F.J. and Withers, P.J. (1996) Development of sulphur deficiency in crops and its treatment. *Fertilizer Society Proceedings* 379, 1–48.

Machet, J.M. and Hebert, J. (1983) Results from six years trials on nitrogen fertilizer supply to sugar beet. *Institut International de Recherches Betteravières Proceedings*, 493–507.

Machet, J.M., Duval, R., Jaouen, V. and Servain, F. (1994) N fertilizer recommendations for sugar beet in France. *Institut International de Recherches Betteravières Proceedings*, 305–319.

Mäck, G. and Tischner, R. (1990) Glutamine synthetase oligomers and isoforms in sugar beet (*Beta vulgaris* L.). *Planta* 181, 10–17.

McLean, E.O. and Watson, M.E. (1985) Soil measurements of plant-available potassium. In: Munson, R.D. (ed.) *Potassium in Agriculture*. American Society of Agronomy, Madison, Wisconsin, pp. 277–308.

McLean, E.O , Hartwig, R.C., Eckert, D.J. and Triplett, G.B. (1983) Basic cation saturation ratios as a basis for fertilizing and liming agronomic crops. II. Field studies. *Agronomy Journal* 75, 635–639.

McMillan, J.A. and Hanley, F. (1936) The effect of sowing fertilizers in contact with the seed of barley and of sugar beet. *Agriculture London* 42, 1205–1211.

MAFF (1976) *Trace Element Deficiencies in Crops*. Advisory Paper 17, London, 34 pp.

MAFF (1998) *The Soil Code*. UK Ministry of Agriculture, Fisheries and Food, London, 66 pp.

MAFF (2000) *Fertilizer Recommendations for Agricultural and Horticultural Crops*. RB209, Ministry of Agriculture, Fisheries and Food, London, 175 pp.

Magdoff, F.R., Ross, D. and Amadon, J. (1984) A soil test for nitrogen availability to corn. *Soil Science Society of America Journal* 48, 1301–1304.

Malnou, C., Jaggard, K.W. and Sparkes, D.C. (2003) Nitrogen fertilizer recommendations for sugar beet: a canopy approach. In: *Proceedings of American Society of Sugar Beet Technologists, 32nd Biennial Meeting, San Antonio* (in press).

Malzer, G.L. and Graff, T. (1995) Impact of turkey manure application on corn production and potential water quality concerns, Westport, MN 1994. In: *Field Research in Soil Science*. Miscellaneous Publication 88–1995, Agricultural Experiment Station, University of Minnesota, St Paul, Minnesota, pp. 121–125.

Mambelli, S.R., Benati, R., Amaducci, M.T. and Venturi, G. (1992) Effects of irrigation on the growth and qualitative characteristics of roots of sugarbeet. *Rivista di Agronomia* 26, 623–632.

Mambelli, S., Dal Rio, M.P., Grandi, S., Amaducci, M.T. and Venturi, G. (1997a) Dynamics of nitrogen uptake and root quality in sugar beet. *Institut International de Recherches Betteravières Proceedings*, 313–319.

Mambelli, S., Dal Rio, M.P. Amaducci, M.T. and Venturi, G. (1997b) Methods of plant analysis to evaluate nitrogen status in sugar beet. *Institut International de Recherches Betteravières Proceedings*, 321–326.

Mann, H.H. and Boyd, D.A. (1958) Some results of an experiment to compare ley and arable rotations at Woburn. *Journal of Agricultural Science, Cambridge* 50, 297–306.

Mann, H.H. and Patterson, H.D. (1963) The Woburn market-garden experiments, Summary 1944–1960. In: *Rothamsted Experimental Station Report for 1962*, Lawes Trust, Harpenden, pp. 186–193.

Mann, J. and Barnes, T.W. (1945) Manuring for the production of sugar beet seed. *Agriculture, London* 52, 400–404.

Mannini, P. and Venturi, G. (1993). Effects of irrigation and fertilization on sugar beet yield and quality in Northern Italy's Po Valley. In: *Institut International de Recherches Betteravières 56 Congress Proceedings*, Brussels, pp. 171–185.

Marcussen, C. (1985) Amino N figures as used in Denmark. *British Sugar Beet Review* 53(4), 46–48.

Marcussen, C. (1991) Placement of NPK to sugar beet. *Institut International de Recherches Betteravières Proceedings*, 401–406.

Marcussen, C. and Smed, E. (1996) The effect of K and Na on sugar beet quality. *Institut International de Recherches Betteravières*, 409–410.

Märländer, B. and Ladewig, E. (1995) Economic optimum N – supply for sugar beet – an analysis of different statistical approaches. *Institut International de Recherches Betteravières Proceedings*, 489–495.

Märländer, B. and Windt, A. (1996) Development of the fibrous root system and its relationship to nutrient uptake and growth of sugar beet. *Institut International de Recherches Betteravières Proceedings*, 187–198.

Marschner, H. (1995) *Mineral Nutrition of Higher Plants*, 2nd edn. Academic Press, London, 889 pp.

Martens, D.C. and Westermann, D.T. (1991) Fertilizer applications for correcting micronutrient deficiencies. In: Luxmoore, R.J. (ed.) *Micronutrients in Agriculture*, 2nd edn. Soil Science Society of America, Madison, Wisconsin, pp. 549–592.

Martens, M. and Oldfield, J.F.T. (1970) Storage of sugar beet in Europe: report of an IIIRB enquiry. *Journal of the International Institute of Sugar Beet Research* 5, 102–127.

Martin, W.P., Taylor, G.S., Engibous, J.C. and Burnett, E. (1952) Soil and crop responses from field applications of soil conditioners. *Soil Science* 73, 455–471.

Mathers, A.C. and Stewart, B.A. (1984) Manure effects on crop yields and soil properties. *Transactions of American Society of Agriculture Engineers* 27, 1022–1026.

Mattingly, G.E.G. (1974) The Woburn organic manuring experiment. I. Design, crop yields and nutrient balance, 1964–72. In: *Rothamsted Experimental Station Report for 1973*, Part 2, Lawes Trust, Harpenden, pp. 98–133.

Mehlich, A. (1957) Aluminum, iron and pH in relation to lime induced manganese deficiencies. *Soil Science Society of America Proceedings* 21, 625–628.

Mellor, J.L., Johnson, H.P.H. and Gardner, R. (1950) Fertilizer placement for sugar beet production. *Proceedings of American Society of Sugar Beet Technologists* 6, 428–435.

Mengel, K. (1994) Scientific improvement by the EUF working group for the importance of soil organic nitrogen to ecological agriculture. *Institut International de Recherches Betteravières Proceedings*, 213–238.

Mengel, K. and Kirkby, E.A. (1978) *Principles of Plant Nutrition*. International Potash Institute, Worblaufen-Bern, Switzerland, 593 pp.

Metochis, C. and Ophanos, P.I. (1988) Irrigation and nitrogen requirements of sugar beet in a Mediterranean environment. *Journal of Agricultural Science, Cambridge* 110, 387–390.

Miles, R.O. (1947) *The Placement of Fertilizers*. Bulletin 4, Jealotts Hill Research Station.

Milford, G.F.J., Biscoe, P.V., Jaggard, K.W., Scott, R.K. and Draycott, A.P. (1980) Physiological potential for increasing yields of sugar beet. In: Hurd, R.G., Biscoe, P.V. and Dennis, C. (eds) *Opportunities for Increasing Crop Yields*. Pitman, London, pp. 71–83.

Milford, G.F.J., Pocock, T.O., Riley, J. and Messem, A.B. (1985) An analysis of leaf growth in sugar beet. III. Leaf expansion in field crops. *Annals of Applied Biology* 106, 187–203.

Milford, G.F.J., Armstrong, M.J., Jarvis, P.J., Houghton, B.J., Bellett-Travers, D.M., Jones, J. and Leigh, R.A. (2000) Effect of potassium fertilizer on the yield, quality and potassium offtake of sugar beet crops grown on soils of different potassium status. *Journal of Agricultural Science, Cambridge* 135, 1–10.

Millar, C.E., Cook, R.L. and Davis, J.F. (1938) The effect of fertilizer placement on the yield and stand of sugar beets at the Michigan Experiment Station in 1938. *Proceedings of Annual Meeting National Joint Committee on Fertilizer Application* 14, 129–131.

Millar, C.E., Cook, R.L. and Davis, J.F. (1940) The effect of fertilizer placement on the yield and stand of sugar beets at the Michigan Experiment Station. *Proceedings of Annual Meeting National Joint Committee on Fertilizer Application* 16, 115–116.

Millar, C.E., Cook, R.L., Rood, P.J. and Robertson, L.S. (1945) A comparison of methods of applying fertilizer plough under versus deep planting time placement for corn and plough under versus row application for corn, sugar beets and potatoes. *Proceedings of Annual Meeting National Joint Committee on Fertilizer Application* 21, 99–104.

Millar, C.E., Robertson, L.S., Cook, R.L., Hansen, C.M., Hulburt, W.C. and Cumings, G.A. (1948) The effect of fertilizer placement on the yield and maturity date of corn at the Michigan Experiment Station in 1948. *Proceedings of Annual Meeting National Joint Committee on Fertilizer Application* 24, 51–54.

Miller, D.E. and Kemper, W.D. (1962) Water stability of aggregates of two soils as influenced by incorporation of alfalfa. *Agronomy Journal* 54, 494–496.

Miller, R.W. and Donahue, R.L. (1990) *Soils: an Introduction to Soils and Plant Growth*, 6th edn. Prentice Hall, Englewood Cliffs, New Jersey, 768 pp.

Miller, S.D. and Dexter, A.G. (1982) No-till crop production in the Red River Valley. *North Dakota Farm Research* 40(2), 3–5.

Mills, H.A. and Jones, J.B., Jr (1996) *Plant Analysis Handbook – A Practical Sampling, Preparation, Analysis and Interpretation Guide*. Micro-Macro Publishing, Athens, Georgia, 422 pp.

Mitchell, J., Gaskell, M., Smith, R., Fouche, C. and Koike, S. (2000) *Soil Management and Soil Quality for Organic Crops*. Publication 7248, University of California, Davis, California.

Mitscherlich, E.A. (1954) *Bodenkunde für Landwirte, Förster und Gärtner*, 7th edn. Parey, Berlin, pp. 259.

Momen, N.M. (1985) Effect of cropping system on soil structure and growth and development of sugar beets (*Beta vulgaris*, L.). PhD thesis, Michigan State University, East Lansing, Michigan.

Mompin, B.A., Lopez, C.T., de Benito, M.A. and Guirao, S.J. (1993) Sugar beet yield response to water and nitrogen in the Duero valley. In: *Institut International de Recherches Betteravières 56 Congress Proceedings*, Brussels, pp. 131–150.

Monteith, J.L. (1978) Reassessment of maximum growth rates for C_3 and C_4 crops. *Experimental Agriculture* 14, 1–5.

Moraghan, J.T. (1972) Water use by sugar beets in a semiarid environment as influenced by population and nitrogen fertilizer. *Agronomy Journal* 64, 759–762.

Moraghan, J.T. (1979) Responses of sugarbeets to potassium fertilizer in the Red River Valley. In: *1978 Sugarbeet Research and Extension Reports* 9. North Dakota State University, Fargo, North Dakota, pp. 139–161.

Moraghan, J.T. (1984) Soil fertility research. In: *Sugarbeet Research and Extension Reports 1983* 14, North Dakota State University, Fargo, North Dakota, pp. 93–110.

Moraghan, J.T. and Ananth, S. (1985) Return of sugar beet tops and the accumulation of certain chemical constituents in soil. *Journal of American Society of Sugar Beet Technologists* 23, 72–79.

Moraghan, J.T. and Cole, D.F. (1978) Lower leaf scorch of sugarbeets resulting from potassium deficiency in the Red River Valley. *Journal of American Society of Sugar Beet Technologists* 20, 133–145.

Moraghan, J.T. and Etchevers, J.D. (1981) Method of phosphorus fertilization for sugar beets in the Red River Valley. *Journal of American Society of Sugar Beet Technologists* 21, 103–111.

Moraghan, J.T., Tiedeman, P. and Torkelson, R. (1973) Sugar beet production in the Red River Valley as affected by population and nitrogen fertilizer. *Journal of American Society of Sugar Beet Technologists* 17, 260–269.

Moraghan, J.T., Smith, L. and Sims, A. (1997) Sugar beet tops and soil nitrogen fertility. *Institut International de Recherches Betteravières Proceedings*, 327–329.

Moraghan, J.T., Sims, A. and Smith, L. (2000) Remote sensing of sugarbeet canopies for improved nitrogen fertilizer recommendations for a subsequent wheat crop. *Communications in Soil Science and Plant Analysis* 31, 827–836.

Morley Davies, W. (1939) Acidity and manganese deficiency problems in connexion with sugar beet growing. *Annals of Applied Biology* 26, 285–392.

Mortvedt, J.J. and Giordano, P.M. (1975) Crop response to manganese sources applied with ortho and polyphosphate fertilizers. *Soil Science Society of America Proceedings* 39, 782–787.

Mortvedt, J.J., Westfall, D.G. and Croissant, R.L. (1996) *Fertilizing Sugar Beets*. Cooperative Extension Bulletin No. 0542, Colorado State University, Denver.

Mostafa, M.A.E. and Ulrich, A. (1974) Calcium uptake by sugar beets relative to concentration and activity of calcium. *Soil Science* 116, 432–436.

Mostafa, M.A.E. and Ulrich, A. (1976a) Absorption, distribution and form of Ca in relation to Ca deficiency (tip burn) of sugarbeets. *Crop Science* 16, 27–30.

Mostafa, M.A.E. and Ulrich, A. (1976b) Interaction of calcium and magnesium in nutrition of intact sugarbeets. *Soil Science* 121, 16–20.
Müller, H.J. (1978) Soluble nitrogen in soil at drilling in relation to soil characteristics and climatological factors. *Institut International de Recherches Betteravières Proceedings*, 45–52.
Mulvaney, R.L. (1996) Nitrogen – inorganic forms. In: Sparks, D.L. (ed.) *Methods of Soil Analysis. Part 3 Chemical Methods*. Soil Science Society of America, Madison, Wisconsin, pp. 1123–1184.
Muro, J., Irigoyen, I. and Lamsfus, C. (1998) Defoliation timing and severity in sugar beet. *Agronomy Journal* 90, 800–804.
Murphy, L.S. (1979) MAP, DAP, poly and rock. In: *North Central Extension–Industry Soil Fertility Workshop*. Potash and Phosphate Institute, Norcross, Georgia, p. 35.
Murphy, L.S. and Smith, G.E. (1967) Nitrate accumulation in forage crops. *Agronomy Journal* 59, 171–174.
Murphy, L.S. and Walsh, L.M. (1972) Correction of micronutrient deficiencies with fertilizers. In: Mortvedt, J.J., Giordano, P.M. and Lindsay, W.L. (eds) *Micronutrients in Agriculture*. Soil Science Society of America, Madison, Wisconsin, pp. 347–388.
Nable, R.O. and Paull, J.G. (1991) Mechanism of genetics of tolerance to boron toxicity in plants. *Current Topics in Plant Biochemical Physiology* 10, 257–273.
Nagarajah, S. and Ulrich, A. (1966) Iron nutrition of the sugar beet plant in relation to growth, mineral balance, and riboflavin formation. *Soil Science* 102, 399–407.
Narayan, D., Chandel, A.S. and Singh, G.R. (1989) Effect of boron fertilization on yield and quality of sugar beet. *Indian Journal of Plant Physiology* 32, 164–168.
Natural Resource Management Laboratories (1994) *Soil Copper Status*. Advisory Note 2.4, Jealott's Hill, 3 pp.
Neeteson, J.J. and Ehlert, P.A.I. (1989) Environmental aspects of applying inorganic fertilizers to sugar beet. *Institut International de Recherches Betteravières Proceedings*, 79–91.
Neeteson, J.J. and Smilde, K.W. (1983) Corrective methods of estimating optimum nitrogen rate for sugar beet based on soil mineral nitrogen at the end of the winter period. *Institut International de Recherches Betteravières Proceedings*, 409–421.
Nelson, J.M. (1969) Effect of row width, plant spacing, nitrogen rate and time of harvest on yield and sucrose content of sugar beets. *Journal of American Society of Sugar Beet Technologists* 15, 509–516.
Nelson, J.M. and Ruppel, E.G. (1970) The effect of manure on sprangling of sugar beet roots. *Journal of American Society of Sugar Beet Technologists* 16, 191–196.
Nelson, S.D. (1998) Krilium: the famous soil conditioner of the 1950s. In: Wallace, A. and Terry, R.E. (eds) *Handbook of Soil Conditioners: Substances that Enhance the Physical Properties of Soil*. Marcel Dekker, New York, pp. 385–398.
Nevins, D.J. and Loomis, R.S. (1970) Nitrogen nutrition and photosynthesis in sugar beet (*Beta vulgaris* L.). *Crop Science* 10, 21–25.
Norvell, W.A. (1991) Reactions of metal chelates in soils and nutrient solutions. In: Luxmore, R.J. (ed.) *Micronutrients in Agriculture*, 2nd edn. Soil Science Society of America, Madison, Wisconsin, pp. 187–227.
Nowicki, A. (1969) The effect of molybdenum upon yield, health and processing quality of sugar beet. *Roczniki Nauk Rolniczych* 95, 55–74.
Nuckols, S.B. (1942) Use of manures for sugar beets. *Proceedings of American Society of Sugar Beet Technologists* 3, 121–137.
Ocio, J.A., Martinez, J. and Brookes, P.C. (1991) Contribution of straw-derived N to total microbial biomass N following incorporation of cereal straw to soil. *Soil Biology and Biochemistry* 23, 655–659.
Ødelein, M. (1963) Long-term field experiments with small applications of boron. *Soil Science* 95, 60–62.
Offiah, O. and Axley, J.H. (1988) Improvement of boron soil test. *Communications in Soil Science and Plant Analysis* 19, 1527–1542.
Ogden, D.B., Finkner, R.F., Olson, R.F. and Hanzas, P.C. (1958) The effect of fertilizer treatment upon three different varieties in the Red River Valley of Minnesota 1. Stand, yield, sugar purity and nonsugars. *Journal of American Society of Sugar Beet Technologists* 10, 265–271.
Olsen, S.R., Gardner, R., Schmehl, W.R., Watanabe, F.S. and Scott, C.O. (1950) Utilization of phosphorus from various fertilizer materials by sugar beets in Colorado. *Proceedings of American Society of Sugar Beet Technologists* 6, 317–331.
Olsen, S.R., Cole, C.V., Watanabe, F.S. and Dean, L.A. (1954) *Estimation of Available Phosphorus in Soils by Extraction with Sodium Bicarbonate*. USDA Circular 939, US Government Printing Office, Washington, DC.

Olsson, R. and Bramstorp, A. (1994) Fate of nitrogen from sugar beet tops. *Institut International de Recherches Betteravières Proceedings*, 189–212.

Overpeck, J.C. and Elcock, H.A. (1937) *Sugar, Beet Seed Production Studies in Southern New Mexico, 1931–36*. Bulletin 252, New Mexico Agriculture Experimental Station, Santa Fe.

Pabin, J., Sienkiewicz, J. and Włodek, S. (1991) Effect of loosening and compacting on soil physical properties and sugar beet yield. *Soil and Tillage Research* 19, 345–350.

Papanicolaou, E.P., Analogides, D.A., Apostolakis, C.G., Skarlou, V. and Nobeli, C. (1982) Effectiveness of P fertilizer placement methods and N fertilizer sources on fertilizer utilization and yield characteristics of sugarbeets. *Journal of American Society of Sugar Beet Technologists* 21, 247–259.

Patterson, H.D. (1960) An experiment on the effects of straw ploughed in or composted in a three-course rotation of crops. *Journal of Agricultural Science, Cambridge* 54, 222–230.

Patterson, H.D. and Watson, D.J. (1960) Farmyard manure and its interactions with fertilizers. In: *Rothamsted Experimental Station Report for 1959*, Lawes Trust, Harpenden, pp. 164–168.

Paull, J.G., Rathjen, A.J. and Cartwright, B. (1988) Genetic control of tolerance to high concentrations of soil boron in wheat. In: *Proceedings of 7th International Wheat Genetics Symposium*. Cambridge, UK, pp. 871–877.

Paulson, K.N. and Kurtz, L.T. (1969) Locus of urease activity in soil. *Soil Science Society of America Proceedings* 33, 897–901.

Pendleton, R.A. (1950) Soil compaction and tillage operation effects on sugar beet root distribution and seed yields. *Proceedings of American Society of Sugar Beet Technologists* 6, 278–285.

Pendleton, R.A. (1954) Forms of nitrogen as related to sugar beet seed production in Oregon. *Proceedings of American Society Sugar Beet Technologists* 8, 140–141.

Pendleton, R.A., Finnell, H.E. and Reimer, F.C. (1950) *Sugar Beet Seed Production in Oregon*. Bulletin 437, Agriculture Experiment Station, Oregon State College.

Penman, H.L. (1952) Experiments on the irrigation of sugar beet. *Journal of Agricultural Science, Cambridge* 42, 286–292.

Penman, H.L. (1962) Woburn irrigation, 1951–59, Parts I, II, III. *Journal of Agricultural Science, Cambridge* 58, 343–348, 349–364, 365–379.

Penman, H.L. (1970) Woburn irrigation, 1960–68, Parts IV, V, VI. *Journal of Agricultural Science, Cambridge* 75, 69–73, 75–88, 89–102.

Peterson, G.A., Anderson, F.N. and Olson, R.A. (1966) A survey of the nutrient status of soils in the North Platte Valley of Nebraska for sugar production. *Journal of American Society of Sugar Beet Technologists* 14, 48–60.

Pfleiderer, U.-E. (1983) Evaluation of the amount of soil nitrogen nitrate in the horizon 0–90 cm from measures obtained at 0–60 cm depth. *Institut International de Recherches Betteravières Proceedings*, 509–516.

Philips, R.E. and Kirkham, D. (1962) Soil compaction in the field and corn growth. *Agronomy Journal* 54, 29–34.

Pidgeon, J.D. (2000) Drought stress, climate change and research strategy for the sugar beet crop. In: *Rothamsted Experimental Station Report for 1999*, Lawes Trust, Harpenden, pp. 36–39.

Pidgeon, J.D., Werker, A.R., Jaggard, K.W., Richter, G.M., Lister, D.H. and Jones, P.D. (2001) Climatic impact on the productivity of sugar beet in Europe, 1961–1995. *Agriculture and Forest Meteorology* 109, 27–37.

Pimpini, F., Giardini, L., Borin, M. and Gianquinto, G. (1992) Effects of poultry manure and mineral fertilizers on the quality of crops. *Journal of Agricultural Science, Cambridge* 118, 215–221.

Pizer, N.H. (1954) Organic matter in some eastern counties soils. *NAAS Quarterly Review* 25, 41–46.

Pizer, N.H., Caldwell, T.H., Burgess, G.R. and Jones J.L.O. (1966) Investigations into copper deficiency in crops in East Anglia. *Journal of Agricultural Science, Cambridge* 66, 303–314.

Pocock, T., Milford, G. and Armstrong, M. (1988) The nitrogen nutrition of sugar beet. *British Sugar Beet Review* 56(3), 41–44.

Pocock, T.O., Milford, G.F.J. and Armstrong, M.J. (1990) Storage root quality in sugar beet in relation to nitrogen uptake. *Journal of Agricultural Science, Cambridge* 115, 355–362.

Potash Development Association (1999) *Potash for Organic Growers*. Leaflet No. 23, Potash Development Association, Laugharne, 12 pp.

Power, R. and Herlihy, M. (1988) Cost reduction in relation to fertilizer usage on the sugar beet crop in Ireland. *Institut International de Recherches Betteravières Proceedings*, 73–82.

Powlson, D.S. (1994) Soil organic nitrogen – its structure, dynamics and role in the nitrogen cycle of agricultural soils. *Institut International de Recherches Betteravières Proceedings*, 155–176.

Powlson, D.S. (1998a) Organic farming. In: *Rothamsted Experimental Station Report for 1997*, Lawes Trust, Harpenden, 34–35.
Powlson, D.S. (1998b) Nitrogen cycling from field to farm. In: *Rothamsted Experimental Station Report for 1997*, Lawes Trust, Harpenden, pp. 35–37.
Powlson, D.S., Jenkinson, D.S., Pruden, G. and Johnston A.E. (1985) The effect of straw incorporation on the uptake of nitrogen by winter wheat. *Journal of Agricultural Science, Cambridge* 36, 29–30.
Price, T.J.A. and Harvey, P.N. (1962) Effect of irrigation on sugar beet and potatoes. *Experimental Husbandry* 7, 1–7.
Prummel, J. (1975) Effect of soil structure on phosphate nutrition of crop plants. *Netherlands Journal of Agricultural Science* 23, 62–68.
Pultz. L.M. (1937) Relation of nitrogen to yield of sugar beet seed and to accompanying changes in the composition of the roots. *Journal of Agricultural Research* 54, 639–654.
Quarles, W. and Grossman, J. (1995) Alternatives to methyl bromide in nurseries – disease suppressive media. *IPM Practice* 17(8), 1–13.
Randall, G.W., Schulte, E.E. and Corey, R.B. (1975) Effect of soil and foliar applied manganese on the micronutrient content and yield of soybeans. *Agronomy Journal* 67, 502–507.
Rasmusson, J. and Wiklund, O. (1960) Characteristics of the technological value of the sugar beet. In: *Proceedings XIth Session of the International Technology Committee on Sucrose 1960*, pp. 13–24.
Rayns, F. (1961) *A Revolution in Arable Farming*. The Lord Hastings Memorial Lecture, Norfolk Agricultural Station.
Rayns, F. and Culpin, S. (1948) Rotation experiments on straw disposal at the Norfolk Agricultural Station. *Journal of the Royal Agricultural Society of England* 109, 128–137.
Rehm, G., Schmitt, M. and Munter, R. (1998) *Fertilizing Soybeans in Minnesota*. Extension Service FS-3813-GO, University of Minnesota, St Paul, Minnesota.
Reisenauer, H.M. (1988) Determination of plant available soil manganese. In: *Manganese in Soils and Plants, Proceedings of an International Symposium, August 1988*. University of Adelaide, Adelaide, pp. 87–98.
Richards, L.A. (1954) *Diagnosis and Improvement of Saline and Alkali Soils*. Handbook No. 60, United States Department of Agriculture, US Government Printing Office, Washington, DC.
Roberts, S., Richards, A.W., Day, M.G. and Weaver, W.H. (1972) Predicting sugar content and petiole nitrate of sugarbeets from soil measurements of nitrate and mineralizable nitrogen. *Journal of American Society of Sugar Beet Technologists* 17, 126–133.
Robertson, L.S. (1952) A study of the effects of seven systems of cropping upon yields and soil structure. *Proceedings of American Society of Sugar Beet Technologists* 7, 255–264.
Robertson, L.S. and Erickson, A.E. (1983) *Till Planting on Ridges*. Extension Bulletin E-1683, Michigan State University, East Lansing, Michigan.
Robertson, L.S. and Lucas, R.E. (1981) *Essential Micronutrients: Manganese*. Cooperative Extension Service Extension Bulletin E-1031, Michigan State University, East Lansing, Michigan.
Robertson, L.S., Cook, R.L., Rood, P.J. and Turk, L.M. (1952) Ten years' results from the Ferden rotation and crop sequence experiment. *Proceedings of American Society of Sugar Beet Technologists* 7, 172–179.
Robertson, L.S., Knezek, B.D. and Belo, J.O. (1975) A survey of Michigan soils as related to possible boron toxicities. *Communications in Soil Science and Plant Analysis* 6, 359–373.
Robertson, L.S., Cook, R.L. and Davis, J.F. (1976a) *The Ferden Farm Report: Part II, Soil Management for Sugar Beets 1940–1970*. Agricultural Experiment Station Research Report 324, Michigan State University, East Lansing, Michigan.
Robertson, L.S., Christenson, D.R. and Warncke, D.D. (1976b) *Essential Secondary Elements: Calcium*. Cooperative Extension Service Extension Bulletin E-996, Michigan State University, East Lansing, Michigan.
Robertson, L.S., Christenson, D.R., Smucker, A.J.M. and Mokma, D.L. (1978) Tillage systems. In: Robertson, L.S. and Frazer, R.D. (eds) *Dry Bean Production – Principles and Practices*. Extension Bulletin E-1251, Michigan State University, East Lansing, Michigan, pp. 78–93.
Robertson, L.S., Christenson, D.R. and Warncke, D.D. (1979a) *Essential Secondary Elements: Magnesium*. Cooperative Extension Service Extension Bulletin E-994, Michigan State University, East Lansing, Michigan.
Robertson, L.S., Erickson, A.E. and Hansen, C.M. (1979b) *Tillage Systems for Michigan Soils and Crops, Part I: Deep, Primary, Supplemental and No-till*. Extension Bulletin E-1041, Michigan State University, East Lansing, Michigan.

Robertson, L.S., Hansen, C.M. and Erickson, A.E. (1979c) *Tillage Systems for Michigan Soils and Crops, Part II: Secondary Tillage and Cultivation.* Extension Bulletin E-1042, Michigan State University, East Lansing, Michigan.

Robins, J.S., Nelson, C.E. and Domingo, C.E. (1956) Some effects of excess water application on utilisation of applied nitrogen by sugar beets. *Journal of American Society of Sugar Beet Technologists* 9, 180–188.

Rodriguez, J. and Tomic, Y.M.T. (1984) Micronutrient availability in andisols and ultisols of Los Lagos region. *Ciencia e Investigacion Agrararia* 11, 169–178.

Romsdal, S.D. and Schmehl, W.R. (1963) The effect of method and rate of phosphate application on yield and quality of sugar beets. *Journal of American Society of Sugar Beet Technologists* 12, 603–607.

Rounds, H.G., Rush, G.E., Oldemeyer, D.L., Parrish, C.P. and Rawlings, F.N. (1958) A study and economic appraisal of the effect of nitrogen fertilization and selected varieties on the production and processing of sugar beets. *Journal of American Society of Sugar Beet Technologists* 10, 97–116.

Roussel, N., van Stallen, R. and Vlassak, K. (1966) Results of two years fertilizer trials with anhydrous ammonia. *Journal of the International Institute of Sugar Beet Research* 2, 35–53.

Rowe, E.A. (1936) A study of heart-rot of young sugar beet plants grown in culture solutions. *Annals of Botany* 50, 735–746.

Rozema, J., De Bruin, J. and Brockman, R.A. (1992) Effect of boron on the growth and mineral economy of some halophytes and non halophytes. *New Phytologist* 121, 249–256.

Russell, E.W. (1956) The effects of very deep ploughing and of subsoiling on crop yields. *Journal of Agricultural Science, Cambridge* 48, 129–144.

Russell, E.W. (1971) Soil structure: its maintenance and improvement. *Journal of Soil Science* 22, 137–151.

Russell, G.E. (1971) Effects on *Myzus persicae* (Sulz) and transmission of beet yellows virus of applying certain trace elements to sugar beet. *Annals of Applied Biology* 68, 67–70.

Russell, R.S. and Goss, M.J. (1974) Physical aspects of soil fertility – the response of roots to mechanical impedance. *Netherlands Journal of Agricultural Science* 22, 305–318.

Ryser, G.K. (1966) A regression study on tare samples of sugar beets in relation to factors influencing productivity and quality. *Journal of American Society of Sugar Beet Technologists* 13, 727–747.

Saive, R. (1987) Solubility of calcium phosphates in the rhizosphere of sugar beet. *Institut International de Recherches Betteravières Proceedings*, 189–201.

Salcedo, I.H., Ellis, B.G. and Lucas, R.E. (1979) Studies in soil manganese: II. Extractable manganese and plant uptake. *Soil Science Society of America Journal* 43, 138–141.

Salter, P.J., Williams, J.B. and Harrison, D.J. (1965) Effects of bulky organic manures on the available water capacity of a fine sandy loam. *Experimental Horticulture* 13, 69–75.

Sanchez, J., Harwood, R., LeCureux, J., Shaw, J., Shaw, M., Smalley, S., Smeenk, J. and Volker, R. (2001) *Integrated Cropping System for corn–sugar beet–dry bean rotation: The experience of the innovative farmers of Michigan.* Extension Bulletin E-2738, Michigan State University, East Lansing, Michigan.

Sands, R. (2002) Should we worry about Silver Y? *British Sugar Beet Review* 69(4), 36–38.

Sanz Saez, A. (1985) Fertilizer trial results with phosphorus and potassium in Central and Northern Spain. *Institut International de Recherches Betteravières Proceedings*, 341–355.

Sauchelli, V. (1969) *Trace Elements in Agriculture.* Van Nostrand Reinhold, New York, 248 pp.

Sawahata, H. and Takase, N. (1966) Studies on germination and early growth of sugar beet in southern Japan. III. Effects of kinds and amounts of nitrogenous fertilizers on germination and early growth of sugar beet. *Research Meeting of Sugar Beet Technologists of the Cooperative of South Japan* 8, 26–30.

SBREC (1995) *Sugar Beet – a Grower's Guide*, 5th edn. Sugar Beet Research and Education Committee of the Ministry of Agriculture, Fisheries and Food, London, 111 pp.

Scherer, H.W. (2001) Sulphur in crop production. *European Journal of Agronomy* 14, 81–111.

Schlegel, A.J., Nelson, D.W. and Sommers, L.E. (1986) Field evaluation of urease inhibitors for corn production. *Agronomy Journal* 78, 1007–1012.

Schmehl, W.R., Olsen, S.R. and Gardner, R. (1952) Effect of type of phosphate material and method of application on phosphate uptake and yield of sugar beets. *Proceedings of American Society of Sugar Beet Technologists* 7, 153–158.

Schmehl, W.R., Olsen, S.R. and Gardner, R. (1954) Effect of method of application on the availability of phosphate for sugar beets. *Proceedings of American Society of Sugar Beet Technologists* 8(2), 363–369.

Schneider, C.L. and Robertson, L.S. (1975) Occurrence of diseases in sugar beet plots in a crop rotation experiment in Saginaw County, Michigan in 1969–70–71. *Plant Disease Reporter* 59, 194–197.

Scott, R.K. (1967) Sugar beet seed production. *Agricultural Progress* 42, 112–118.

Scott, R.K. (1968) Sugar beet growing in Europe and North America. *Journal of the International Institute of Sugar Beet Research* 3, 53–84.

Scott, R.K. (1969) The effect of sowing and harvesting dates, plant population and fertilizers on seed yield and quality of direct-drilled sugar beet seed crops. *Journal of Agricultural Science, Cambridge* 73, 373–385.

Scott, R.K. and Jaggard, K.W. (1993) Crop physiology and agronomy. In: Cooke, D.A. and Scott, R.K. (eds) *The Sugar Beet Crop*. Chapman & Hall, London, pp. 179–237.

Scott, R.K. and Jaggard, K.W. (2000) Impact of weather, agronomy and breeding on yields of sugar beet grown in UK since 1970. *Journal of Agricultural Science, Cambridge* 134, 341–352.

Scott, R.K., English, S.D., Wood, D.W. and Unsworth, M.H. (1973) Yield of sugar beet in relation to weather and length of growing season. *Journal of Agricultural Science, Cambridge* 81, 339–347.

Scott, T.W. and Erickson, A.E. (1964) Effect of aeration and mechanical impedance on the root development of alfalfa, sugar beets and tomatoes. *Agronomy Journal* 56, 575–576.

Sexton, J.J. (1996) Sulphur survey of the sugar beet crop in England 1995. *British Sugar Beet Review* 64(2), 50–53.

Shephard, L., Lawton, K. and Davis, J.F. (1960) The effectiveness of various manganese materials in supplying manganese to crops. *Soil Science Society of America Proceedings* 24, 218–221.

Shepherd, M.A., Lord, E.I., Mitchell, R. and Groves, S.J. (1997) Sugar beet – nitrate leaching after harvest and consequences for longer-term losses. *Institut International de Recherches Betteravières Proceedings*, 347–351.

Shierlaw, J. and Alston, A.M. (1984) Effect of soil compaction on root growth and uptake of phosphorus. *Plant and Soil* 77, 15–28.

Shock, C.C., Seddigh, J., Saunders, L.D., Stieber, T.D. and Miller, J.G. (2000) Sugarbeet nitrogen uptake and performance following heavily fertilized onion. *Agronomy Journal* 92, 10–15.

Shoemaker, H.E., McLean, E.O. and Pratt, P.F. (1961) Buffer methods of determining lime requirement of soils with appreciable amounts of extractable aluminium. *Soil Science of America Proceedings* 25, 274–277.

Short, J.L. (1973) Straw disposal trials at the experimental husbandry farms. *Experimental Husbandry* 25, 103–136.

Shotton, F.E. (1962) Placement of fertilizer for sugar beet. *Experimental Husbandry* 7, 8–16.

Siegenthaler, A. (1987) Influence of phosphorus fertilizer on yield and sugar content of sugar beet in field experiments. *Institut International de Recherches Betteravières Proceedings*, 271–279.

Silverberg, J., Young, R.D. and Hoffmeister, G. (1972) Preparation of fertilizers containing micronutrients. In: Mortvedt, J.J., Giordano, P.M. and Lindsay, W.L. (eds) *Micronutrients in Agriculture*. Soil Science Society of American, Madison, Wisconsin, pp. 431–458.

Silverbush, M. and Barber, S.A. (1983) Sensitivity of simulated phosphorus uptake to parameters used by a mechanistic–mathematical model. *Plant and Soil* 74, 93–100.

Simon, M., Roussel, N. and Vanstallen, R. (1966) Potassium in the fertilising of sugar beet. *Potassium Symposium* 1966, 61–87.

Sims, A.L. and Smith, L.J. (2001) Early growth response of sugarbeet to fertilizer phosphorus in phosphorus deficient soils of the Red River Valley. *Journal of American Society of Sugar Beet Technologists* 38, 1–17.

Sipitanos, K.M. and Ulrich, A. (1969) Phosphorus nutrition of sugarbeet seedlings. *Journal of American Society of Sugar Beet Technologists* 15, 332–346.

Skaggs, R.W. and van Schilfgaarde, J. (1999) *Agricultural Drainage*. American Society of Agronomy, Madison, Wisconsin, 1328 pp.

Smilde, K.W. (1968) Manganous oxide (MnO) as a fertilizer for controlling manganese deficiency in oats. *Netherlands Journal of Agricultural Science* 16, 197–203.

Smilde, K.W. (1970) Soil analysis as a basis for boron fertilisation of sugar beets. *Zeitschrift fuer Pflanzenernachrunk Enähr und Bodenkunde* 125, 130–143.

Smith, C.H. (1954) Influence of Krilium soil conditioner on sugar content of beets. *Soil Science Society of Sugar Beet Technologists* 8, 143–146.

Smith, L.J. (1982) The effects of nitrogen carrier, time of application and placement on sugarbeet yield and quality. In: *1981 Sugarbeet Research and Extension Reports* 12. North Dakota State University, Fargo, North Dakota, pp. 108–110.

Smith, L.J. (1997) 1996 and 3-year summary of grid soil testing and variable rate fertilization on sugarbeet yield, quality and profitability. In: *1996 Sugarbeet Research and Extension Reports* 27. North Dakota State University, Fargo, North Dakota, pp. 83–87.

Smith, L.J., Cattanach, A.W. and Lamb, J.A. (1990) Uniform versus variable in-row spacing of sugarbeet. In: *1989 Sugarbeet Research and Extension Reports* 20. North Dakota State University, Fargo, North Dakota, pp. 151–156.

Smucker, A.J.M. and Leep, R.H. (1975) Influence of peroxide and polyelectrolyte treatments on germination, seedling emergence and yield. *Proceedings of the Association of Official Analysts* 65, 147–153.

Smucker, A.J.M., Adams, M.W., Christenson, D.R. and Hogaboam, G.J. (1978) Inhibition of dry bean and sugar beet production by soil physical stresses. In: *Saginaw Valley Bean and Beet Research Farm Report.* Michigan State University, East Lansing, Michigan, pp. 68–95.

Sneddon, J.L. (1963) Sugar beet seed production experiments. *Journal of the National Institute of Agricultural Botany* 9, 333–345.

Snyder, F.W. (1959) Effect of nitrogen on yield and subsequent germinability of sugar beet seed. *Journal of American Society of Sugar Beet Technologists* 10, 439–443.

Soane, G.C., Godwin, R.J. and Spoor, G. (1986) Influence of deep loosening techniques and subsequent wheel traffic on soil structure. *Soil and Tillage Research* 8, 231–237.

Soine, O.C. (1967) The effect of simulated hail damage on sugar beet. *Journal of American Society of Sugar Beet Technologists* 14, 424–432.

Sommers, L.E. (1977) Chemical composition of sewage sludges and analysis of their potential use as fertilizers. *Journal of Environmental Quality* 6, 225–232.

Sonneveld, C. and van Dijk, P.R. (1982) The effectiveness of some washing procedures on the removal of contaminates from plant tissues of glass-house crops. *Communications in Soil Science and Plant Analysis* 13, 487–496.

Sørensen, C. (1960) The influence of nutrition on the nitrogenous constituents of plants. II. Field experiments with heavy dressings of nitrogen to fodder sugar beets. *Acta Agriculturae Scandinavica* 10, 17–32.

Sørensen, C. (1962) The influence of nutrition on the nitrogenous constituents of plants. III. Nitrate tests and yield structure of fodder sugar beet leaves. *Acta Agriculturae Scandinavica* 12, 106–124.

Spoor, G. (1979) Soil management and the sugar beet crop. *British Sugar Beet Review* 47(4), 23–24.

Steven, M.D. (2001) Crop monitoring by remote sensing. *Society of Chemical Industry Proceedings,* London, 9–10.

Steven, M.D., Biscoe, P.V., Jaggard, K.W. and Paruntu, J. (1986) Foliage cover and radiation interception. *Field Crops Research* 13, 75–87.

Stevenson, F.J. (1996) Nitrogen – organic forms. In: *Methods of Soil Analysis. Part 3 Chemical Methods.* Soil Science Society of America, Madison, Wisconsin, pp. 1185–1200.

Stockinger, K.R., MacKenzie, A.J. and Cary, E.E. (1963) Yield and quality of sugar beets as affected by cropping systems. *Journal of American Society of Sugar Beet Technologists* 12, 492–496.

Stoker, G.L. and Tolman, B. (1941) Boron deficiency relations in sugar beets grown for seed in Oregon. *Journal of the American Society of Agronomy* 33, 657–665.

Stolzy, L.H., Letey, J., Szuszkiewicz, T.E. and Lunt, O.R. (1961) Root growth and diffusion rates as functions of oxygen concentration. *Soil Science Society of America Proceedings* 25, 436–467.

Stout, B.A., Snyder, F.W. and Carleton, W.M. (1956) The effect of soil moisture and compaction on sugar beet emergence. *Journal of American Society of Sugar Beet Technologists* 9, 277–283.

Stout, M. (1961) A new look at some nitrogen relationships affecting the quality of sugar beets. *Journal of American Society of Sugar Beet Technologists* 11, 388–398.

Stout, M. and Smith, C.H. (1950) Studies on the respiration of sugar beets as affected by bruising, by mechanical harvesting, severing into top and bottom halves, chemical treatment, nutrition and variety. *Proceedings of American Society of Sugar Beet Technologists* 6, 670–679.

Subbarao, G.V., Wheeler, R.M., Stutte, G.W. and Levine, L.H. (1999) How far can sodium substitute for potassium in red beet? *Journal of Plant Nutrition* 22(11), 1745–1761.

Swennson, C. (2002) Effect of manure handling system, N fertilizer use and area of sugar beet on N surpluses from dairy farms in Southern Sweden. *Journal of Agricultural Science, Cambridge* 138, 403–413.

Swift, E.L. (1980) Hillsboro deep freeze beet storage project. In: *1979 Sugarbeet Research and Extension Reports* 10. North Dakota State University, Fargo, North Dakota, pp. 215–216.

Syers, J.K. (1998) Soil and plant potassium in agriculture. *Fertilizer Society Proceedings* 411, 1–32.

Syers, J.K., Skinner, R.J. and Curtin, D. (1987) Soil and fertilizer sulphur in UK agriculture. *Fertilizer Society Proceedings* 264, 1–43.

Sylvester-Bradley, R. and Shepherd, M.A. (1997) Effects of interposing sugar beet on the nitrogen responses of successive wheat crops. *Journal of the Science of Food and Agriculture* 74, 323–330.

Tandon, P.K. (1979) Relative susceptibility of four varieties of sugar beet to boron deficiency. *Indian Journal of Agricultural Research* 13, 55–56.

Tariq, M., Khattak, J.K. and Sarwar, G. (1993) Effect of boron on the yield and quality of sugar beet in Peshawar Valley. *Scientific Khyber* 6, 97–106.
Taylor, H.M. and Bruce, R.R. (1968) Effect of soil strength on root growth and crop yield in the Southern United States. In: *9th International Congress of Soil Science Society Transactions*, Vol. 1, pp. 803–811.
Terman, G.L. (1971) *Phosphate Fertilizer Sources: Agronomic Effectiveness in Relation to Chemical and Physical Properties*. Fertiliser Society, London, 39 pp.
Terry, N. (1977) Photosynthesis, growth and the role of chloride. *Plant Physiology* 60, 69–75.
Terry, N. and Ulrich, A (1973) Effects of phosphorus deficiency on the photosynthesis and respiration of leaves in sugar beet. *Plant Physiology* 51, 43–47.
Terry, N., Evans, P.S. and Thomas, D.E. (1975) Manganese toxicity effects on leaf cell multiplication and expansion and on dry matter in sugar beets. *Crop Science* 15, 205–208.
Thielebein, M. (1960) Postulates for germination of sugar beet seed in the field. *Zucker* 13, 539–545.
Thomsen, I.K. and Christensen, B.T. (1996) Availability to subsequent crops and leaching of nitrogen ^{15}N-labelled sugar beet tops and oilseed rape residues. *Journal of Agricultural Science, Cambridge* 126, 191–199.
Thorne, G.N. and Watson, D.J. (1956) Field experiments on uptake of nitrogen from leaf sprays by sugar beet. *Journal of Agricultural Science, Cambridge* 47, 12–22.
Thorud, D.B. and Frissell, S.S. (1976) *Time Changes in Soil Density Following Compaction Under an Oak Forest*. Minnesota Forest Research Notes 257, University of Minnesota, Minneapolis, 4 pp.
Tiarks, A.E., Mazurak, A.P. and Chesin, L. (1974) Physical and chemical properties of soil associated with heavy applications of manure from cattle feedlots. *Soil Science Society of America Proceedings* 38, 826–830.
Tills, A.R. and Alloway, B.J. (1981) Subclinical copper deficiency in crops on the Breckland in East Anglia. *Journal of Agricultural Science, Cambridge* 97, 473–476.
Tinker, P.B.H. (1965) The effects of nitrogen, potassium and sodium fertilizers on sugar beet. *Journal of Agricultural Science, Cambridge* 65, 207–212.
Tinker, P.B.H. (1967a) The relationship of sodium in the soil to uptake of sodium by sugar beet in the greenhouse and to yield responses in the field. *Proceedings of International Soil Science Society*, 223–231.
Tinker, P.B.H. (1967b) The effects of magnesium sulphate on sugar beet yields and its interactions with other fertilizers. *Journal of Agricultural Science, Cambridge* 68, 205–212.
Tinker, P.B.H. (1970) *How Long does Applied Sodium Remain in the Soil?* Information 115, Nitrate Corporation of Chile, London.
Tinker, P.B.H. (2001) Organic farming – nutrient management and productivity. *International Fertilizer Society Proceedings* 471, 1–24.
Tinker, P.B.H., Jones, M.D. and Durall, D.M. (1992) A functional comparison of ecto- and endomycorrhizas. In: Read, D.J., Lewis, D.H., Fitter, A.H. and Alexander, I.J. (eds) *Mycorrhizas in Ecosystems*. CAB International, Wallingford, pp. 303–310.
Tolman, B. (1943) *Sugar beet Seed Production in Southern Utah, with Special Reference to Factors Affecting Yield and Reproductive Development*. Technical Bulletin 845, United States Department of Agriculture.
Trist, P.J.O. and Boyd, D.A. (1966) The Saxmundham rotation experiments: Rotation I, Rotation II, 1899–1952. *Journal of Agricultural Science, Cambridge* 66, 327–336, 337–339.
Trouse, A.C., Jr (1985) Development of controlled traffic concept. In: *International Conference on Soil Dynamics Proceedings* 5. Auburn University, Auburn, Alabama, pp. 1112–1119.
Tugnoli, V. and Bettini, S. (2000) Nitrogen fertilizers in sugar beet spring sowing: use of SPAD optical instrument. *Institut International de Recherches Betteravières Proceedings*, 419–424.
Tzilivakis, J., Jaggard, K., Lewis, K.A., May, M. and Warner, D.J. (2002) Environmental assessment of sugar beet production. *Institut International de Recherches Betteravières Proceedings*, 13–33.
Ulrich, A. (1948) Plant analysis as a guide to the nutrition of sugar beets in California. *Proceedings of American Society of Sugar Beet Technologists* 5, 364–377.
Ulrich, A. (1950) Critical nitrate levels of sugar beets estimated from analysis of petioles and blades, with special reference to yields and sucrose concentration. *Soil Science* 69, 291–309.
Ulrich, A. (1961) Plant analysis in sugar beet nutrition. In: *Plant Analysis and Fertilizer Problems*. American Institute of Biological Science, Washington, DC, pp. 190–211.
Ulrich, A. (1964) The relative constancy of the critical nitrogen concentration of sugar beet plants. In: *Plant Analysis and Fertiliser Problems*, IV Colloquium, University of California, Davis, pp. 371–391.
Ulrich, A. and Hills, F.J. (1952) Petiole sampling of sugar beet leaves in relation to their nitrogen, phosphorus, potassium and sodium status. *Proceedings of American Society of Sugar Beet Technologists* 7, 32–45.

Ulrich, A. and Hills, F.J (1969) *Sugarbeet Nutrient Deficiency Symptoms – A Color Atlas and Chemical Guide.* Division of Agriculture Sciences, University of California, Davis, California.

Ulrich, A. and Mostafa, M.A.E. (1976) Calcium nutrition of the sugarbeet. *Communication in Soil Science and Plant Analysis* 7, 483–495.

Ulrich, A. and Ohki, K. (1956a) Hydrogen ion effects on early growth of sugar beet plants in culture solution. *Journal of American Society of Sugar Beet Technologists* 9, 265–274.

Ulrich, A. and Ohki, K. (1956b) Chlorine, bromine and sodium as nutrients for sugar beet plants. *Plant Physiology* 31, 181–191.

Ulrich, A., Ririe, D., Hills, F.J., George, A.G. and Morse, M.D. (1959) *Plant Analysis a Guide for Sugar Beet Fertilisation. Analytical Methods for Use in Plant Analysis.* Bulletin 766, Californian Agricultural Experimental Station, Davis, California.

Unger, P.W. and Stewart, B.A. (1974) Feedlot waste effects on soil conditions and water evaporation. *Soil Science Society of America Proceedings* 38, 954–957.

Urbano, P., Arroyo, J.M., Conde, J.R., Rojo, C. and Gonzalez, F. (1992) Sugarbeet irrigation: trials in the Duero valley. *Agricultural Review Agropecuaria* 61, 380–385.

USEPA (1994) *A Plain English Guide to the EPA Part 503 Biosolids Rule.* EPA/832/R-93/003, United States Environmental Protection Agency, Washington, DC, 155 pp.

van Bavel, C.H.M. (1949) Mean weight-diameter of soil aggregates as a statistical index of aggregation. *Soil Science Society of America Proceedings* 14, 20–23.

van Burg, P.F.J., van Brakel, G.D. and Schepers, J.H. (1967) *The Agricultural Value of Anhydrous Ammonia on Arable Land 1963–1966.* Technical Bulletin of The Netherlands No. 3, Amsterdam.

Vandergeten, J.P. and Vanstallen, M. (1991) Influence of localized application of rational N dose rates on yield and technical quality of sugar beet. *Institut International de Recherches Betteravières Proceedings,* 297–318.

Vandergeten, J.P., Duval, R. and Vereerstraeten, R. (1997) Placement of reduced quantities of nitrogen fertilizer in sugar beet cultivation. In: *Proceedings of Institut International de Recherches Betteravières Congress,* Brussels, pp. 63–90.

van Egmond, F. (1979) Fate of calcium in the sugar beet plant. *Communications in Soil Science and Plant Analysis* 10, 311–323.

van Luit, B. and Smilde, K.W. (1969) Boron fertilisation of sugar beets based on soil analysis. *Rapport Insituut voor Bodemvruchtbaarheed Haren* 9, 1–48.

van Ouwerkerk, C. (1968) Two model experiments on the durability of subsoil compaction. *Netherlands Journal of Agricultural Science* 16, 204–210.

van Schilfgaarde, J. (1974) *Drainage for Agriculture.* American Society of Agronomy, Madison, Wisconsin, 700 pp.

van Schreven, D.A. (1936) Copper deficiency in sugar beets. *Phytopathology* 26, 1106–1117.

Vanstallen, R. and Vandergeten, J.P. (1987) Quantitative and qualitative study of nutrient elements in crop residues in the course of long term trials. *Institut International de Recherches Betteravières Proceedings,* 25–41.

Varsa, E.C. (1970) The use of soil and petiole tests for detecting residual nitrogen and for predicting responses of sugar beets (*Beta vulgaris*) to nitrogen. PhD thesis, Michigan State University, East Lansing, Michigan.

Vepraskas, M.J. (1988) A method to estimate the probability that sub-soiling will increase crop yields. *Soil Science Society of America Journal* 52, 229–232.

Vepraskas, M.J. and Miner, G.S. (1986) Effects of sub-soiling and mechanical impedance on tobacco root growth. *Soil Science Society of America Journal* 50, 423–427.

Vepraskas, M.J., Miner, G.S. and Peedin, G.F. (1986) Relationships of dense tillage pans, soil properties and sub-soiling to tobacco root growth. *Soil Science Society of America Journal* 50, 1541–1546.

Vereerstraeten, R., Hofman, G. and Vandergeten, J.P. (1997) Study of mineral nitrogen dynamics in soil and possible adjustments of recommendations. *Institut International de Recherches Betteravières Proceedings,* 379–385.

Viets, F.G., Jr and Robertson, L.S. (1971) Secondary nutrients and micronutrients. In: Johnson, R.T., Alexander, J.T., Rush, G.E. and Hawkes, G.R. (eds) *Advances in Sugar Beet Production: Principles and Practices.* Iowa State University Press, Ames, Iowa, pp. 171–187.

Vitosh, M.L., Warncke, D.D. and Lucas, R.E. (1994) *Secondary and Micronutrients for Vegetables and Field Crops.* Extension Bulletin E-486, Michigan State University Extension Service, East Lansing, Michigan, 18 pp.

Vitosh, M.L., Warncke, D.D. and Lucas, R.E. (1998) *Secondary- and Micro-nutrients.* Bulletin E-486, Michigan State University Extension Service, East Lansing, Michigan.

von Lüdecke, H. and Nitzsche, M. (1967) Influence of excessive amounts of mineral fertilizers on yield and quality of sugar beets. *Zucker* 20, 461–466.

von Müller, K., Niemann, A. and Werner, W. (1962) Influence of nitrogen : potassium ratio on yield and quality of sugar beet. *Zucker* 15, 142–147.

Voorhees, W.B. (1983) Relative effectiveness of tillage and natural forces in alleviating wheel-induced soil compaction. *Soil Science Society of America Journal* 47, 129–133.

Voorhees, W.B., Farrell, D.E. and Larson, W.E. (1975) Soil strength and aeration effects on root elongation. *Soil Science Society of America Journal* 39, 948–953.

Voorhees, W.B., Senst, C.G. and Nelson, W.W. (1978) Compaction and soil structure modification by wheel traffic in the northern corn belt. *Soil Science Society of America Journal* 42, 344–349.

Voorhees, W.B., Johnson, J.F., Randall, G.W. and Nelson, W.W. (1989) Corn growth and yield as affected by surface and subsoil compaction. *Agronomy Journal* 81, 294–303.

Voss, J. (1996) Input and offtake of nitrogen, phosphorus and potassium in cropping systems with potato as a main crop and sugar beet and spring wheat as subsidiary crops. *European Journal of Agronomy* 5, 105–114.

Voth, R.D. (1977) Effect of boron, manganese and fertilizers on yield, quality and nutrition of sugar beets (*Beta vulgaris,* L.). PhD thesis, Michigan State University, East Lansing, Michigan.

Voth, R.D. and Christenson, D.R. (1980a) Yield, quality and tissue N and Mn levels as affected by interactions between applied N and Mn. *Journal of American Society of Sugar Beet Technologists* 20, 544–552.

Voth, R.D. and Christenson, D.R. (1980b) Effect of fertilizer reaction and placement on availability of manganese. *Agronomy Journal* 72, 769–773.

Voth, R.D., Reisen, J. and Christenson, D.R. (1979) *Effect of Applied Boron on Yield of Sugarbeets.* Research Report 376, Michigan State University Agricultural Experiment Station, East Lansing, Michigan.

Wadman, W.P. and Ehlert, P.A.I. (1989) Environmental effects of organic manures in sugar beet production. *Institut International de Recherches Betteravières Proceedings,* 93–101.

Wadsworth, G.A. and Webber, J. (1980) Deposition of minerals and trace elements in rainfall. In: *Inorganic Pollution and Agriculture.* MAFF Reference Book 326, HMSO, London, pp. 47–55.

Wallace, A. and Terry, R.E. (1998) *Handbook of Soil Conditioners: Substances that Enhance the Physical Properties of Soil.* Marcel Dekker, New York, 596 pp.

Wallace, A. and Wallace G. (1986) Effects of soil conditioners on emergence and growth of tomato, cotton and lettuce seedlings. *Soil Science* 141, 313–316.

Wallace, T. (1945) Some aspects of mineral deficiencies in farm crops. *Agricultural Progress* 20, 20–25.

Walsh, T. (1970) Towards efficiency in the use of our soils. *Scientific Proceedings of the Dublin Society* 2, 285–327.

Walters, D.T. and Malzer, G.L. (1990a) Nitrogen management and nitrification inhibitor effects on nitrogen-15 urea: I. Yield and fertilizer use efficiency. *Soil Science Society of America Journal* 54, 115–122.

Walters, D.T. and Malzer, G.L. (1990b) Nitrogen management and nitrification inhibitor effects on nitrogen-15 urea: II. Nitrogen leaching and balance. *Soil Science Society of America Journal* 54, 122–130.

Walther, U. (1983) Does the N-min method assist in the aim to accurately fertilize sugar beet? *Institut International de Recherches Betteravières Proceedings,* 521–532.

Warburg, O. and Lüttgens, W. (1946) Photochemiche Reduction des Chinons in grünen Zellen und Granual. *Biochemia* 11, 321–322.

Warncke, D.D. and Brown, J.R. (1998) Potassium and other basic cations. In: Brown, J.R. (ed.) *Recommended Chemical Soil Test Procedures for the North Central Region.* North Central Regional Research Publication No. 221. University of Missouri, Columbia, Missouri.

Warren, R.G. and Cooke, G.W. (1962) Comparisons between methods of measuring soluble phosphorus and potassium in soils used for fertilizer experiments on sugar beet. *Journal of Agricultural Science, Cambridge* 59, 269–274.

Warren, R.G. and Johnston, A.E. (1960) The exhaustion land site. In: *Rothamsted Experimental Station Report for 1959,* Lawes Trust, Harpenden, pp. 230–239.

Warren, R.G. and Johnston, A.E. (1961) Soil organic matter and organic manures. In: *Rothamsted Experimental Station Report for 1960,* Lawes Trust, Harpenden, pp. 43–48.

Warren, R.G. and Johnston, A.E. (1962a) Barnfield. In: *Rothamsted Experimental Station Report for 1961,* Lawes Trust, Harpenden, pp. 58–59.

Warren, R.G. and Johnston, A.E. (1962b) Barnfield. In: *Rothamsted Experimental Station Report for 1961,* Lawes Trust, Harpenden, pp. 227–247.

Warren, R.G., Johnston, A.E. and Penny, A. (1962) The value of residues of PK fertilizers in soils: continuous barley site at Woburn. In: *Rothamsted Experimental Station Report for 1961*, Lawes Trust, Harpenden, pp. 58–59.

Watson, D.J. (1947) Comparative physiological studies on the growth of field crops. Parts I and II. *Annals of Botany* 11, 41–76, 375–407.

Watson, D.J. (1952) Physiological basis of variation in yield. *Advances in Agronomy* 4, 101–145.

Watson, D.J. and Russell, E.J. (1943) *et seq.* The Rothamsted experiments on mangolds, 1872–1940. Parts I, II, III, IV(i), IV(ii). *Empire Journal of Experimental Agriculture* 9, 49–64; 9, 65–77; 13, 62–79; 14, 49–56; 14, 57–70.

Watson, M.E. and Brown J.R. (1998) pH and lime requirement. In: Brown, J.R. (ed.) *Recommended Chemical Soil Test Procedures for the North Central Region*. SB 1001, Missouri Agricultural Experiment Station, University of Missouri, Columbia, Missouri.

Weatherly, A.B. and Dane, J.H. (1979) Effect of tillage on soil-water movement during corn growth. *Soil Science Society of America Journal* 43, 1222–1225.

Weaver, J.E. (1926) *Root Development of Field Crops*. McGraw-Hill, London.

Webb, B.C., Harrison, C.M. and Dexter, S.T. (1960) The growth of sugar beets in sand cultures fertilized solely with several green manures. *Quarterly Bulletin Michigan State University Agricultural Experiment Station* 43, 367–374.

Webb, J., Sylvester-Bradley, R. and Seeney, F.M. (1997) Effects of site and season on the fate of nitrogen residues from root crops grown on sandy soils. *Journal of Agricultural Science, Cambridge* 128, 445–460.

Welkie, G.W. and Miller, G.W. (1989) Sugar beet responses to iron nutrition and stress. *Journal of Plant Nutrition* 12(8), 1041–1054.

Wendenburg, C. (1996) Determination of N uptake of sugar beet plants as the basis of fertilizer supply. *Institut International de Recherches Betteravières Proceedings*, 303–307.

Werner, D., Werner, B. and Herzau, M. (1998) Effects of mechanical stress and regenerative soil processes on the structure of conventionally and conservationally tilled loess soil. In: *Advances in Sugar Beet Research*. Monograph Vol. 1, Institut International de Recherches Betteravières, Brussels, pp. 21–34.

Westerman, D.T. and Crothers, S.E. (1980) Measuring soil nitrogen mineralization under field conditions. *Agronomy Journal* 72, 1009–1012.

Westerman, D.T., Leggett, G.E. and Carter, J.N. (1977) Phosphorus fertilization of sugarbeets. *Journal of American Society of Sugar Beet Technologists* 19, 262–269.

Whisler, F.D., Engle, C.F. and Baughman, N.M. (1965) *The Effect of Soil Compaction on Nitrogen Transformations in the Soil*. Bulletin 516 T, West Virginia University Agricultural Experimental Station, Morgantown, West Virginia.

White, R.P. and Doll, E.C. (1971) *Phosphorus and Potassium Fertilizers Increase Soil Test Levels*. Research Report 127, Michigan Agricultural Experiment Station, Michigan State University, East Lansing, Michigan.

White, T.L. (1959) Petiole analysis as a guide to the manuring of sugar beet. *Plant and Soil* 11, 78–86.

Whitehead, D.C. (1963) Some aspects of the influence of organic matter on soil fertility. *Soils Fertility, Harpenden* 26, 217–223.

Whitehead, D.C. (2000) *Nutrient Elements in Grassland*. CAB International, Wallingford, UK, 369 pp.

Widdowson, F.V. (1974) Results from experiments measuring the residues of nitrogen fertilizer given for sugar beet, and of ploughed-in sugar beet tops, on the yield of following barley. *Journal of Agricultural Science, Cambridge* 83, 415–421.

Widdowson, F.V. and Penny, A. (1967) Results of an experiment at Woburn testing farmyard manure and N, P and K fertilizers on five arable crops and a long ley. I. Yields. *Journal of Agricultural Science, Cambridge* 68, 95–102.

Widdowson, F.V. and Penny, A. (1979) Results from the Woburn reference experiment. In: *Rothamsted Experimental Station Report for 1978*, Part 2, Lawes Trust, Harpenden, pp. 67–82.

Wiedemann, H. (1994) Soil analysis and N fertilizer recommendations for sugar beet in southern Germany and Austria. *Institut International de Recherches Betteravières Proceedings*, 335–341.

Wiersma, D. and Mortland, M.M. (1953) Response of sugar beets to peroxide fertilizations and its relationship to oxygen diffusion. *Soil Science* 75, 355–360.

Wiersum, L.K. (1962) Uptake of nitrogen and phosphorus in relation to soil structure and nutrient mobility. *Plant and Soil* 16, 62–70.

Wiklicky, L.M., Nemeth, K. and Recke, H. (1983) Assessment of nitrogen fertilizer requirements for sugar beet by means of EUF. *Institut International de Recherches Betteravières Proceedings*, 533–543.

Williams, R.J.B. (1971) Relationships between the composition of soils and physical measurements made on them. In: *Rothamsted Experimental Station Report for 1970*, Part 2, Lawes Trust, Harpenden, pp. 5–35.

Williams, R.J.B. and Cooke, G.W. (1971) Results of the Rotation I experiment at Saxmundham, 1964–69. In: *Rothamsted Experimental Station Report for 1970*, Part 2, Lawes Trust, Harpenden, pp. 68–97.

Winder, T.L. and Nishio, J.N. (1995) Early iron deficiency stress response in leaves of sugar beet. *Plant Physiology* 108, 1487–1494.

Winner, C. (1966) Düngung überdüngung und qualität der Zuckerrübe. In: *Potassium Symposium 1966*, International Potash Institute, Bern, pp. 89–106.

Winner, C. (1993) History of the crop. In: Cooke, D.A. and Scott R.K. (eds) *The Sugar Beet Crop*. Chapman & Hall, London, pp. 6–7.

Winter, S.R. (1975) Influence of nitrogen placement and source on surfact nitrate accumulation and sugar-beet production. *Journal of American Society of Sugar Beet Technologists* 18, 343–348.

Winter, S.R. (1983) Efficient deep tillage for sugarbeets on Pullman clay loam. *Journal of American Society of Sugar Beet Technologists* 22, 29–34.

Wood, G. (2002) Calibrating airborne digital photography to determine variable application rates of fertilizer nitrogen for optimising canopy size in winter wheat. In: *Society of the Chemical Industry, Agriculture Group, Proceedings, January 2002*, SCI, London.

Woolley, D.G. and Bennett, W.H. (1962) Effect of soil moisture, nitrogen fertilisation, variety and harvest date on root yields and sucrose content of sugar beets. *Journal of American Society of Sugar Beet Technologists* 12, 233–237.

Wright, M.J. and Davidson, K.K. (1964) Nitrate accumulation in crops and nitrate poisoning of animals. *Advances in Agronomy* 16, 197–247.

Xu, C. (1991) Tillage and row spacing effects on development and growth of dry beans (*Phaseolus vulgaris* L.) and sugar beet (*Beta vulgaris* L.) on a Parkhill loam soil. MS thesis, Michigan State University, East Lansing, Michigan.

Yackle, G.A. and Cruse, R.M. (1983) Corn plant residue age and placement effects upon early corn growth. *Canadian Journal of Plant Science* 63, 871–877.

Yates, F. and Patterson, H.D. (1958) A note on the six-course rotation experiments at Rothamsted and Woburn. *Journal of Agricultural Science, Cambridge* 50, 102–109.

Yerokun, O.A. and Christenson, D.R. (1989) Use of urea and ammonium phosphates as starter fertilizers. *Journal of Fertilizer Issues* 6, 12–16.

Yerokun, O.A. and Christenson, D.R. (1990) Relating high soil test phosphorus concentrations to plant phosphorus uptake. *Soil Science Society of America Journal* 54, 796–799.

Yonts, C.D. and Smith, J.A. (1997) Effects of plant population and row width on yield of sugar beet. *Journal of Sugar Beet Research* 34, 21–30.

Young, R.D. and Davis, C.H. (1980) Phosphate fertilizers and process technology. In: Khasawneh, F.E., Sample, E.C. and Kamprath, E.J. (eds) *The Role of Phosphorus in Agriculture*. American Society of Agronomy, Madison, Wisconsin, pp. 195–226.

Glossary

Adjusted tonnage: weight of roots in tonnes adjusted to sugar percentage of 16.

Ammonium fixation: fixation of ammonium ions by the mineral fraction of soil in forms that cannot be replaced by a neutral potassium salt solution.

Banding or band application: method of fertilizer or other agrochemical application above, below or alongside the sown seed row.

Biological nitrogen fixation: conversion of molecular nitrogen gas (N_2) to ammonia and subsequently to organic forms utilizable in biological processes.

Biuret ($H_2NCONHCONH_2$): product formed at high temperature during manufacturing of urea, toxic to plants.

Broadcast or broadcast application: application of solid or liquid fertilizer or other agrochemical on the soil surface.

Cation exchange capacity (CEC): potential of a soil to adsorb cations, often expressed in mg 100 g^{-1} soil.

Chelates: organic chemicals with two or more functional groups that can bind with metals to form a ring structure.

Critical nutrient concentration: concentration below which yield, quality or performance are less than optimum.

Denitrification: reduction by microbial activity of nitrogen oxides to molecular nitrogen or nitrogen oxides of lower oxidation state.

Fertilizer: any organic or inorganic material of natural or synthetic origin (other than liming materials) added to soil to supply one or more nutrients essential for growth of plants.

Immobilization: conversion of an element from inorganic to organic form in microbial or plant tissue, such as when mineral nitrogen is taken up by plants.

Juice purity: concentration of sugar percentage of total soluble solids in fresh roots.

Labile: readily available (nutrient) to plants.

Leaching: removal of soluble constituents (usually from soil) by water moving downwards.

Leaf area index (LAI): area of leaf on a plant divided by the area of land occupied by the plant.

Lime, agricultural: a soil amendment containing calcium carbonate, magnesium carbonate and other materials, used to neutralize soil acidity and furnish calcium and magnesium for plant growth.

Macronutrient: plant nutrient found at relatively high concentration ($>$ 500 mg kg^{-1}) in plants. Usually refers to nitrogen, phosphorus and potassium, but includes calcium, magnesium, sulphur and sodium.

Marc: water-insoluble part of the sugar beet root.

Micronutrient: plant nutrient found in relatively small concentrations (< 100 mg kg^{-1}) in plants. Usually refers to boron, chlorine, copper, iron, manganese, molybdenum and zinc. However, in some plants it may include nickel and cobalt.

Mineralization: conversion of an element from the organic to the inorganic state, mainly as a result of microbial activity.

Molar: the concentration resulting from 1 mole (see below) dissolved in 1 l of solution.

Mole: the amount of a substance that has a weight in grams numerically equal to the molecular weight of the substance.

Nitrification: the biological oxidation of NH_4^+ first to nitrite and then to nitrate.

Organic farming: crop-production system that reduces, avoids or largely excludes the use of synthetically compounded fertilizers, pesticides, growth regulators and livestock feed additives.

Pesticide: chemical designed to kill pests (weeds, insects, mites, nematodes, fungi, rodents, algae, bacteria), including many natural chemicals to trap or confuse insects.

Saline soil: soil containing sufficient soluble salt to adversely affect the growth of crops. Electrical conductivity is greater than 4 dS m^{-1}.

Slurry: mixture of dung and urine with variable amount of water from housed livestock.

Sodic soil: soil containing sufficient exchangeable sodium to adversely affect crop production and soil structure.

Soil-analysis units: concentrations of elements in volumes of soil (mg l^{-1}) or weight of soil (mg kg^{-1}).

Soil pH: degree of acidity or alkalinity of soil. Descriptive terms associated with pH ranges: extremely acid, < 4.5; very strongly acid, 4.5–5.0; strongly acid, 5.1–5.5; moderately acid, 5.6–6.0; slightly acid, 6.1–6.5; neutral, 6.6–7.3; slightly alkaline, 7.4–7.8; moderately alkaline, 7.9–8.4; strongly alkaline, 8.5–9.0; and very strongly alkaline, > 9.1.

Soil-test critical concentration: the concentration of an extractable nutrient above which crop response to the added nutrient would not be expected.

Sucrose: disaccharide composed of one glucose linked to one fructose molecule.

Sucrose/sugar: the terms are used interchangeably for the sweetener extracted from either sugar beet or sugar cane. Chemically, sucrose is the disaccharide α-D-glucopyranosyl-β-fructofuranoside. Sucrose is a disaccharide composed of one glucose and one fructose molecule.

Sugar: constituent extracted from either sugar beet or cane, being almost entirely sucrose.

Sugar percentage: concentration of sugar in fresh roots, expressed as a percentage.

Tillage terms:

Conservation – tillage or tillage/planting sequence that leaves 30% or greater cover of crop residue on soil.

No till – procedure whereby a crop is planted directly into soil with no primary or secondary tillage since the harvest of the previous crop.

Primary – tillage at any time constituting initial, major soil-manipulation operation, most commonly ploughing.

Secondary – any of a group of separate or distinct operations, following primary tillage, often to produce a seedbed.

Reduced – system in which the total number of operations is reduced from the normal number practised.

Ridge – system in which ridges are formed and the crop is sown on the ridge.

Tops: sugar beet leaves, petioles and crown, cut at the level of the lowest leaf scar.

Volatilization: escape of nutrients, mainly nitrogen as ammonia, in a gaseous state from the soil.

Appendix A
Phosphorus and Potassium in Crop Material

		Phosphorus (P_2O_5)	Potassium (K_2O)
		kg t^{-1} of fresh material	
Cereals	Grain only (all cereals)	7.8	5.6
	Grain and straw		
	Winter wheat/barley[a]	8.6	11.8
	Spring wheat/barley[a]	8.8	13.7
	Winter/spring oats[a]	8.8	17.3
Oilseed rape	Seed only	14.0	11.0
	Seed and straw[a]	15.1	17.5
Peas	Dried	8.8	10.0
	Vining	1.7	3.2
Field beans		11.0	12.0
Potatoes		1.0	5.8
Sugar beet	Roots only	0.8	1.5
	Roots and tops	1.9	7.3
Grass	Fresh grass (15–20% DM)	1.4	4.8
	Silage (25% DM)	1.7	6.0
	Silage (30% DM	2.1	7.2
	Hay (86% DM)	5.9	18.0
Kale		1.2	5.0
Maize	Silage (30% DM)	1.4	4.4
Swedes	Roots only	0.7	2.4
Broad beans		1.6	3.6
French beans		1.0	2.4
Beetroot		1.0	4.5
Cabbage		0.9	3.6
Carrots		0.7	3.0
Cauliflowers		1.4	4.8
Onions	Bulbs only	0.7	1.8
Sprouts	Buttons	2.6	6.3
	Stems	2.1	7.2
Bulbs		2.4	6.3

[a]Offtake values are per tonne of grain or seed removed but include nutrients in straw.
DM, dry matter.

Example

Winter wheat yields 10 t ha^{-1} of grain. The straw is baled and removed from the field.

Phosphorus offtake = $10 \times 8.6 = 86$ kg P_2O_5 ha^{-1}
Potassium offtake = $10 \times 11.8 = 118$ kg K_2O ha^{-1}

Reproduced with permission of the Potash Development Association.

Appendix B
Conversion Factors

To convert	To	Multiply by	Reciprocal
Inches (in)	Centimetres (cm)	2.54	0.394
Yards (yd)	Metres (m)	0.914	1.09
Square yards (yd^2)	Square metres (m^2)	0.836	1.20
Acres (4840 yd^2)	Hectares (ha)	0.405	2.47
Pounds (lb)	Kilograms (kg)	0.454	2.20
UK tons (2240 lb)	Tonnes (t)	1.02	0.984
Short tons (2000 lb)	Tonnes (t)	0.911	1.10
lb $acre^{-1}$	kg ha^{-1}	1.12	0.892
mmho cm^{-1}	dS m^{-1}	1	1
mg kg^{-1}	p.p.m	1	1
P_2O_5	P	0.436	2.29
K_2O	K	0.830	1.20
CaO	Ca	0.715	1.40
MgO	Mg	0.603	1.66
Na_2O	Na	0.742	1.35
NaCl	Na	0.393	2.70
SO_3	S	0.4	2.5
B_2O_3	B	0.31	3.2

Index